MAP OF BORNEO

Orchids Of Borneo
Vol. 3

Jejewoodia jiewhoei (J.J. Wood & Shim) Szlach. *(Photo: C.L. Chan)*

ORCHIDS OF BORNEO

VOL. 3. DENDROBIUM, DENDROCHILUM AND OTHERS

By J.J. WOOD

Series Editor
P.J. Cribb

Series Co-ordinator
C.L. Chan & A. Lamb

The Sabah Society
Kota Kinabalu

in association with

Royal Botanic Gardens
Kew, England

1997

Published by

THE SABAH SOCIETY,
P.O. Box 10547,
88806 Kota Kinabalu, Sabah, Malaysia.

in association with

THE BENTHAM-MOXON TRUST,
The Royal Botanic Gardens,
Kew, Richmond, Surrey TW9 3AB, England.

ORCHIDS OF BORNEO VOL. 3
DENDROBIUM, DENDROCHILUM AND OTHERS
BY J.J. Wood

Series Editor: P.J. Cribb
Series Co-ordinator: C.L. Chan & A. Lamb

First published 1997

Perpustakaan Negara Malaysia Cataloguing-in-Publication Data

Dendrobium, Dendrochilum and others / J.J. Wood.
(Orchids of Borneo ; v. 3)
Bibliography; p.230
includes index
ISBN 967-99947-5-9
1. Orchids —Borneo. 2. Orchids—Borneo—Identification.
I. Title. II. Series.
584.15095983

Printed and bound in Malaysia.

CONTENTS

LIST OF COLOUR PLATES

(taxon number in this account given in brackets)

ACKNOWLEDGEMENTS

I would like to acknowledge the continuing support and enthusiasm for this project expressed by Professor Grenville Lucas, former Keeper of the Herbarium at Kew and Dr Ti Teow Chuan, President and Chan Chew Lun, Vice President of the Sabah Society and Co-ordinator of this project for their support. The enthusiasm and support provided by the present Keeper, Professor Simon Owens, is greatly appreciated. A special thank you is also extended to Dr Phillip Cribb, Assistant Keeper of the Herbarium at Kew, for his work as the series editor and for his useful comments and continuing commitment to the project.

Thanks are also due to Anthony Lamb without whose exhaustive knowledge of orchids in Borneo and enthusiasm, this project would never have been undertaken.

The skill of several artists has succeeded in portraying the beauty of the one hundred varied species depicted in this volume without sacrificing scientific accuracy. I would like to thank Eleanor Catherine, Lucy F.L. Liew, Chin Wan Wai, C.L. Chan, Maureen Church, Judi Stone, Susanna Stuart-Smith, Mair Swann, Sarah Thomas and Jaap Vermeulen for their invaluable contribution.

Excellent colour slides were kindly provided by C.L. Chan, Todd Barkman, Reed Beaman, Sheila Collenette, Jim Comber, Phillip Cribb, John Dransfield, Peter Jongejan, Anthony Lamb, Ed de Vogel, Louis Vogelpoel, Yii Puan Ching, Ronald Zabeau, and the Royal Botanic Garden, Edinburgh.

I would like to express my thank for computer assistance provided by Chua Kok Hian. The tedious work of seeing the book through the press by C.L. Chan and maintaining such a high quality of production throughout the series is especially acknowledged. The invaluable financial contribution provided by Mr Tan Jiew Hoe (Singapore), without which this volume would not have been published, is gratefully acknowledged.

Bulbophyllum mutabile (Blume) Lindl. var. *obesum* J.J. Verm. *(Photo: C.L. Chan)*

CHAPTER 1

INTRODUCTION

Two large and important genera are given centre stage in this volume: *Dendrobium* (Tribe Dendrobieae), with some 143 named species in sixteen sections, is the largest orchid genus after *Bulbophyllum*, on the island of Borneo; *Dendrochilum* (Tribe Coelogyneae), represented by 88 currently accepted species, is a characteristic element in the montane orchid flora of Borneo. The Bornean species are currently under revision at Kew.

Dendrobium, with perhaps as many as 1000 species, is the second largest orchid genus in the Old World after *Bulbophyllum*. *Dendrobium*, however, contains far more species of horticultural merit than *Bulbophyllum* and consequently a wider range can be found in collections. Occurring in mainland Asia, the Malay Archipelago, New Guinea, the Pacific islands and Australia, *Dendrobium* is less widely distributed than *Bulbophyllum*. *Bulbophyllum* is, in addition to the above, well represented in Africa, Madagascar and the Mascarene Islands, and a few species are also found in the New World.

Dendrobium as a genus is still poorly understood and no complete account of it has been produced since that of Fritz Kraenzlin in 1910 which appeared in Engler's monumental Das Pflanzenreich. This, however, has been less influential than the treatment of Rudolf Schlechter (1912) who revised the Papua New Guinea (formerly German New Guinea) species and produced a sectional treatment that is still widely followed. Three of Schlechter's sections containing Bornean species, out of 41 listed, have been recognised at generic level for some time. These are *Desmotrichum* (now *Flickingeria*), *Diplocaulobium* and *Sarcopodium* (now *Epigeneium*), with fourteen, two and ten species in Borneo respectively. A recent attempt by Brieger (1981) to further divide *Dendrobium* into a number of smaller genera based on established sections has not gained recognition by most other workers in the field. A key to the Bornean sections is provided in Chapter 2.

Only 30 species of *Dendrobium* occurring in Borneo are shared with Thailand, which has 147 native species, while 39 are also found in Sumatra where 109 species have been recorded. Further study of critical sections such as *Aporum*, *Calcarifera* and *Rhopalanthe* is required and will probably demonstrate that several of the Bornean taxa are conspecific.

The Bornean figure of 143 species in sixteen sections of *Dendrobium* is low when compared with New Guinea which can boast some 700 named species in 26 sections. This high figure for New Guinea will inevitably drop, however, with further study and when synonymy is taken into account. Three distinctive sections centred on New Guinea, each containing many beautiful species, are notably absent from Borneo. These are *Latouria*,

Oxyglossum and *Spatulata* (syn. *Ceratobium*), all having long-lived flowers of various unusual hues. Similarly, important sections from mainland Asia, notably *Callista* and *Stachyobium*, which require a more seasonal climate, are also absent. Another section of great horticultural value, *Dendrobium* (syn. *Eugenanthe*), also centred on mainland Asia, is represented by only four species in Borneo including the widespread *D. anosmum* Lindl. and *D. heterocarpum* Lindl. This is in striking contrast with the 37 species of this section recorded from Thailand. The distinctive section *Grastidium*, which has short-lived flowers, is only represented in Borneo by seven species, whereas in New Guinea it is the largest section with 125 named species.

Fourteen sections are represented in both Borneo and Thailand and fifteen are shared with Sumatra. Only thirteen of the 26 sections occurring in New Guinea, however, are found in Borneo. Section *Distichophyllum*, though, has diversified in Borneo more than elsewhere with twenty named species and several more awaiting description. The curious brown-flowered *D. piranha* C.L. Chan & P.J. Cribb, described in volume one, belongs in this section.

Several Bornean species of horticultural merit are found in section *Formosae* (formerly *Nigrohirsutae*), notably *D. lowii* Lindl. and *D. spectatissimum* Rchb.f. This is a large, mainly mainland Asiatic section of between 30 and 35 species which extends east to the Philippines. It is distinguished by the dark brown or blackish hairs on the leaf sheaths and elsewhere, and by the large often white flowers usually borne near the stem apex. Twelve species have been recorded from Borneo.

Another attractive section with 21 (twelve endemic) named species from Borneo, is *Calcarifera* which was, until recently, included by many in the predominantly New Guinean *Pedilonum*. The recently described *D. cymboglossum* J.J. Wood & A. Lamb is featured on the front of the dust jacket of this volume.

Dendrobium seldom occurs above an elevation of 1800 metres in Borneo. Its niche here is often filled by species of *Epigeneium*. Only 6 of the 61 species of *Dendrobium* recorded from Mt. Kinabalu, for example, are found at high elevation. Four of these belong to section *Rhopalanthe* which includes the well known and widespread Pigeon orchid, *D. crumenatum* Sw. Most Bornean species of *Dendrobium* (Wood 1990) occur in hill and lower montane forest between about 600 and 1600 metres elevation. The paucity of high-elevation species is in striking contrast to New Guinea which, of course, has a considerably larger land area above 2400 metres. This is mirrored by the notable absence of sections such as *Calyptrochilus* (except *D. erosum* (Blume) Lindl.), *Latouria*, *Oxyglossum* and *Pedilonum* (except *D. secundum* (Blume) Lindl.), which form an important and colourful element in the montane flora of New Guinea. One of the most attractive Bornean members of section *Rhopalanthe* is *D. cinnabarinum* Rchb.f. which is recorded between 700 and 1900 metres at various localities, but curiously has never been observed on Mt. Kinabalu. There is considerable variation in the form and colour of the flowers and a study of this variation would be interesting. It seems likely that the distinctive bright orange-flowered var. *angustitepalum* Carr should be recognised at specific level.

The scope in the genus *Dendrobium* for growers is immense with so many diverse species from which to choose. It is to be hoped that some of the more desirable Bornean species, currently under threat in the wild, can be successfully propagated in large numbers and made available to a wider public.

Dendrochilum is known to growers from only a handful of species which are mostly native to the Philippines. Most species are restricted to the cool, humid conditions of montane forests which explains the high diversity found in Sumatra, Borneo and the Philippines. There is a corresponding paucity of species in areas lacking high mountains, eg. Peninsular Malaysia, and those which experience a distinct dry season, eg. East Java and the Lesser Sunda Islands. A total of 88 named species have been recorded from Borneo. The geographical isolation of populations, which is an unavoidable consequence of strictly montane preferences, probably explains the extraordinary high percentage of endemism.

Dendrochilum is closely allied to *Pholidota* and can be distinguished from it by several characters, the most important of which are the four entire pollinia and the often somewhat concave but usually not saccate lip. *Pholidota* has four porate pollinia and a distinctly saccate lip.

A recent revised subdivision of *Dendrochilum* proposed by Pedersen, Wood & Comber (1996) follows the traditional concept of recognising four subgenera. The subdivision of three of the subgenera, however, includes some radical changes. Of the thirteen sections recognised, nine are described as new, one is a formerly proposed section which is redefined and validly published, while only the three sections carrying the autonyms are accepted from previous subdivisions. In general the classification is expanded when compared to any of the earlier systems. Three of the four subgenera are represented in Borneo, viz. *Acoridium*, *Dendrochilum* and *Platyclinis*. A key to these and the sections within them is provided in chapter three.

The distribution of *Dendrochilum* is of the Indo-Malesian pattern as defined by van Steenis (1979). The subgenera and sections of the genus occupy variously delimited areas within the total geographic range.

Subgenus *Acoridium* comprises four sections, two of which are endemic to the Philippines which is the centre of diversity. Section *Acoridium* is represented in Borneo by two species, while a third is the type of the monotypic endemic section *Falsiloba*.

Subgenus *Dendrochilum*, although relatively widespread, has its centre of diversity in Sumatra where fourteen of the 23 species occur. Eight species are recorded from Borneo. Diversity decreases rapidly further east, with only the widespread *D. pallidiflavens* Blume occurring in the Philippines.

Subgenus *Platyclinis* comprises seven sections with very restricted to very wide distributions. Four sections occur in Borneo including *Cruciformia* and *Mammosa* which are both endemic. Section *Eurybrachium*, with a total of 70 species, is distributed in

Borneo, Java, the Lesser Sunda Islands, the Philippines, Sulawesi, and Sumatra, the centre of diversity, where 30 species occur. Section *Platyclinis* has its primary centre of diversity in Borneo where 50 species are found, but is also well represented in Sumatra and the Philippines.

A provisional artificial key to all of the Bornean species in sections *Acoridium*, *Eurybrachium* and *Platyclinis* is provided in Chapter 4. Further work on species delimitation in subgenus *Dendrochilum* is necessary before a workable key for these can be constructed.

CHAPTER 2

ARTIFICIAL KEY TO THE SECTIONS OF *DENDROBIUM* IN BORNEO

The separation of *Dendrobium* into sections is at best highly artificial and is largely based on characters that are not always distinctive or exclusive. A certain level of personal knowledge and field experience of the genus is clearly advantageous when attempting to understand sectional delimitation. Despite this, it is hoped that the key reproduced below will be of some practical use.

1. Plants small and tufted in habit. Stems 3–10 cm long, composed of very few internodes, the uppermost swollen, bearing 2 fleshy, sheathless, opposite apical leaves between which appear one-flowered inflorescences successively or 2–3 at the same time .... **Bolbidium**

1. Plants without this combination of characters 2

2. Leaves without basal sheaths. Flowers solitary, ephemeral. Lip with a circular blade above a linear claw, concave .... **Euphlebium**

2. Leaves with distinct basal sheaths, often enclosing most of the internodes. Flowers solitary to many, ephemeral or long-lived. Lip variously shaped 3

3. Lip with a fleshy, flange-like basal appendage. Inflorescences short, with papery floral bracts .... **Amblyanthus**

3. Plants without this combination of characters 4

4. Stems with at least some of the internodes fleshy and swollen 5

4. Stems hard, thin and wiry, the internodes not fleshy and swollen 15

5. Stems with only a few fleshy and swollen internodes just above the base, the upper portion of the stem thin and wiry. Leaves sometimes terete .... **Rhopalanthe**

5. Stems with internodes uniformly cylindrical, fleshy and swollen, or sometimes fusiform with internodes at the middle portion of the stem fleshier than those above and below. Leaves dorsiventral, never terete 6

6. Leaf sheaths and sometimes other parts of the plant covered with dark brown or blackish hairs. Flowers usually borne near the stem apex, often large, white, with coloured markings on the lip, rarely yellow .... **Formosae**

6. Dark brown or blackish hairs absent 7

7. Mentum ± same length or longer than the dorsal sepal, usually narrow 8

7. Mentum usually much shorter than the dorsal sepal, saccate 12

8. Lip distinctly 3-lobed, margins long-hairy .... **Breviflores**

8. Lip usually entire, margins glabrous 9

9. Mentum parallel to the pedicel with ovary, always distinctly longer than the dorsal sepal 10

9. Mentum divergent from the pedicel with ovary, usually ± equal in length to the dorsal sepal, but sometimes longer 11

10. Apex of the lip incurved and fimbriate .... **Calyptrochilus**

10. Apex of the lip flat, entire .... **Pedilonum**

11. Lip with distinct, though sometimes small, side lobes, fleshy, stiff, often adpressed to column, epichile with keels or warts. Flowers always borne on leafy stems .... **Distichophyllum**

11. Lip entire, usually broadened abruptly above a narrow claw, often with a projection of some type on the upper surface of the claw, blade without keels or warts. Flowers borne on leafy or leafless stems .... **Calcarifera**

12. Lip entire, lacking side lobes .... **Dendrobium** (***D. anosmum, D. heterocarpum and D. stuartii***)

12. Lip with side lobes which are sometimes small or indistinct 13

13. Lip margins long-hairy. Flowers borne on leafless stems .... **Breviflores**

13. Lip margins not long-hairy. Flowers borne on leafy stems, or on both leafy and leafless stems 14

14. Flowers about 4–4.5 cm in diameter, white, with a lilac or purple central patch on lip. Side lobes of lip indistinct, margins serrated, the epichile without keels or warts .... **Dendrobium** (***D. tetrachromum***)

14. Flowers lacking this combination of characters .... **Distichophyllum**

15. Leaves laterally compressed, broadest at base 16

15. Leaves dorsiventral or terete 17

16. Lip with a small conical projection beneath the apex .... **Oxystophyllum**

16. Lip without an apical projection .... **Aporum**

17. Leaves terete, ± as wide as they are thick .... **Strongyle**

17. Leaves dorsiventral 18

18. Leaf sheaths hairy. Base of lip connate with sides of the column-foot, forming a spur .... **Conostalix**

18. Leaf sheaths glabrous. Base of lip free, not forming a spur with the column-foot .... **Grastidium**

CHAPTER 3

ARTIFICIAL KEY TO SUBGENERA AND SECTIONS OF *DENDROCHILUM* IN BORNEO

1. Column sometimes with an apical hood but never distinctly prolonged beyond the anther; stelidia absent (**subgenus Acoridium**) 2

1. Column distally prolonged into an apical wing or hood which distinctly exceeds the anther; stelidia present or absent 3

2. Column distinctly shorter than the dorsal sepal. Lip usually sessile (rarely shortly clawed), entire to distinctly 3-lobed but never with two prominent, wing-like keels reminiscent of side lobes .... **section Acoridium**

2. Column subequal to the dorsal sepal. Lip distinctly clawed, without side lobes but proximally with two prominent, suberect, wing-like keels reminiscent of side lobes .... **section Falsiloba** (*D. pandurichilum* only)

3. Inflorescences heteranthous. Rhizome elongate, creeping. Lip entire (often more or less pandurate). Seed filiform; central testa cells nearly isodiametric, with the junctions of the anticlinal walls conspicuously protruding .... **subgenus Dendrochilum**

3. Inflorescences synanthous (to nearly hysteranthous). Rhizome short or elongate (if elongate usually pendent). Lip entire or 3-lobed. Seed fusiform; central testa cells prosenchymatic, with the junctions of the anticlinal walls not protruding (**subgenus Platyclinis**) 4

4. Lip divided into a distinctly saccate hypochile and a flat epichile by two free, prominent calli; firmly attached to the column. Column strongly incurved; short foot present .... **section Mammosa**

4. Lip not divided into a hypochile and epichile by two prominent calli; firmly or elastically attached to the column. Column straight to somewhat (rarely strongly) incurved; foot present or absent 5

5. Lip more or less distinctly cruciform; firmly attached to the column. Column slender (relatively stout in *D. haslamii*); foot absent; stelidia always basal and subequal to the column proper, usually narrowly subspathulate, never absent .... **section Cruciformia**

5. Lip entire to variously 3-lobed but never cruciform; firmly or elastically attached to the column. Column stout or slender, foot present or absent; stelidia very variable, sometimes absent 6

6. Lip firmly attached to the column, entire to obscurely 3-lobed (rarely distinctly 3-lobed). Column comparatively stout and straight; foot absent .... **section Eurybrachium**

6. Lip elastically attached to the column, entire to distinctly 3-lobed. Column comparatively slender (stout in *D. glumaceum*), more or less incurved; foot present, usually short .... **section Platyclinis**

CHAPTER 4

KEYS TO SPECIES OF *DENDROCHILUM* IN SECTIONS OF SUBGENERA *ACORIDIUM* AND *PLATYCLINIS* IN BORNEO

The three keys provided below are provisional and will appear, with revisions where necessary, in an account of the Bornean species currently under preparation. Subgenus *Dendrochilum* and subgenus *Platyclinis* section *Mammosa* are omitted as further study of species delimitation is necessary. The monotypic section *Falsiloba* is omitted, but appears in the subgeneric and sectional key in Chapter 3.

Artificial key to species of subgenus *Acoridium* section *Acoridium*

1. Pseudobulbs 2.8–3 cm long. Inflorescences with flowers borne 2–2.5 mm apart. Lip with auriculate side lobes, the disc with a fleshy, transverse basal ridge. Sepals 4 mm long .... ***D. auriculilobum*** *J.J. Wood*

2. Pseudobulbs 10–14.5 cm long. Inflorescences with flowers borne 1 mm apart. Lip entire, triangular-ovate to obliquely subquadrate, the disc lacking basal ridge. Sepals 2–2.5 mm long .... ***D. hologyne*** *Carr*

Artificial key to species of subgenus *Platyclinis* section *Eurybrachium*

1. Lip broader than long 2

1. Lip as long as or longer than broad 5

2. Lip strongly concave, strongly cupulate, with 2 small rounded and separate central calli, and a separate basal ridge. Pseudobulbs borne up to 2.2 cm apart on rhizome. Leaves generally 4–10 × 0.4–0.5 cm; petioles 0.1–0.5 cm long .... ***D. cupulatum*** *J.J. Wood*

2. Lip flatter, never strongly cupulate, sometimes concave at base, calli much larger and keel-like, often united and forming a horseshoe-shaped structure. Pseudobulbs borne closer together on rhizome, often caespitose. Leaves larger, generally 8–23 × 0.5–1.2 cm; petioles 0.8–4 cm long 3

3. Margin of lip shortly and irregularly fimbriate. Lip sometimes shallowly three-lobed. Stelidia shorter than the column. Petals denticulate .... ***D. corrugatum*** *(Ridl.) J.J.Sm.* (syn. *D. fimbriatum* Ames)

3. Margin of lip entire. Lip never shallowly lobed. Stelidia as long as or slightly longer than the column. Petals entire 4

4. Lip shallowly retuse, c. 0.18 × 0.27 cm. Pseudobulbs 1.5–2.5 cm long. Petals porrect, never twisted and aligned 90° from vertical. Stelidia glabrous, dark reddish-brown .... ***D. scriptum*** *Carr*

4. Lip subapiculate, c. 0.30 × 0.44 cm. Pseudobulbs generally longer, up to 4.5 cm long. Petals sometimes twisted and aligned up to 90° from vertical. Stelidia minutely papillose, ochre, tipped with dark ochre-brown .... ***D. transversum*** *Carr*

5. Flowers minute, the sepals 0.1–0.15 × 0.075–0.08 cm. Lip *c.* 0.05–0.07 × 0.05 cm, cuneate-cucullate, entire 6

5. Flowers larger, the sepals 0.3–0.7 × 0.1–0.44 cm. Lip flat, ovate to oblong–obovate and entire, oblong to oblong-elliptic or narrowly elliptic and shallowly lobed, or sharply deflexed near base, oblong-lanceolate, acuminate, obscurely auriculate at base, or shallowly cymbiform and obscurely lobed 7

6. Lip apex retuse .... ***D. microscopicum*** *J.J. Sm.*

6. Lip apex acutely apiculate .... ***D. minimiflorum*** *Carr*

7. Petals twisted and aligned 90° from vertical. Lip entire, longer than broad 8

7. Petals porrect to ascending, not twisted and aligned 90° from vertical. Lip entire, auriculate at base, or obscurely lobed 11

8. Flowers relatively large, the sepals 0.78 × 0.4–0.44 cm, the lip 0.64 × 0.58 cm .... ***D. alpinum*** *Carr*

8. Flowers smaller, the sepals 0.3–6 × 0.1–0.3 cm, the lip 0.25–0.35 × 0.15–0.35 cm 9

9. Lip as long as broad, nearly circular or cordate in outline, sometimes cymbiform. Callus consisting of two broader than high, very fleshy keels which touch but are not united at the base and are distinctly horseshoe-shaped .... ***D. pseudoscriptum*** *T.J. Barkman & J.J. Wood*

9. Lip generally somewhat longer than broad, oblong-obovate to obcuneate or ovate to oblong. Callus keels higher and united at the base by a transverse ridge, less obviously horseshoe-shaped 10

10. Lip oblong-obovate to obcuneate, shortly apiculate, margins not recurved, *c.* 0.3 × 0.15 cm .... ***D. alatum*** *Ames*

10. Lip ovate or oblong, minutely apiculate, margins sometimes strongly recurved in the upper half, *c.* 0.26–0.35 × 0.27–0.30 cm .... ***D. pterogyne*** *Carr*

11. Stelidia normally absent. Exclusively epilithic. Flowers creamy-white with a pink pedicel with ovary and column .... ***D. stachyodes*** *(Ridl.) J.J.Sm.*

11. Stelidia present. Not this combination of characters 12

12. Lip sharply deflexed near base, otherwise flat, oblong-lanceolate, acute to acuminate, obscurely auriculate and somewhat erose at base .... ***D. lewisii*** *J.J. Wood*

12. Lip not sharply deflexed near base. Not this combination of characters 13

13. Stelidia very slender, linear, subacute, reaching only as far as rostellum. Lip flat, mid-lobe ovate or transversely elliptic, side lobes short, but distinct, erose .... ***D. acuiferum*** *Carr*

13. Stelidia oblong, rounded to truncate, as long as or slightly shorter than column. Lip shallowly cymbiform, entire or obscurely lobed, side lobes, when present, not erose 14

14. Plant small, up to *c.* 10 cm high. Leaves 0.3–0.5 cm wide. Flowers salmon-pink, the sepals 0.3 × 0.1 cm. Lip elongate-acuminate .... ***D. joclemensii*** *Ames*

14. Plant large, up to 28 cm high. Leaves 0.8–2 cm wide. Flowers translucent pale lemon-yellow, the lip chocolate brown, sepals 0.55–0.6 × 0.2–0.25 cm. Lip obtuse to subacute .... ***D. trusmadiense*** *J.J. Wood*

Artificial key to species of subgenus *Platyclinis* section *Cruciformia*

1. Mature leaf-blades 5.3–5.7 cm wide, elliptic. Inflorescence rather rigid, the rachis fleshy, terete. Disc of lip bearing a fleshy transverse basal callus and two separate crest-like keels above at either side which, although lying very close to, are not united with the basal callus .... ***D. hosei*** *J.J. Wood*

1. Mature leaf-blades much narrower, rarely up to 3.5 cm wide, narrowly linear, ligulate or elliptic. Inflorescence rarely rigid, (if so then rachis thin and quadrangular), often arcuate and rather graceful, the rachis narrow, often quadrangular. Disc of lip bearing two keels united at base to form a U-shaped structure 2

2. Mature leaf-blades (0.7–)0.9–2.6 (rarely to 3.5) cm wide 3

2. Mature leaf-blades 0.1–0.6 cm wide (usually between 0.1 and 0.4 cm wide) 5

3. Leaf-blades oblong-elliptic, obtuse, 1.8–2.6 cm wide, with finely reticulate venation .... ***D. exasperatum*** *Ames*

3. Leaf-blades linear-lanceolate to elliptic, acute to acuminate, 0.7–2.6(–3.5) cm wide, with or without finely reticulate venation 4

4. Inflorescence densely many-flowered, each flower borne 2–3 mm apart on rachis. Flowers small, 4–5 mm across, transparent yellow, the disc with a brown, yellow-brown or citron blotch around the distal part of the keels, often extending to around base of the lobules, or reduced to two separate blotches, one at base of each mid-lobe side lobule. Pseudobulbs closely caespitose, without small transverse wrinkles. Leaves without finely reticulate venation, containing sparsely or densely distributed crystalline calcium oxalate bodies (clearly visible in herbarium material), generally linear-lanceolate or oblong-lanceolate, 9–25 × 0.7–1.8 cm. Lip appearing 5-lobed, the side lobes reduced, the mid-lobe subulate, the side lobules variable in length and shape, usually oblong-falcate, obtuse, retrorse, ascending. Stelidia equalling or longer than the column apex .... ***D. gibbsiae*** *Rolfe*

4. Inflorescence laxly few- to several-flowered, each flower borne up to 5 mm apart on rachis. Flowers relatively large, 1 cm across, flesh-pink to pinkish-brown, the lip with darker keels, without a darker central blotch. Pseudobulbs less closely caespitose, often borne 1–2 cm apart on rhizome, with small transverse wrinkles (especially apparent on herbarium material). Leaves with fine reticulate venation, generally shorter and broader, often around 2 cm wide (occasionally up to 3.5 cm wide), calcium oxalate bodies absent. Lip pandurate, tridentate, the side lobes absent, the mid-lobe with 3 acute lobules. Stelidia slightly shorter than or equalling the column apex .... ***D. grandiflorum*** *(Ridl.) J.J. Sm.*

5. Side lobes of lip distinct, obtuse or acute to acuminate, often falcate 6

5. Side lobes of lip absent, or obscure and rounded, never falcate 7

6. Lip 1.5–1.6 mm long, the mid-lobe transversely oblong to narrowly hastate, the side lobes small, keel-like, obtuse to subacute, somewhat falcate. Sepals and petals 2–3 mm long .... ***D. dolichobrachium*** *(Schltr.) Schltr.*

6. Lip 3.8–4 mm long, the mid-lobe broadly hastate, the side lobes well developed, triangular, acute. Sepals and petals 4–4.1 mm long .... ***D. hastilobum*** *J.J. Wood*

7. Lip cruciform. Stelidia shorter than column apex or, if longer, then the lip apex longly subulate-acuminate. Pedicel with ovary usually 0.5–1(–2) mm long. Disc usually with a dark purple-brown central blotch, apex of keels dark purple-brown .... ***D. cruciforme*** *J.J. Wood*

7. Lip pandurate, or expanded above the middle into large lanceolate or oblong, falcate, acute or obtuse divaricate lobules. Pedicel with ovary 3 mm long. Disc and keels without purple-brown pigmentation 8

8. Lip pandurate, not expanded above the middle into divaricate lobules. Stelidia 2–2.1 mm long, ligulate, decurved at the apex, hamate. Flowers pale green .... ***D. devogelii*** *J.J. Wood*

8. Lip expanded above the middle into large lanceolate or oblong, falcate, acute or obtuse divaricate lobules, or rarely lobules almost straight to subfalcate, brown or deep yellow. Sepals and petals obtuse or subacute, strongly reflexed. Stelidia equalling or slightly longer than column apex, not hamate .... ***D. haslamii*** *Ames*

Artificial key to species of subgenus *Platyclinis* section *Platyclinis*

1. Leaves, cataphylls, inflorescences and sometimes sepals covered with a black or brownish, finely setose indumentum. Lip never 3-lobed, margin lacerate, serrulate to obscurely erose 50

1. Leaves, cataphylls and inflorescences lacking a black or brownish, finely setose indumentum (vegetative parts and sometimes the sepals, petals and ovary provided with very minute, scattered trichomes). Lip entire to distinctly 3-lobed 2

2. Lip entire, or obscurely 3-lobed (margins sometimes erose or toothed), side lobes (when present) less than 2 mm long (rarely up to 2 mm long in forms of *D. dewindtianum* var. *dewindtianum*) 3

2. Lip distinctly 3-lobed, 2 mm or more long, but usually longer (sometimes only 1.5 mm long in *D. crassilabium*) 28

3. Lip with a papillose-hairy disc and ridges. Column with a bilobed flange at base of stigmatic cavity .... ***D. papillilabium*** *J.J. Wood*

3. Lip glabrous or minutely papillose, never papillose-hairy. Column lacking a bilobed flange on the stigma 4

4. Lip entire, the side lobes absent, margins not erose or toothed 5

4. Lip obscurely 3-lobed, the side lobes often rudimentary; or side lobes absent, but lower margins of lip erose or toothed 11

5. Pseudobulbs borne 1–4.5 cm apart on rhizome. Rhizome up to 40 cm long. Plants often terrestrial. Flowers white, sometimes flushed pale yellowish-green, or yellowish-cream, fragrant .... ***D. simplex*** *J.J. Sm.* (see also couplet 16)

5. Pseudobulbs crowded on rhizome, at most 0.8 cm apart, usually much less. Rhizome shortly creeping, usually less than 3 cm long, rarely up to 15 cm long. Plants usually epiphytic. Flowers variously coloured, unscented or fragrant 6

6. Peduncle strongly flattened. Flowers bright yellow-green or citron-yellow, with a dark brown or maroon-purple lip and pink stelidia. Sepals and petals with revolute margins. Lip strongly coiled-up distally .... ***D. planiscapum*** *Carr*

6. Peduncle terete. Flowers coloured otherwise. Sepals and petals with flat margins. Lip not strongly coiled-up distally 7

7. Leaf-blades 10.5–21.5 cm long, linear-ligulate. Sepals and petals 6–7.1 mm long, narrowly lanceolate, acuminate. Lip surface minutely papillose. Stelidia longer than the column apex .... ***D. tenuitepalum*** *J.J. Wood*

7. Leaf-blades 3–8 cm long, narrowly linear to elliptic. Sepals usually shorter (if as long as *D. tenuitepalum*, then sepals acute, petals spathulate and lip obtuse). Lip surface glabrous. Stelidia equal to or shorter than the column apex 8

8. Flowers yellow, flushed plum-purple. Stelidia basal, shorter than the column, not reaching the stigma, obtusely truncate. Rhizome up to 15 cm or more long. Pseudobulbs 6–8 mm apart .... ***D. suratii*** *J.J. Wood*

8. Flowers cream, white or green. Stelidia subapical, either borne opposite the stigma, between it and the apical hood, or basal and almost equal to the column apex. Rhizome abbreviated, rarely up to c. 6 cm long. Pseudobulbs crowded, at most 3–4 mm apart 9

9. Leaves narrowly linear, 2–3(–4) mm wide. Stelidia basal, almost equal to the column apex, obtuse to acute. Sepals 2–2.3 mm long. Lip 1.3 mm long. Rhizome up to 6 cm long. Pseudobulbs 3–4 mm apart .... ***D. integrilabium*** *Carr*

9. Leaves elliptic to ligulate, (5–)7–12(–15) mm wide. Stelidia subapical to median, almost equal to the column apex, acuminate. Sepals 5–8 mm long. Lip around 3 mm long. Rhizome abbreviated. Pseudobulbs crowded 10

10. Flowers white with a brown lip and column, sweetly scented. Rachis 3–4.5 cm long. Petals 2.5–3 mm long. Lip oblanceolate, narrowest at base, broadest towards apex .... ***D. globigerum*** *(Ridl.) J.J. Sm.*

10. Flowers green, unscented. Rachis 8–18 cm long. Petals 6–6.8 mm long. Lip oblong-pandurate, constricted at middle, not broadest towards the apex .... ***D. lumakuense*** *J.J. Wood*

11. Plants loose in habit, the rhizomes elongated, pseudobulbs spaced 1–6.5 cm apart, sometimes contiguous 12

11. Plants caespitose, forming tight clumps, the rhizomes abbreviated, pseudobulbs usually crowded together, never more than 1 cm apart 17

12. Pseudobulbs (1.5–) 5–16 (–22) cm long 13

12. Pseudobulbs 0.8–1.5 cm long 16

13. Lip lanceolate, acuminate, simple, margins minutely erose, fimbriate to denticulate, not strongly decurved above the base or about the middle, weakly attached to the column-foot, versatile 14

13. Lip spathulate-elliptic with very small side lobes, or subentire, broadly ovate-triangular, with obscure, erect, rounded side lobes, strongly decurved above the base or about the middle, less mobile or immobile 15

14. Leaves 0.6–1 (–1.5) cm wide, linear-ligulate or narrowly elliptic. Rhizome cataphylls without dark green to brown apical margins. Lip simple, widest across the middle .... ***D. lancilabium*** *Ames*

14. Leaves much wider, usually 2.5–4.5 cm wide, broadly elliptic. Rhizome cataphylls with dark green to brown apical margins. Lip subentire, with obscure side lobes, widest towards base .... ***D. subintegrum*** *Ames*

15. Stelidia arising from just below the middle of the column, almost equalling apical hood, subacute or obtuse, never dilated, obliquely obtuse and spathulate-truncate at apex. Lip mobile, spathulate-elliptic, with irregularly serrulate decurved margins, the side lobes very small, triangular and tooth like, the disc with a fleshy M-shaped basal callus .... ***D. gracilipes*** *Carr*

15. Stelidia basal, longer than apical hood, apex dilated, obliquely obtuse, spathulate-truncate. Lip immobile, subentire, mid-lobe broadly ovate-triangular, margins undulate and not decurved, side lobes obscure, erect, rounded, the disc with 2 prominent fleshy keels adnate above the base to the margin of the side lobes .... ***D. longipes*** *J.J. Sm.*

16. Lip oblong-ligulate to subpandurate, narrowest at middle, the proximal margins irregularly toothed, 3–3.5 mm long. Stelidia subulate, hamate, decurved. Rhizome much less than 40 cm long. Leaves linear-ligulate, 0.3–0.4 cm wide .... ***D. hamatum*** *Schltr.*

16. Lip semiorbicular-ovate, with obscure auriculate side lobes, versatile. Stelidia porrect, neither hamate nor decurved. Rhizome often up to 40 cm long. Leaves narrowly elliptic to ovate-elliptic, (0.5–)0.8–1.7 cm wide .... ***D. simplex*** *J.J. Sm.*

17. Leaves narrowly linear to linear, grass-like, the blade (6–)10–18 × 0.2–0.5 cm 18

17. Leaves broader; if linear, then shorter or wider 20

18. Pseudobulbs borne 5–7 mm apart on rhizome. Leaves rigid and coriaceous, generally 3–4 mm wide. Rachis 12–16 cm long. Flowers cream. Stelidia lanceolate, slightly exceeding column apex .... ***D. mucronatum*** *J.J. Sm.*

18 Pseudobulbs closely aggregated on rhizome. Leaves, inflorescence and flowers lacking this combination of characters 19

19. Sepals and petals 4–4.3 mm long. Lip broadest across mid-lobe, 3 mm long, the side lobes minutely erose, the mid-lobe elliptic. Stelidia arising just below stigma at the middle of the column, subulate, reaching or almost reaching rostellum. Flowers lemon-yellow .... ***D. graminoides*** *Carr*

19. Sepals and petals 2–2.5 mm long. Lip broadest at base, 1.5 mm long, the side lobes obscure and rounded, the mid-lobe triangular, acute to acuminate. Stelidia basal, linear, obtuse, slightly shorter than the apical hood. Flowers greenish-yellow .... ***D. sublobatum*** *Carr*

20. Leaves linear-subspathulate, distinctly broadest towards apex, obtuse and retuse. Lip simple, central margins slightly irregularly toothed .... ***D. johannis-winkleri*** *J.J. Sm.*

20. Leaves otherwise, either of equal width or broadest at middle. Lip simple or obscurely lobed 21

21. Plants very small, 5 cm or less high 22

21. Plants larger, 8–45 cm high 23

22. Leaves linear (margins often revolute in dried material), *c.* 2 mm wide. Flowers pale salmon-pink with a yellow lip. Sepals and petals minutely papillose at base. Lip *c.* 4.3 × 3.7 mm. Stelidia cuneate, rounded and subtruncate .... ***D. dulitense*** *Carr*

22. Leaves narrowly elliptic, with very minute black, spot-like trichomes, particularly on abaxial surface, 3–4 mm wide. Flowers pale green, lip brownish-ochre. Sepals and petals glabrous. Lip 2.8 × 1.2–1.3 mm. Stelidia ligulate, acute .... ***D. ochrolabium*** *J.J. Wood*

23. Leaves linear to linear-ligulate, 3–6 mm wide 24

23. Leaves narrowly elliptic to oblong-elliptic, 0.8–3.5 cm wide 25

24. Lip cuneate-obovate, widest at apex, with tiny subulate side lobes, the proximal margins and apex minutely erose. Flowers green or greenish-yellow with brown patch on lip .... ***D. dewindtianum*** *W.W. Sm.* var. ***sarawakense*** *Carr*

24. Lip elliptic, acute, simple, erose towards base. Flowers greenish-yellow or bright yellow, the lip yellow, brown centrally, the callus red .... ***D. galbanum*** *J.J. Wood*

25. Leaf-blade 8–30 × 2–3.8 cm, obtuse or acute. Flowers cream or white, rarely greenish-white. Floral bracts conspicuous, glumaceous, spreading, (5–)7–12 mm long. Lip obscurely lobed, not sharply deflexed above base .... ***D. havilandii*** *Pfitzer*

25. Not this combination of characters 26

26. Leaves thick and fleshy, narrowly linear-elliptic, acute, sulcate above, curved, the blade 2–7 × 0.3–0.6 cm. Flowers salmon-pink to pale brownish, the lip white or dull yellow, red and rather fleshy at centre .... ***D. pachyphyllum*** *J.J. Wood & A. Lamb*

26. Leaves thin-textured, often coriaceous, never thick and fleshy, variously shaped. Flowers variously coloured 27

27. Stelidia slightly shorter than apical hood. Sepals and petals lemon-yellow or greenish-yellow, rarely whitish, centre of the lip often suffused bright green, the keels usually bright green. Floral bracts 3–4 mm long ....
.... ***D. dewindtianum*** *W.W. Sm.* var. ***dewindtianum***

27. Stelidia a little longer than apical hood. Sepals, petals and lip pale green. Floral bracts 5–7 mm long .... ***D. geesinkii*** *J.J. Wood*

28. Flowers white, lip with a yellowish-orange or brownish blotch, non-resupinate. Rachis stiffly erect to porrect. Leaves 0.4–1.1 cm wide. Side lobes of lip entire, never erose or toothed .... ***D. muluense*** *J.J. Wood*

28. Not this combination of characters 29

29. Flowering plants small, 15(–20) cm or less high (usually much less). Leaves narrowly linear-ligulate to linear-lanceolate 30

29. Flowering plants much larger. Leaves lanceolate, oblong-elliptic to elliptic 33

30. Flowers brownish-salmon with brownish olive-green column. Lip mid-lobe thick and fleshy, oblong-spathulate and shallowly retuse. Sepals obtuse ....
.... ***D. crassilabium*** *J.J. Wood*

30. Flowers greenish, greenish-cream, pale yellow to pale ochre, or salmon-pink to buff. Lip mid-lobe thin, not fleshy, broadly elliptic or broadly ovate from a cuneate base, obtuse, truncate or acute. Sepals acute to acuminate 31

31. Stelidia borne at each side of or just below stigmatic cavity, lanceolate, falcate, usually rather obtuse. Mid-lobe of lip acute .... ***D. tenompokense*** *Carr*

31. Stelidia borne at or just above base of column, subulate, acicular. Mid-lobe of lip obtuse to truncate 32

32. Flowers salmon-pink to buff, or cream with an orange to pale ochre central area on sepals, petals and lip. Petals elliptic, obtuse or mucronate. Mid-lobe of lip broadly elliptic, obtuse .... ***D. magaense*** *J.J. Wood*

32. Flowers yellow, yellowish-green or pale green, lip with 2 pale brown longitudinal central streaks. Petals acuminate. Mid-lobe of lip broadly ovate to obscurely 6-angled, truncate .... ***D. subulibrachium*** *J.J. Sm.*

33. Floral bracts large, ovate-elliptic, cymbiform, 8–13 mm long, almost concealing the flowers. Flowers translucent greenish-yellow with brown areas on lip, the column cream, orange at base. Stelidia subulate, borne just below stigmatic cavity, barely reaching the rostellum .... ***D. imbricatum*** *Ames*

33. Floral bracts much smaller, never almost concealing the flowers. Flowers lacking this combination of characters 34

34. Column distinctly papillose, stelidia arising from just below the stigmatic cavity, short, falcate, papillose. Sepals and petals pale green, tipped pink to dull orange-yellow, the lip and the column orange to salmon-pink or reddish. Lip *Coelogyne*-like, posterior margin of side lobes erect, roundly dilate below apex, recurved towards apex and with a short transverse fold; mid-lobe abruptly and strongly recurved, ovate, subacute, fleshy, erose, provided with 3 inconspicuous broad rounded keels; disc with a V-shaped keel between side lobes, with a longer, thinner keel either side. Leaves oblong-elliptic, margins slightly recurved, 9–15 × 3.3–4 cm .... ***D. anomalum*** *Carr*

34. Plants lacking this combination of characters 35

35. Stelidia toothed or bifid. Apical hood of column ovate, dilated or prominently cucullate, often semi-rotund, entire or minutely denticulate 36

35. Stelidia entire. Apical hood of column variously shaped 37

36. Rhizome pendulous, slender, up to 18 cm long, bearing distinctly spotted cataphylls. Pseudobulbs narrowly cylindrical or narrowly fusiform, 2–3 mm wide. Lateral sepals reflexed. Side lobes of lip triangular, acute; mid-lobe obovate. Stelidia arising just above the base of the column, ligulate, the apex bifid. Column 3.5 mm long. Disc of lip with 3 fleshy papillose keels. Rostellum small, ovate, *c.* 0.2 mm wide ....
.... ***D. imitator*** *J.J. Wood*

36. Rhizome horizontal, robust, usually rather short, cataphylls not distinctly spotted. Pseudobulbs ampulliform or oblong-avoid, 0.5–1 cm wide. Lateral sepals spreading. Side lobes of lip auriculate, obtuse to acute; mid-lobe oblong to oblong-ovate, obtuse

to shortly acuminate-cuspidate. Stelidia basal, often decurrent up to the middle of the column, small, wing-like and variously toothed, usually bifurcate, or reduced to 2 narrowly triangular teeth, 1 large, 1 small. Column 2–2.5 mm long, sometimes dorsally carinate. Disc of lip with short and rounded, hard and fleshy keels, curved near the base, their forward ends depressed and converging, often touching. Rostellum large, quadrate, truncate, up to 1.6 mm wide .... ***D. kingii*** *(Hook.f.) J.J. Sm.*

37. Leaves ovate to oblong-oblanceolate, obtuse, 3.5–10 × 0.9–3 cm, petiole 2–5 mm long. Inflorescence from apex of nearly mature pseudobulb; peduncle 3–5.5 cm long, filiform, with several adpressed bracts at junction with rachis; rachis 20–35 cm long, pendulous, with up to 80 or more flowers. Flowers yellowish-green; lip pale greenish-white, the mid-lobe yellowish-green with 2 converging brown streaks. Lip stipitate to column-foot by a narrow claw, mid-lobe oblong-oblanceolate or oblong above a cuneate base, shortly apiculate, *c.* 2.3 × 0.2–0.5 mm, disc with 2 papillose keels .... ***D. angustilobum*** *Carr*

37. Not this combination of characters 38

38. Side lobes of lip irregularly laciniate, mid-lobe strongly recurved, ovate-elliptic, rounded or subacute, minutely papillose. Disc of lip with 2 flange-like basal keels, each *c.* 0.4 mm wide, which curve toward the middle and meet. Stelidia borne either side of stigmatic cavity, shorter than apical hood. Sepals and petals creamy-white; lip sometimes very pale green, with yellowish-cinnamon or ochre keels .... ***D. lacinilobum*** *J.J. Wood & A. Lamb*

38. Not this combination of characters 39

39. Lip entirely dark chocolate-brown or reddish-brown 40

39. Lip never entirely brown, but often with brown markings 41

40. Side lobes of lip distinctly erose-fimbriate, rounded. Disc of lip with 2 fleshy keels, connected at base by a transverse thickening with a prominent median vein between. Petals minutely denticulate or erose. Sepals and petals yellow-green or lemon-yellow .... ***D. kamborangense*** *Ames*

40. Side lobes of lip entire or a little uneven, never erose-fimbriate, narrowly triangular-subulate, acute. Disc of lip with 2 main fleshy keels, not united at base, and 2 shorter keels in between higher up. Petals entire. Sepals and petals pale green suffused brown, cinnamon to orange-brown or reddish-brown .... ***D. oxylobum*** *Schltr.*

41. Leaves with a tough, coriaceous texture, rigid when dry .... ***D. crassifolium*** *Ames*

41. Leaves thin-textured, often chartaceous 42

42. Flowers white, lip orange-yellow, fragrant. Side lobes of lip falcate, acute, erect, mid-lobe ovate-orbicular, recurved. Disc of lip 2-keeled. Column with tridentate apical hood .... ***D. glumaceum*** *Lindl.*

42. Flowers lacking this combination of characters 43

43. Leaves ligulate, blade 12–22 × 0.8-1.2 cm, gradually attenuated into a slender petiole. Petals linear-oblong, acute, *c.* 0.7–0.8 mm wide. Sepals and petals pale yellow or greenish yellow, lip pale yellow or orange with whitish margins; keels and elevated median vein bright salmon-pink; column and anther-cap salmon-pink, the stelidia and apical hood whitish. Stelidia arising from middle of the column just below stigmatic cavity, equalling apical hood .... ***D. angustipetalum*** *Ames*

43. Leaves broader, blade (1–)1.5–6.5 cm wide, not ligulate. Not this combination of characters 44

44. Leaves (22–)25–60 cm long, variable in width, usually between 3.5 and 6.5 cm wide. Pseudobulbs narrowly conical, 3–10 × 2–3 cm wide at base. Disc of lip with 2 small basal keels. Side lobes of lip short, tooth-like, acute, falcate, margins entire; mid-lobe ovate-rhomboid. Sepals and petals yellow or yellowish-green, tipped ochre; lip green with 2 sepia brown streaks or a sepia brown central area .... ***D. longifolium*** *Rchb. f.*

44. Leaves shorter, between 2.3 and 18 cm long. Not this combination of characters 45

45. Sepals and petals semi-translucent cream, suffused pale salmon; lip cream, the base and keels pale salmon; column pale salmon. Plants often pendulous. Mid-lobe of lip obovate or suborbicular, obtuse, often as long as wide. Disc of lip with 2 prominent papillose keels, incurved and touching at base. Petals erose. Stelidia lanceolate, shortly acuminate .... ***D. lacteum*** *Carr*

45. Sepals and petals pale green, greenish-cream to yellow; lip similar, with 2 brown streaks. Plants lacking this combination of characters 46

46. Leaves obtuse, oblong-elliptic. Inflorescence densely many flowered. Mid-lobe of lip broadly elliptic, narrowed at base, shortly acuminate; side lobes entire. Column with an obscurely 5-toothed apical hood .... ***D. tardum*** *J.J. Sm.*

46. Leaves acute, linear-lanceolate to elliptic-lanceolate. Inflorescence rather laxly to subdensely many flowered. Mid-lobe of lip spathulate to suborbicular, rhomboid, or oblong-ovate to ovate-elliptic; margin of side lobes usually uneven to serrate. Column apical hood rounded, entire, or very obscurely lobed, never 5-toothed 47

47. Rachis with a solitary basal non-floriferous bract .... ***D. gracile*** *(Hook. f.) J.J. Sm.*

47. Rachis with a group of 3–7 adpressed, closely imbricate, glumaceous basal non-floriferous bracts 48

48. Sepals and petals papillate on adaxial surface. Stelidia spathulate, rounded to truncate, only reaching to base of stigmatic cavity. Lip 2.5–2.6 mm long ....
.... **D. papillitepalum** *J.J. Wood*

48. Sepals and petals glabrous. Stelidia linear to linear-subulate, acute, slightly exceeding anther-cap. Lip 3.5–7.7 mm long 49

49. Sepals and petals 4–4.5 mm long. Lip c. 3.5 mm long. Pseudobulbs usually 5.5–9 × 0.3–0.7 cm, pencil-like. Leaves 2–4.5 cm wide, petiole 0.4–1 cm long. Side lobes of lip reaching to about the middle of the mid-lobe, narrowly triangular with a setaceous tip; mid-lobe 2 mm wide .... **D. longirachis** *Ames*

49. Sepals and petals 8–9 mm long. Lip 7.5–7.7 mm long. Pseudobulbs 2–5 × 0.5–0.7 cm. Leaves 1–2.7(–3) cm wide, petiole 1.2–2.4 cm long. Side lobes of lip reaching to only just beyond the base of the mid-lobe; mid-lobe 3.4–3.5 mm wide ....
.... **D. lyriforme** *J.J. Sm.*

50. Leaf-blades (6.5–)9–17(–25) × 0.7–2.5(–3.5) cm. Inflorescence lax, 2– to 7(–12)-flowered. Sepals 7–10 × 3–3.5 mm. Petals 6–7.5 × 2.5–3.2 mm. Lip 5–6 × 2.5–4.1 mm. Column apex entire. Stelidia triangular, acute, falcate, equalling or slightly shorter than rostellum .... **D. pubescens** *L.O. Williams*

50. Leaf-blades 3–5.4 × 0.8–1 cm. Inflorescences rather dense, 15- to 17-flowered. Sepals 5.3–5.5 × 1.3–1.7 mm. Petals 4.5 × 1.5 mm. Lip 3.4 × 2.4 mm. Column apex bidentate. Stelidia linear-subulate, the apex filiform, slightly longer than column apex
.... **D. vestitum** *J.J. Sm.*

Please note that, although *D. lyriforme* is recognised at specific level here, study of recently acquired material suggests that the solitary non-floriferous bract used to distinguish *D. gracile sensu stricto* from *D. lyriforme* is not always constant and that the rather distinctive *D. lyriforme* may perhaps be better treated at varietal rank under *D. gracile* using additional floral characters. A fuller study based on a wider range of material of the *D. gracile* complex *sensu lato* is underway. The results will appear in a forthcoming revision of Bornean *Dendrochilum* currently under preparation.

Opposite: *Dendrobium cinnabarinum* Rchb.f. Sabah, Maliau Basin. *Photo: A. Lamb.*

CHAPTER 5

DESCRIPTIONS AND FIGURES

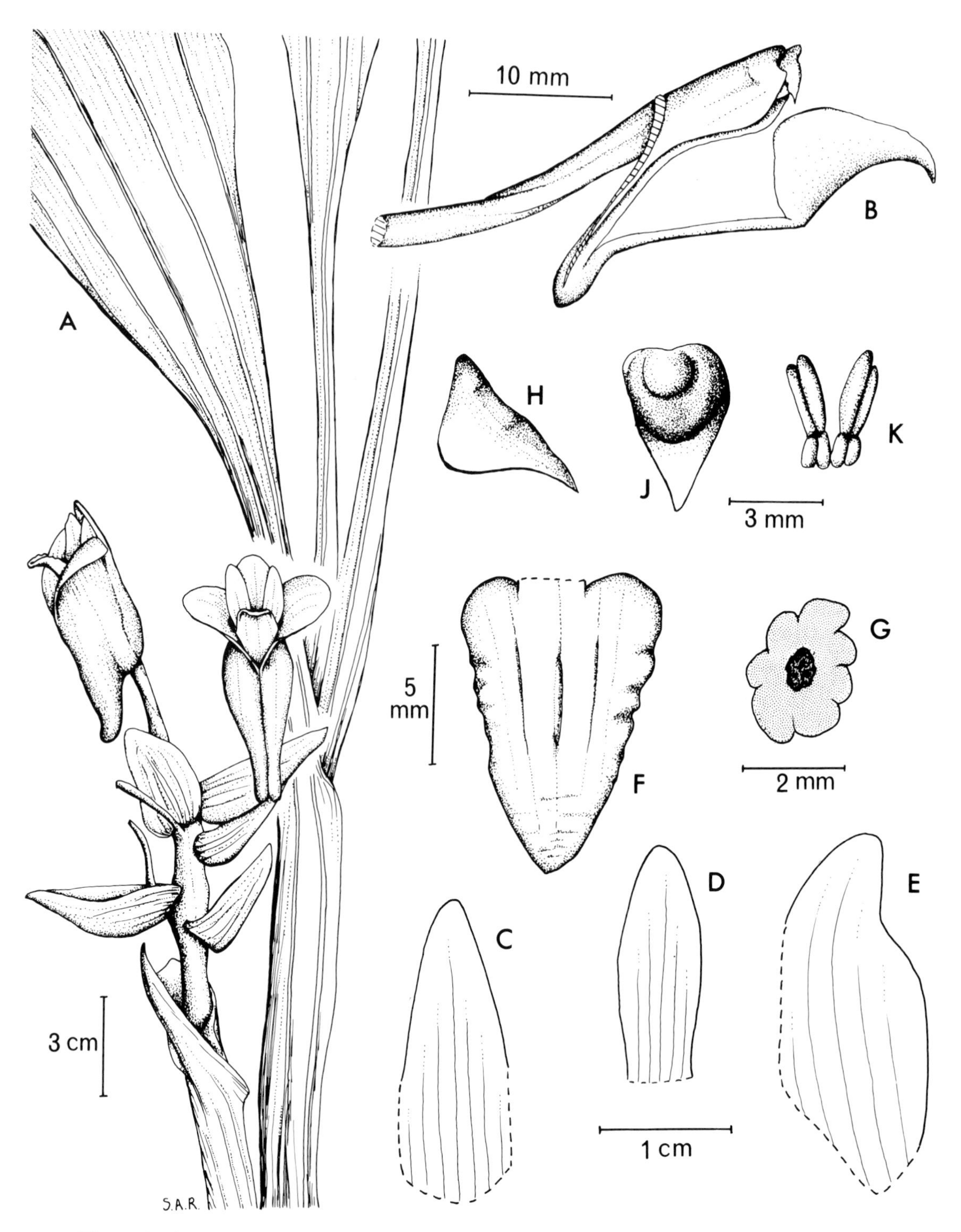

Figure 1. Acanthephippium eburneum Kraenzl. - A: habit. - B: pedicel with ovary, column and lip, side view. - C: dorsal sepal. - D: petal. - E: lateral sepal. - F: lip, front view. - G: ovary, transverse section. - H: anther-cap, side view. - J: anther-cap, back view. - K: pollinia. All drawn from *Ridley* s.n. (type of *A. lycaste* Ridl.) by Sarah Thomas (neé Robbins).

1. ACANTHEPHIPPIUM EBURNEUM Kraenzl.

Acanthephippium eburneum Kraenzl. in Gard. Chron. ser. 3, 20: 266 (1896). Type: Cult. *Wolter*, Magdeburg-Wilhemstadt (holotype B, probably lost).

Acanthephippium lycaste Ridl. in Sarawak Mus. Journ. 1, 2: 35 (1912). Type: Borneo, Sarawak, Kuching, June 1911, *Ridley* s.n. (holotype SING, isotype K).

Erect ***terrestrial*** to about 70 cm high. ***Roots*** few, to 6 cm long, 3–4 mm in diameter. ***Pseudobulb*** 5–15 cm long, 1 cm in diameter, 2- to 3-leaved, ovoid to fusiform, with 4–6 nodes, more or less covered by persistent sheath fibres. ***Leaves*** 30–40 × 12–14 cm, elliptic to broadly elliptic, acute to apiculate, with 5 prominent veins on underside; petiole 8–11 cm long, sheathing, ribbed. ***Inflorescence*** lateral, arising from basal node on new growth, 3- to 4-flowered; peduncle 5–10 cm long, 5 mm in diameter, stout, bearing 2–3 sterile bracts 1–2.5 cm long; rachis 3 cm long; floral bracts 2.1–2.5 × 1.7–2.4 cm, cymbiform, ovate-lanceolate, acute, foliaceous. ***Flowers*** fleshy, urceolate, white with dark crimson-red markings on the inside of the sepals and petals; lip white with yellow and crimson-red markings; column white, yellow at base, with crimson-red markings. ***Pedicel*** with ***ovary*** 3–4 cm long, grooved, sparsely and finely pubescent. ***Dorsal sepal*** 2.4–2.5 × 1–1.1 cm, oblong-lanceolate, rounded, obtuse, adnate to lateral sepals for the basal third, 5- to 7-veined. ***Lateral sepals*** 2.7–3 × 1.3–1.5 cm, obliquely oblong, apex rounded and strongly recurved, 6- to 7-veined. ***Mentum*** slightly shorter than or as long as pedicel with ovary, obtuse, geniculate, spur-like, becoming less geniculate as flower matures. ***Petals*** 1.8 × 0.6 cm, lanceolate, obtuse, fleshy, free, mid-vein adnate to margin of dorsal sepal and lateral sepal in basal half, 3-veined. ***Lip*** 1.2–1.4 × 0.8–0.9 cm, entire or obscurely 3-lobed, fleshy with thickened margins; disc of one low, indistinct median keel extending along the central third of the lip, sometimes with a slight thickening along lateral veins. ***Column*** 1–1.2 cm long, 4–5 mm in diameter, straight; stigmatic cavity *c.* 5 mm long, broadly obovate to cordate; anther-cap 4.6 × 3.4 mm, with a rounded attachment on the adaxial side, front margin drawn out, acuminate; pollinia eight, unequal. Plate 1A.

HABITAT AND ECOLOGY: Unknown.

DISTRIBUTION IN BORNEO: SARAWAK: Kuching area.

GENERAL DISTRIBUTION: Sumatra and Borneo.

DERIVATION OF NAME: The generic name is derived from the Greek *akantha*, thorn, and *ephippion*, a saddle, referring to the lip which has toothed crests and resembles a saddle. The specific epithet is derived from the Latin *eburneus*, ivory white, in reference to the flower colour.

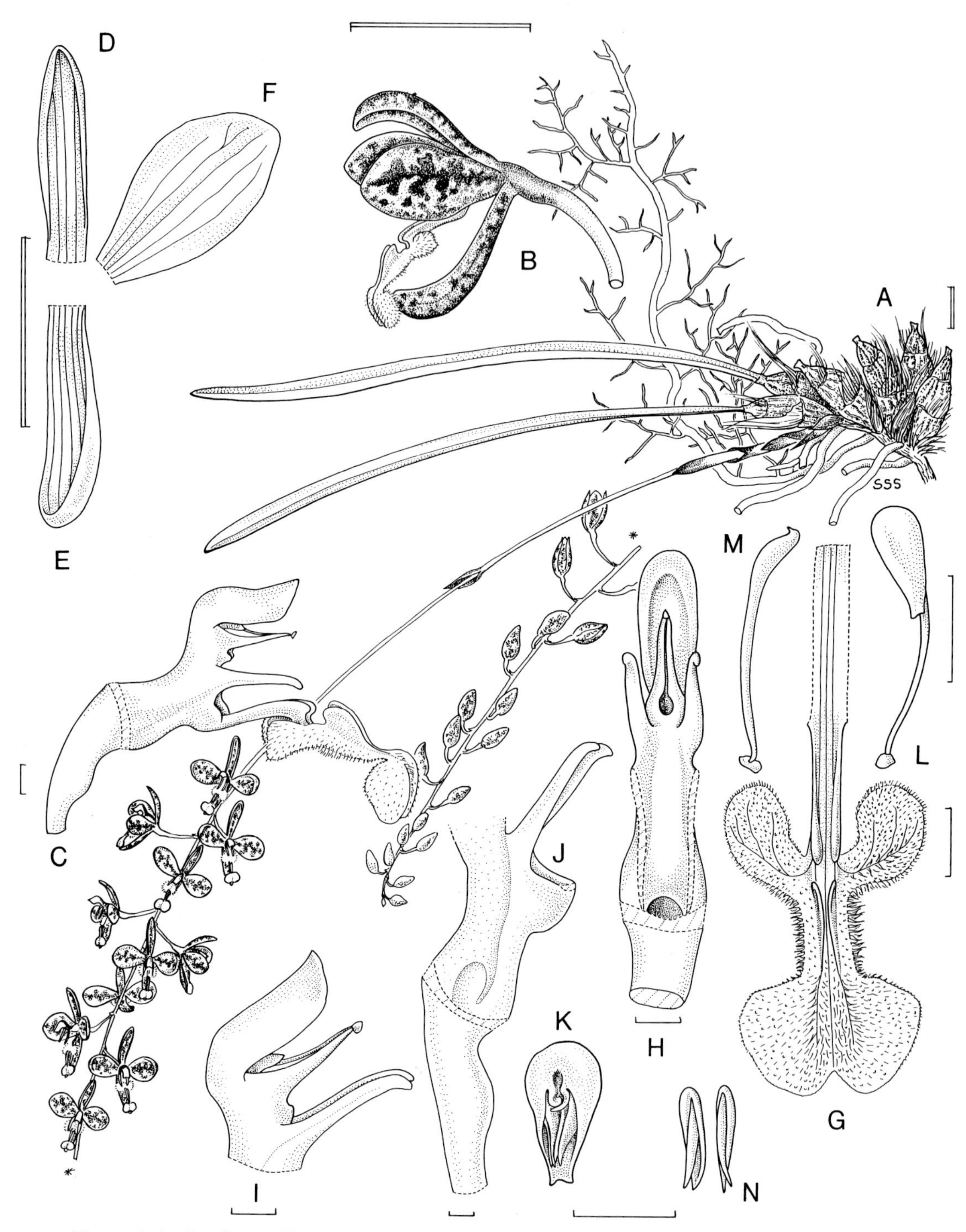

Figure 2. Acriopsis gracilis Minderh. & de Vogel - A: habit. - B: flower, side view. - C: pedicel with ovary, lip and column, side view. - D: dorsal sepal. - E: synsepal. - F: petal. - G: lip, flattened. - H: upper portion of ovary and column, front view. - I: upper portion of column, side view. - J: pedicel with ovary and lower portion of column, side view. - K: anther-cap with pollinia, front view. - L: pollinarium, front view. - M: stipes and viscidium, side view. - N: pollinia. A (habit) drawn from *Lamb* AL 1110/89 and B–N from *Sands et al.* 3772 by Susanna Stuart-Smith. Scale: single bar = 1 mm; double bar = 1 cm.

2. ACRIOPSIS GRACILIS Minderh. & de Vogel

Acriopsis gracilis *Minderh. & de Vogel* in Orchid Monogr. 1: 7, fig. 4, plate 1a (1986). Type: Borneo, Sook, *Lamb* s.n. (holotype L).

Clump-forming ***epiphyte. Main roots*** 1.5–3 mm in diameter, branching, smooth; catch-roots 0.2–1 mm in diameter, much branched, smooth. ***Rhizome*** up to 10 cm long, 3.5–5 mm in diameter, with about 3 internodes 2.5–3 mm long between each pseudobulb; scales 1–4 cm long, grey, often finely speckled blackish, usually becoming fibrous. ***Pseudobulbs*** *c.* 1 cm apart, 2–3 × 1–1.3 cm, with *c.* 4 internodes, wrinkled, olive-green. **Leaves** 2 or 3 per pseudobulb; blade 11–23.5 × 0.5–0.7 cm, linear, obtuse; petiole 3–6 mm long. ***Inflorescence*** borne at base of new pseudobulb, simple, decurved to pendulous, laxly many-flowered, 3.5–4 cm wide; peduncle 10–25 cm long, dark green, with 3 or 4 purple sheaths 5–7 mm long at base and a solitary purple sterile bract 5–6 mm long *c.* halfway up; rachis 10–17 cm long, dark green; floral bracts 1–1.2 × 1 mm, acute, pink or purple. ***Flowers*** borne 2–19 mm apart, *c.* 15 mm in diameter; sepals and petals greenish or translucent yellow, with purple blotches, spots and bars; lip white with pale mauve, purple or pinkish keels; column pale green, dark purple dorsally; stelidia purplish red, with greenish ochre or yellow tips; anther-cap pale yellowish. ***Pedicel*** with ***ovary*** 8–11 mm long. ***Dorsal sepal*** 7.8–10 × 2–2.6 mm, narrowly oblong-elliptic, acute, concave, cymbiform. ***Lateral sepals*** fused to form a synsepalum, 8.3–1 × 2–3.1 mm, oblong-elliptic, obtuse, concave, cymbiform. ***Petals*** 7.9–8 × 3.5–4 mm, obovate, rounded. ***Lip*** 3-lobed, ± panduriform; hypochile with basal margins adnate for 2.5 mm long to column, 2.5 mm long, centrally constricted, apically widened, the basal outgrowth *c.* 0.9 mm up column, *c.* 1 mm long, deflexed, the free part 2.5 × 0.4 mm, narrowly linear, patent, ± semi-orbicular in diameter, with two small apical knobs; epichile 4.6 × 3.8 mm, sigmoid, the side lobes 1.6 × 1.4 mm, obliquely obovate, reflexed, rounded, pilose, the terminal lobe 2 × 0.5 mm, broadly spathulate, narrowed at base, with hairy margin, apical part 2.2 × 3.6 mm, ± semi-orbicular, centrally convex, concave near margin, emarginate, the keels on base of terminal lobe 1.2 × 1 mm, reclined, rounded. ***Column*** 4 mm long, sigmoid; stelidia 2.6 mm long, slightly swollen at apex, decurved; hood 2.6–2.8 mm long, entire, reflexed; rostellum subulate; anther-cap narrowly pyriform; pollinia four.

HABITAT AND ECOLOGY: Open stunted "kerangas" forest on podsolic soils; hill forest. Alt. 300 to 600 m. Flowering observed in February, March, April, November and December.

DISTRIBUTION IN BORNEO: SABAH: Nabawan area; Tenom District.

GENERAL DISTRIBUTION: Endemic to Borneo.

NOTE: The main attachment roots are thick, but bear numerous thin, ascending, extensively branched catch-roots which trap and extract nutrients from leaf litter falling from the canopy above. Similar catch-roots are found in, e.g., *Grammatophyllum* and *Porphyroglottis*, both found in similar nutrient-poor habitats.

DERIVATION OF NAME: The generic name is derived from the Greek *akris*, a locust, and *opsis*, appearance, from the supposed resemblance of the column to a locust.

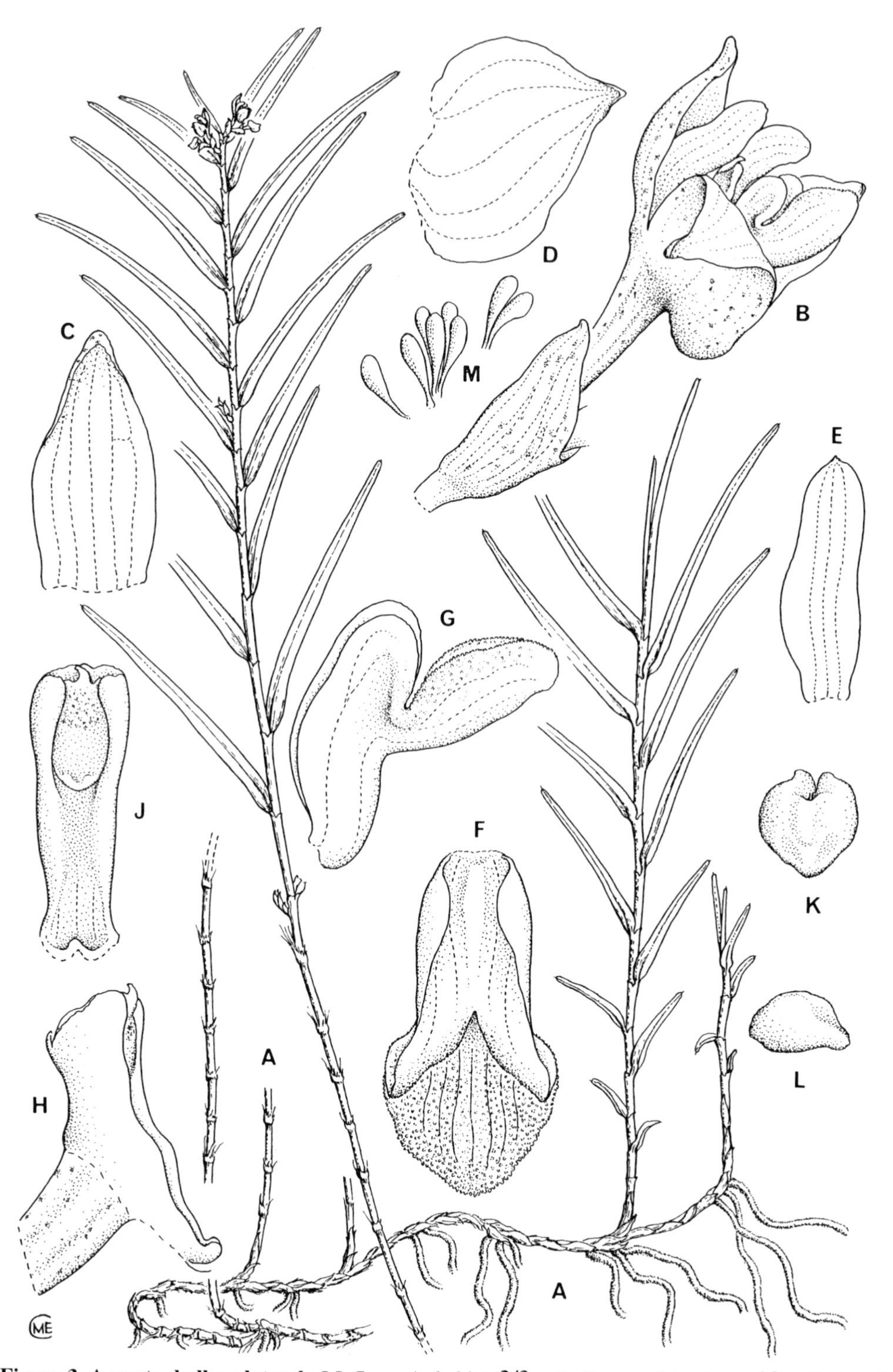

Figure 3. Agrostophyllum laterale J.J. Sm. - A: habit ×2/3. - B: flower, side view ×10. - C: dorsal sepal ×12. - D: lateral sepal ×12. - E: petal ×12. - F: lip, front view ×16.- G: lip, side view ×16. - H: pedicel with ovary and column, side view ×14. - J: column, front view ×14. - K: anther-cap, back view ×14. - L: anther-cap, side view ×14. - M: pollinia ×14. All drawn from *Hansen* 40 (holotype) by Maureen Church.

3. AGROSTOPHYLLUM LATERALE J.J. Sm.

Agrostophyllum laterale *J.J. Sm.* in Bull. Jard. Bot. Buitenzorg, ser. 2, 13: 50(1914). Type: Borneo, Kalimantan Timur, Mt. Labang, *Amdjah* 154 (holotype BO)

Poaephyllum hansenii J.J. Wood in Kew Bull. 39(1): 88, fig. 10 (1984). Type: Borneo, Sarawak, Mt. Mulu National Park, confluence of Sungai Lansat and Sungai Tutuh, *Hansen* 40 (holotype C, isotype K).

Epiphyte. Rhizome slender, covered in greyish-brown imbricate sheaths, internodes 2.5–5 cm long. ***Stems*** 32–38 cm long, slender, wiry, leafy. ***Leaves*** 2–7 × 0.2–0.36 cm, linear-ligulate, equally or unequally obtusely bilobed with a seta in the apical sinus; sheaths 1 cm long. ***Inflorescences*** lateral and terminal, 1- to 3-flowered; peduncle 2–4 mm long, enclosed in 2–3 dark brown, ovate, acute, prominently veined, imbricate bracts 2–2.5 mm long. ***Flowers*** c. 4.3 mm across, whitish or pale yellowish brown, column yellowish, flushed purple. ***Pedicel*** with ***ovary*** 4 mm long, sparsely ramentaceous. ***Sepals*** sparsely ramentaceous. ***Dorsal sepal*** 3–3.5 × 1–1.7 mm, oblong-elliptic, obtuse, cucullate, concave, 4- to 5-veined. ***Lateral sepals*** 2.5–3 × 2–2.5 mm, obliquely ovate-triangular, acute, 5-veined. ***Mentum*** 1 mm long, rounded. ***Petals*** 2.3–3 × 0.7–1 mm, oblique-oblong, obtuse, convex, entire, 3-veined. ***Lip*** erect, deeply 3-lobed, 2.7–3.5 × 1 mm, recurved, fleshy, slightly bigibbous at base; side lobes oblong, rounded, erect; mid-lobe 1.6 mm wide, ovate to suborbicular, obtuse, minutely papillose; callus a fleshy transverse ridge lying across the side lobes. ***Column*** including foot 3 mm long; stelidia obtuse and somewhat truncate; anther-cap minute, ovate, cucullate, retuse; pollinia eight, compressed, clavate, unequal. Plate 1B.

HABITAT AND ECOLOGY: Steep river banks in lowland alluvial forest; riparian forest. Alt. lowlands to 1000 m. Flowering observed in February, July and September.

DISTRIBUTION IN BORNEO: KALIMANTAN TIMUR: Long Ampung area; Mt. Labang. SARAWAK: Mt. Mulu National Park; Sungai Menalio, Bario.

GENERAL DISTRIBUTION: Endemic to Borneo.

DERIVATION OF NAME: The generic name is derived from the Greek *agrostis,* grass, and *phyllon*, leaf, referring to the grass-like leaves of most species. The specific epithet is derived from the Latin *lateralis,* lateral, borne from the side, referring to the origin of the majority of inflorescences in this species.

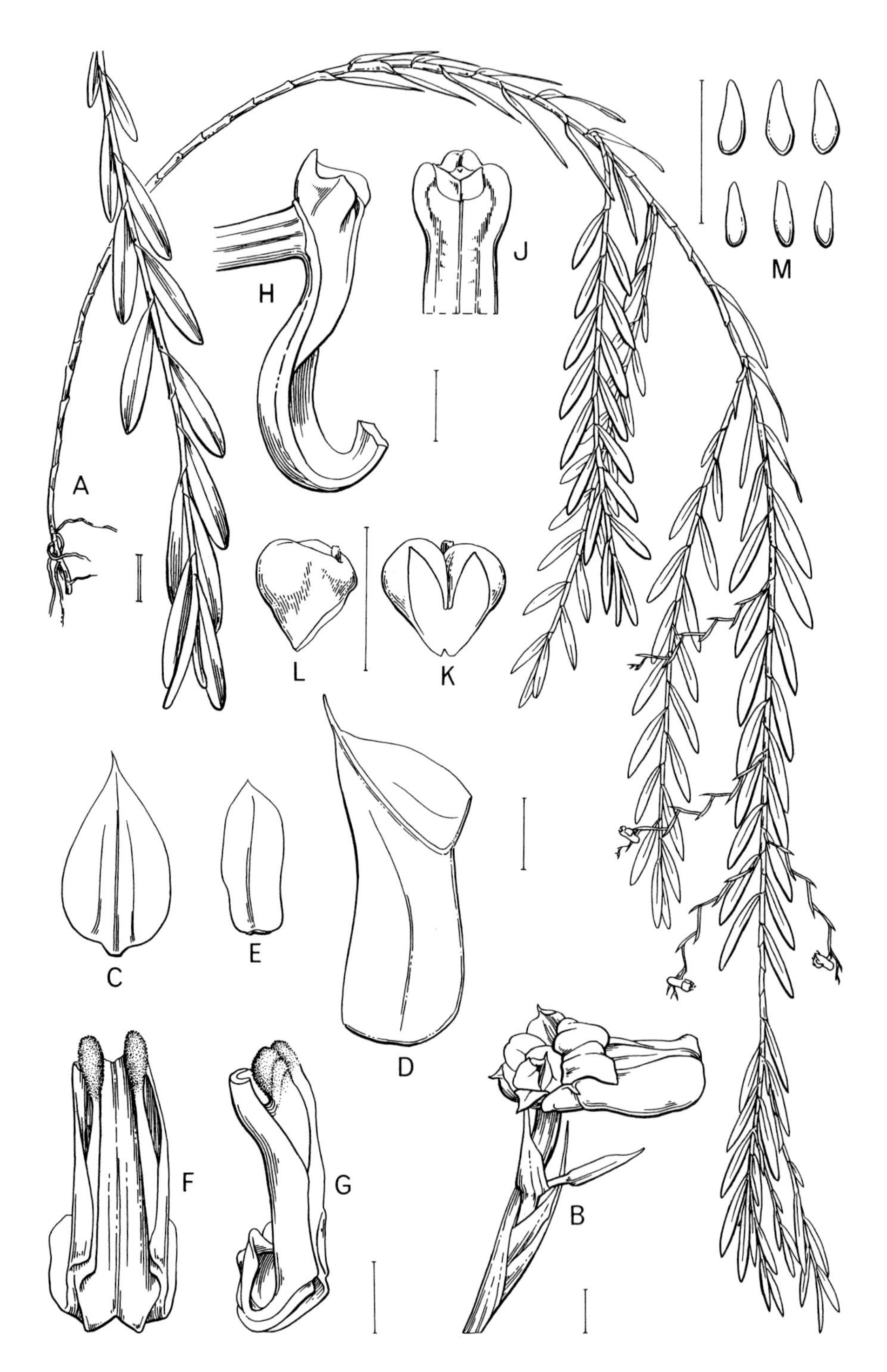

Figure 4. Appendicula fractiflexa J.J. Wood. - A: habit. - B: flower and part of rachis. - C: dorsal sepal. - D: lateral sepal. - E: petal. - F: lip, front view. - G: lip, side view. - H: column and pedicel with ovary, side view. - J: column, front view. - K: anther-cap, front view. - L: anther-cap, side view. - M: pollinia. All drawn from *Wood* 827 (holotype) by Eleanor Catherine. Scale: single bar = 1 mm; double bar = 1 cm.

4. APPENDICULA FRACTIFLEXA J.J. Wood

Appendicula fractiflexa *J.J. Wood* in Wood, Beaman & Beaman, Plants of Mt. Kinabalu, 2, Orchids: 89, fig. 7 (1993). Type: Borneo, Sabah, Mt. Kinabalu, East Mesilau/Menteki Rivers, 31 May 1988, *Wood* 827 (holotype K).

Pendulous ***epiphyte***. ***Stems*** to 70 cm long, much branched, slender, flexuous, branches 10–35 cm long, internodes 5–8 mm long. ***Leaves*** 1–2.1(–2.5) × 0.2–0.4(–0.7) cm, those on lower part of stem usually broader, those on flowering branches narrower, narrowly oblong or narrowly obovate, apex unequally retuse, with a mucro in the sinus, sheaths 5–8 mm long. ***Inflorescences*** borne from middle or upper part of stem, one flower open at a time; peduncle 1.5–3.5 cm long, strongly fractiflex; sterile bracts 4–5.5 mm long, acicular; rachises usually 2, emerging from several imbricate, acicular bracts 3–4 mm long, each rachis 3–4 mm long, fractiflex; floral bracts 1–2 mm long, triangular-ovate, acuminate, concave. ***Flowers*** non-resupinate, pale lilac to purple, with a cream lip. ***Pedicel*** with ***ovary*** 1.8–2 mm long. ***Sepals*** with prominent and slightly raised median vein. ***Dorsal sepal*** 2.5 × 1.5 mm, ovate, slightly aristate, concave. ***Lateral sepals*** 2.5 × 1.5 mm, obliquely triangular-ovate, aristate, fused to column-foot to form a rounded ***mentum*** 3 mm long. ***Petals*** 1.8 × 0.8 mm, oblong, apiculate. ***Lip*** 4.5 × 1.9 mm, entire, oblong, retuse, apex sharply decurved, margin below deflexed apex erect; disc provided with a minutely papillose, concave, bilobed basal appendage which is continued as two smooth fleshy keels which terminate *c.* 1 mm below the apex; a small obscure low subapical keel is also present. ***Column*** 0.8 mm long, foot 3 mm long, fleshy and sulcate above, apex incurved; anther-cap 0.8 × 0.8 mm, ovate, cucullate; pollinia six.

HABITAT AND ECOLOGY: Lower montane forest with *Drimys*, rattans, etc.; mixed hill forest on poor sandy soils with sandstone outcrops; epiphytic on mossy tree boles associated with filmy ferns, aroids, etc. Alt. 1200 to 1700 m. Flowering observed in February and May.

DISTRIBUTION IN BORNEO: SABAH: Mt. Kinabalu; Maliau Basin.

GENERAL DISTRIBUTION: Endemic to Borneo.

DERIVATION OF NAME: The generic name is Latin for a little appendix, describing the appendiculate calli on the lip. The specific epithet is derived from the Latin *fractiflexus*, zigzag, in reference to the inflorescence.

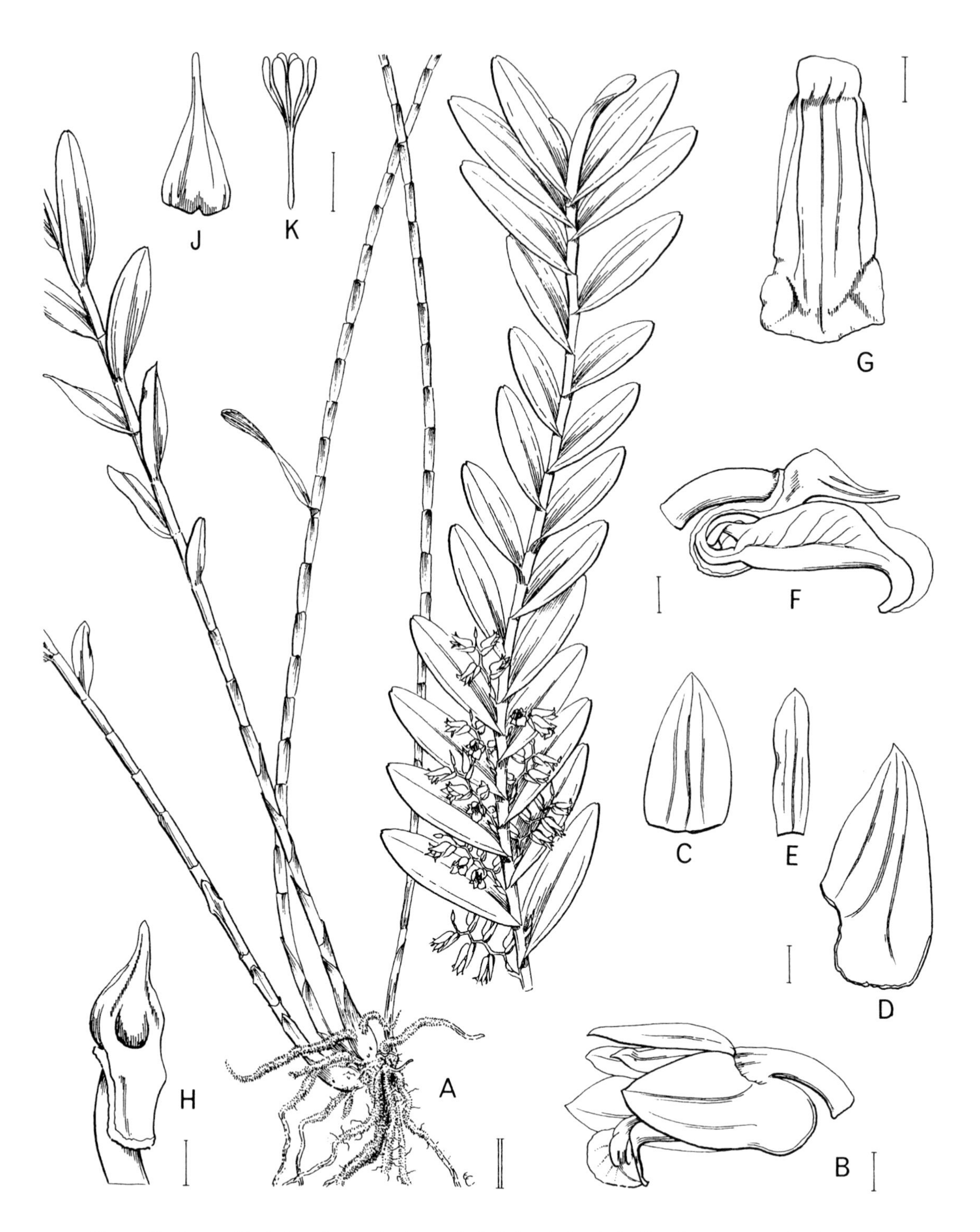

Figure 5. Appendicula longirostrata Ames & C. Schweinf. - A: habit. - B: flower, side view. - C: dorsal sepal. - D: lateral sepal. - E: petal. - F: pedicel with ovary, column and lip, side view. - G: lip, front view. - H: column, front view. - J: anther-cap, back view. - K: pollinia. A (habit) drawn from *Beaman* 10496a and *Wood* 792, and B-K from *Bailes* & *Cribb* 619 by Eleanor Catherine. Scale: single bar = 1 mm; double bar = 1 cm.

5. APPENDICULA LONGIROSTRATA Ames & C. Schweinf.

Appendicula longirostrata *Ames & C. Schweinf.*, Orchidaceae 6: 149 (1921). Type: Borneo, Sabah, Mt. Kinabalu, Marai Parai Spur, 2 December 1915, *J. Clemens* 387 (holotype AMES, isotype K).

Epiphyte or ***terrestrial***. ***Stems*** suberect, to 60 or more cm high, wiry, concealed by leaf sheaths. ***Leaves*** variable in size, 2–5 × 0.8–1.3 cm, elliptic to oblong-elliptic, apex obtusely unequally bilobed, complicate, with a mucro in the sinus, distichous, spreading, sheaths striate-veined, loose, often flushed crimson. ***Inflorescences*** 6–1.5 cm long, 3- to 6-flowered, lateral, single or several in a fascicle, borne at middle of internodes, base concealed by scarious sheaths; rachis zigzag, spreading to ascending or reflexed; floral bracts *c.* 6 mm long, ovate, acuminate, strongly concave, distichous, scarious, nervose. ***Flowers*** creamy white; sepals tipped yellow-green in bud. ***Pedicel*** with ***ovary*** 2–3 mm long. ***Dorsal sepal*** *c.* 4.5 × 2.5 mm, ovate-elliptic, acuminate to complicate, concave, carinate. ***Lateral sepals*** *c.* 4.6 × 3.5–4 mm, obliquely triangular-ovate, long-acuminate to complicate-mucronate at apex, carinate, forming a deeply saccate ***mentum***. ***Petals*** *c.* 4.5 × 1.25 mm, linear, acute or rounded at apex. ***Lip*** folded against the column in natural position, *c.* 6 × 3.5–4 mm, spathulate; hypochile saccate-conduplicate, when flattened suborbicular with the anterior margins continued into centre of epichile; claw short, decurved; epichile transversely oval, somewhat retuse. ***Column*** very short; rostellum conspicuous, longly attenuate, bidentate, 2.5 mm long; foot deeply unciform. Plate 1C & D.

HABITAT AND ECOLOGY: Oak-laurel forest; lower montane forest; secondary forest; rocky roadside banks, with *Arundina*, ferns, etc., sometimes on ultramafic substrate. Alt. 1250 to 2700 m. Flowering observed in May, June, July, August and December.

DISTRIBUTION IN BORNEO: BRUNEI: Mt. Pagon. SABAH: Mt. Kinabalu; Mt. Alab; between Mt. Alab and Mt. Emas; Kimanis road. SARAWAK: Mt. Penrissen.

GENERAL DISTRIBUTION: Endemic to Borneo.

DERIVATION OF NAME: The specific epithet is derived from the Latin *longus*, long, and *rostratus*, beaked, referring to the conspicuous, attenuate rostellum.

Figure 6. Appendicula uncata Ridl. subsp. **sarawakensis** J.J. Wood. - A: habit ×2/3. - B: bud ×8. - C: flower, side view ×8. - D: dorsal sepal ×10. - E: lateral sepal ×10. - F: petal ×10. - G: lip, flattened ×14. - H: lip, side view ×14. - J: column, side view ×14. - K: upper part of column with anther-cap, side view ×20. - L: upper part of column, front view ×14. - M: anther-cap, back view. All drawn from *Nielsen* 387 (holotype) by Maureen Church.

6. APPENDICULA UNCATA Ridl. subsp. SARAWAKENSIS J.J. Wood

Appendicula uncata *Ridl.* subsp. **sarawakensis** *J.J. Wood* in Kew Bull. 39(1): 92, fig. 13 (1984). Type: Borneo, Sarawak, Mt. Mulu National Park, plateau NW of Melinau Gorge, *Nielsen* 387 (holotype AAU).

Epiphyte 10–18(–26) cm high. ***Leaves*** 3–6 × 0.5–1 cm, oblong-lanceolate or narrowly elliptic, minutely obliquely bilobed, with a seta in the apical sinus. ***Inflorescences*** 1–3.5 cm long, mostly terminal, occasionally also lateral, pendent, solitary or in two's or three's; rachis and bracts minutely ramentaceous; floral bracts 2–3 mm long, ovate, acute, reflexed. ***Flowers*** white, with two purple lines on lip. ***Pedicel*** with ***ovary*** 2 mm long. ***Dorsal sepal*** 3 × 1.5 mm, ovate, cucullate, obtuse. ***Lateral sepals*** 3 × 3 mm, obliquely triangular-ovate, acute. ***Mentum*** 2 mm long, obtuse. ***Petals*** 2.8 × 1 mm, oblong-spathulate, obtuse. ***Lip*** 3.5 × 2 mm, entire, oblong-ovate, obtuse, concave below, with U-shaped basal callus, strongly recurved. **Column** *c.* 2 mm long; foot 2 mm long; anther-cap *c.* 1.5 mm long, triangular-ovate, emarginate, acuminate.

HABITAT AND ECOLOGY: In western Sarawak recorded as an epiphyte on root pneumatophores and in Mt. Mulu National Park from sclerophyllous ridge forest and "kerangas" vegetation; mixed hill forest on ultramafic substrate. Alt. sea-level to 700 m. Flowering observed in January, February, March, May and November.

DISTRIBUTION IN BORNEO: BRUNEI: Belait District. SABAH: Mt. Nicola. SARAWAK: Bako National Park; Baram District; Kuching area; Mt. Matang; Mt. Mulu National Park; Simanggang District.

GENERAL DISTRIBUTION: Endemic to Borneo.

NOTE: *A. uncata* subsp. *sarawakensis* differs from subsp. *uncata*, described from Selangor in Peninsular Malaysia, in having shorter stems, smaller leaves and flowers with shorter sepals and petals and a narrower lip.

DERIVATION OF NAME: The specific epithet is derived from the Latin *uncatus*, hooked, bent inwards, probably referring to the incurved column-foot. The subspecific epithet refers to Sarawak from where the majority of collections have originated.

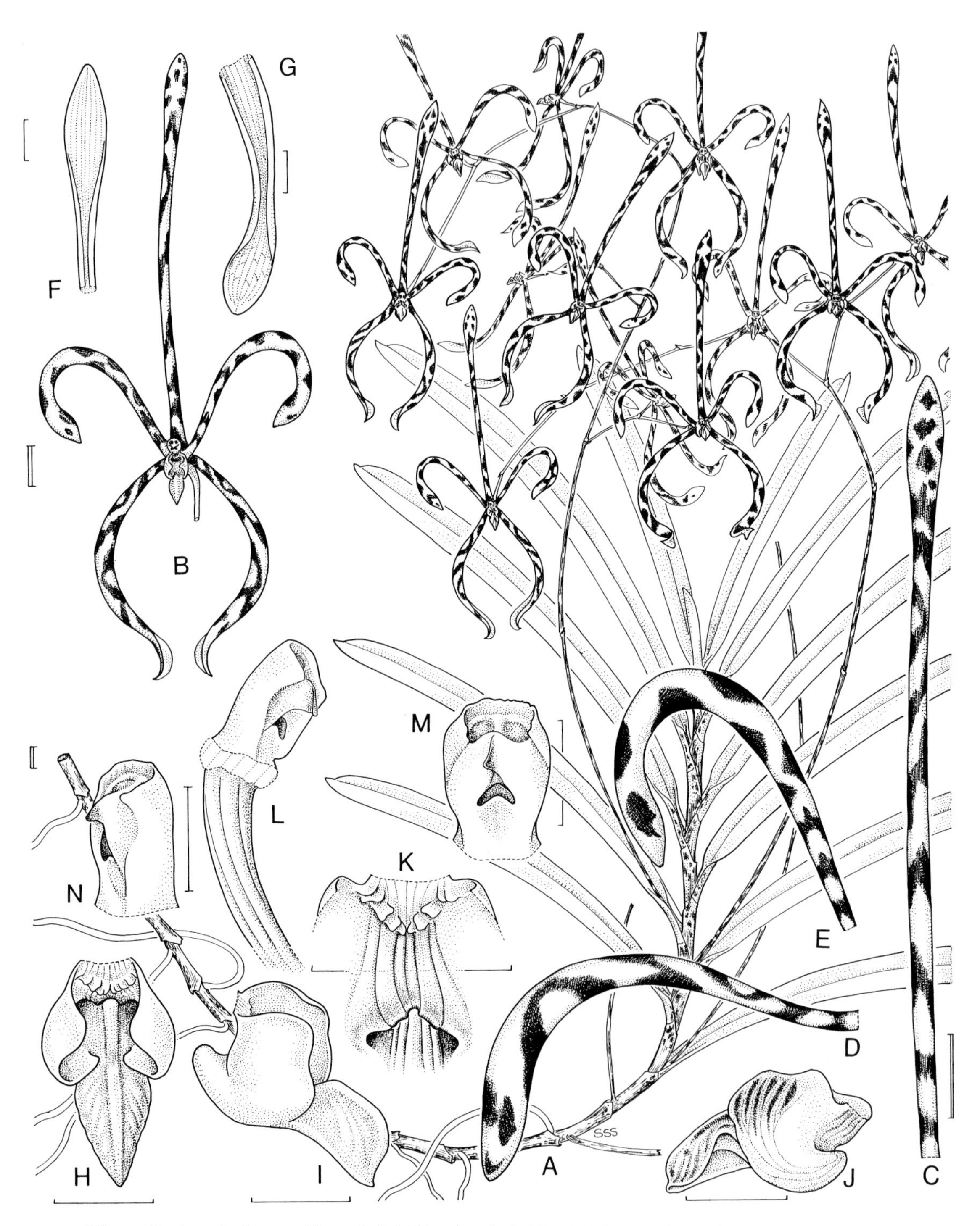

Figure 7. Arachnis grandisepala J.J. Wood. - A: habit. - B: flower, front view. - C: dorsal sepal. - D: lateral sepal. - E: petal. - F: apex of dorsal sepal, reverse. - G: apex of lateral sepal, reverse. - H: lip, front view. - I & J: lip, side views. - K: base of lip. - L: pedicel with ovary and column, side view. - M: column, front view. - N: column, side view. All drawn from *Lamb* 105 in SAN 91558 (holotype) by Susanna Stuart-Smith. Scale: single bar = 5 mm; double bar = 1 cm.

7. ARACHNIS GRANDISEPALA J.J. Wood

Arachnis grandisepala *J.J. Wood* in Orchid Rev. 89 (1050): 113, fig. 95 (1981). Type: Borneo, Sabah, Crocker Range, between Keningau and Kimanis, south of Kota Kinabalu, October 1979, *Lamb* 105 in SAN 91558 (holotype K, isotype SAN).

Semi-pendulous scrambling ***epiphyte*** to 100 cm or so high including inflorescence. ***Stem*** to 50 cm long, simple or branching, rooting and clothed in old grey leaf sheaths below. ***Leaves*** coriaceous; blade 19–23 × 1.5–2 cm, ligulate, unequally bilobed, articulated at a sheathing base 3–4 cm long. ***Inflorescences*** with a few branches, 8- to 10-flowered, lax, porrect; peduncle 25–28 cm long, 3 mm in diameter, reddish-brown, with 2 small pale brown, scarious sheaths at base and 2 remote, ovate, obtuse, scarious sterile bracts 5 mm long above; rachis 17–28 cm long, 1–2 mm in diameter, reddish-brown or green; floral bracts 3–4 mm long, ovate, obtuse, scarious, persistent, yellowish-green. ***Flowers*** large; pedicel with ovary cream; sepals and petals pale yellow or cream with irregular reddish brown blotches and bands; lip pale yellowish cream, its side lobes with brownish red striations, its mid-lobe with brownish orange markings, its disc with orange proximal and red distal keels, the spur cream; column cream with fine purple-brown spots at base. ***Pedicel*** with ***ovary*** 1.5–2 cm long, narrow. ***Dorsal sepal*** 10.2 cm long, 3 mm wide below, 4.5 mm wide at apex, erect, narrowly linear-ligulate, apex dilated, spathulate, obtuse, margins revolute in middle part. ***Lateral sepals*** 5.5 cm long, 5–6 mm wide at broadest point, ligulate, strongly falcate-decurved, obtuse, margins revolute in upper part. ***Petals*** 4.5–5.5 cm long, 4–5 mm wide at broadest point, ligulate, falcate, obtuse, margins slightly revolute along entire length. ***Lip*** mobile, 3-lobed, fleshy, glabrous, 1.1 cm long, 5–6 mm wide across side lobes; side lobes 4–5 mm high, rounded, obtuse, convex, margins incurved and somewhat divergent; mid-lobe 5 mm long, 3.5 mm wide at base, triangular-ovate, obtuse, decurved, with a median keel; disc with four short low ribs and with eight imbricate plate-like outgrowths at base; spur 1.5–2 mm long, conical, obtuse, upcurved. ***Column*** 6–7 × 5 mm, thick, fleshy, obovate-oblong; foot very short; anther-cap lost; pollinia four, in two pairs. Plate 1E & F.

HABITAT AND ECOLOGY: Lower montane forest, recorded as epiphytic 30 metres above ground level. Alt. 900 m. Flowering observed in October.

DISTRIBUTION IN BORNEO: SABAH: Crocker Range.

GENERAL DISTRIBUTION: Endemic to Borneo.

DERIVATION OF NAME: The generic name is derived from the Greek *arachne*, spider, and refers to the spidery appearance of the flowers. The specific epithet describes the grand, showy sepals.

Figure 8. Arachnis longisepala (J.J. Wood) Shim & A. Lamb. - A: habit. - B: flower. - C: dorsal sepal. - D: lateral sepal. - E: petal. - F: apex of dorsal sepal, reverse. - G: pedicel with ovary, lip and column, side view. - H: lip, front view. - I: close-up view of area surrounding spur entrance. - J: column, front view. - K: anther-cap with pollinarium, front and back views. - L: pollinarium. A (habit) drawn from *Cockburn* in SAN 84949 (holotype) and B–L from *Bailes & Cribb* 654 by Susanna Stuart-Smith. Scale: single bar = 5 mm; double bar = 1 cm.

8. ARACHNIS LONGISEPALA (J.J. Wood) Shim & A. Lamb

Arachnis longisepala (*J.J. Wood*) *Shim & A. Lamb* in Orchid Digest 46: 178 (1982). Type: Borneo, Sabah, Lahad Datu District, Mt. Tribulation, near Segama, *Cockburn* in SAN 84949 (holotype K, isotypes L, SAN).

Arachnis calcarata Holttum subsp. *longisepala* J.J. Wood in Orchid Rev. 89(1050): 113, fig. 96 (1981).

Semi-pendulous ***epiphyte. Stem*** branching, covered in disintegrating grey leaf sheaths, total length not recorded, 6 mm in diameter above. ***Leaves*** 21–28 × 1.5–1.7 cm, ligulate, unequally obtusely bilobed, coriaceous, articulated to a sheathing base to 3 cm long. ***Inflorescences*** with a few branches, few-flowered, lax; peduncle *c.* 28 cm long, 3 mm in diameter; sterile bracts 5 mm long, usually 2, remote; rachis *c.* 24 cm long, 2 mm in diameter; floral bracts 5 mm long, ovate, obtuse. ***Flowers*** unscented; dorsal sepal uniformly magenta to purple; lateral sepals and petals uniformly deep orange; mid-lobe of lip lemon-yellow; anther-cap yellow spotted magenta. ***Pedicel*** with ***ovary*** 1.5 cm long, narrow. ***Dorsal sepal*** 6.5–7 cm long, 1 mm wide below, 5 mm wide above, erect, narrowly linear-ligulate, obtuse. ***Lateral sepals*** 3–3.4 cm long, 2 mm wide below, 5 mm wide above, ligulate, falcate, obtuse. ***Petals*** 3.3–3.4 cm wide, 2 mm wide below, 4 mm wide above, ligulate, subfalcate distally, obtuse. ***Lip*** mobile, 3-lobed, fleshy, 1.7 cm long, 6 mm wide across side lobes; side lobes 7 mm long, 5 mm high, rounded, margins incurved proximally, somewhat divergent; mid-lobe 1 cm long, 2.5 cm wide at base, subulate, acute, carinate above and below, sharply inflexed at middle, glabrous; disc at mouth of spur with a cushion of papillose hairs at base above which is a minutely papillose, fleshy area giving rise to three keels, the median extending to apex of mid-lobe, the two lateral much shorter; spur 2.3 mm long, saccate, rounded. ***Column*** 8 × 5.5 mm, straight, slightly winged; anther-cap 4.5 × 4 mm, ovate, retuse. Plate 2A.

HABITAT AND ECOLOGY: Hill forest on ultramafic substrate; epiphytic in shade low down on trees, often not far above the ground. Alt. 600 to 800 m. Flowering observed in August.

DISTRIBUTION IN BORNEO: SABAH: Mt. Tribulation.

GENERAL DISTRIBUTION: Endemic to Borneo.

DERIVATION OF NAME: The specific epithet describes long sepals.

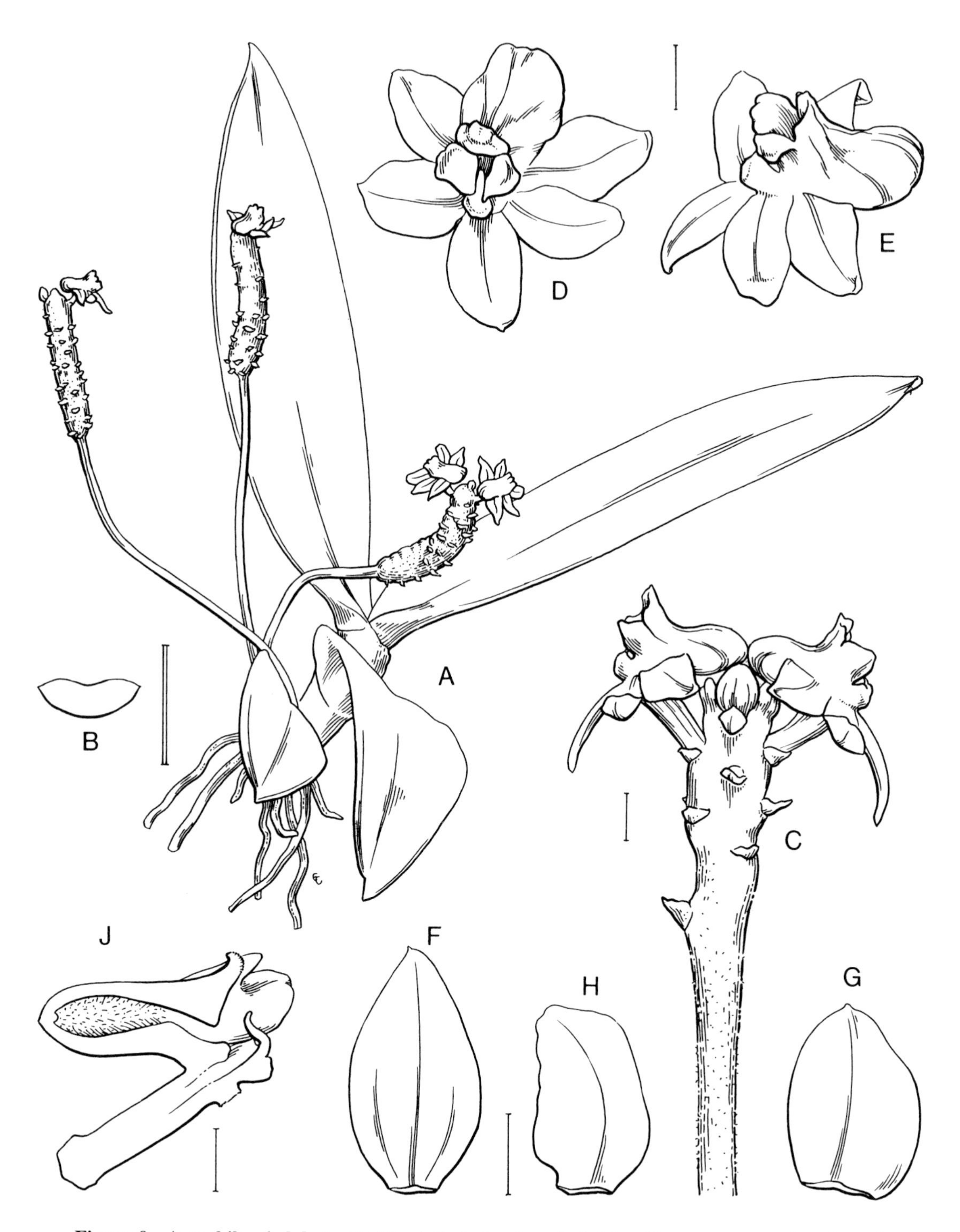

Figure 9. Ascochilopsis lobata J.J. Wood & A. Lamb. - A: habit. - B: transverse section through leaf. - C: inflorescence. - D: flower, front view. - E: flower, side view. - F: dorsal sepal. - G: lateral sepal. - H: petal. - J: longitudinal section through pedicel with ovary, lip and column. All drawn from *Lamb & Surat* in *Lamb* AL 1252/90 (holotype) by Eleanor Catherine. Scale: single bar = 1 mm; double bar = 1 cm.

9. ASCOCHILOPSIS LOBATA J.J. Wood & A. Lamb

Ascochilopsis lobata *J.J. Wood & A. Lamb* in Wood & Cribb, Checklist Orchids Borneo: 348, fig. 37, plate 1E (1994). Type: Borneo, Sabah, Tenom District, Paling Paling Hills, Ulu Batu Tiningkang, April 1990, *Lamb & Surat* in *Lamb* AL 1252/90 (holotype K, spirit material only).

Epiphyte. Stems 1–2.5 cm long. ***Leaves*** 4–7 per stem, 2–5 × 0.8–1.3 cm, fleshy, up to 2 or 3 mm thick, narrowly elliptic, ligulate, gradually narrowed at base, apex minutely unequally bilobed, the lobules acute. ***Inflorescences*** erect or porrect, bearing many tiny flowers in succession, 1 or 2 open at a time; peduncle 1.5–2.3 cm long, scabrous-hairy, greenish-purple; rachis 0.5–1.5(–2.5) cm long, thickened, clavate, minutely scaberulus, green; floral bracts 0.4 mm long, triangular-ovate, acute, fawn-coloured. ***Flowers*** 4–5 mm long, 5 mm across; pedicel with ovary pale yellow, flushed green; sepals and petals translucent yellow, whitish at base; lip white with a reddish brown patch on upper margins of side lobes; column white; anther-cap yellow. ***Pedicel*** with ***ovary*** 2 mm long. ***Sepals*** and ***petals*** spreading. ***Dorsal sepal*** 2 × 1.2 mm, oblong-elliptic, acute. ***Lateral sepals*** 2.2 × 1.5 mm, ovate-elliptic, subacute, slightly asymmetrical. ***Petals*** 1.8 × 1 mm, oblong-elliptic, obtuse. ***Lip*** 3-lobed, *c.* 3 mm long, fleshy, immobile; mid-lobe very small, 0.4–0.5 mm long, trilobulate, the side lobules rounded, mid-lobule triangular; side lobes *c.* 1 mm long, oblong, apex very shallowly unequally retuse, longer than broad; spur 1.9–2 × 1.5 mm, saccate, obtuse, transversely flattened. ***Column*** 1 mm long; foot absent; rostellum elongate, apex upcurved; pollinia not seen. Plate 2B.

HABITAT AND ECOLOGY: Epiphytic on branches of dipterocarp trees in mixed hill-dipterocarp forest on sandstone ridges. Alt. 700 to 800 m. Flowering observed in April.

DISTRIBUTION IN BORNEO: SABAH: Tenom District.

GENERAL DISTRIBUTION: Endemic to Borneo.

NOTE: *A. lobata* is distinguished from *A. myosurus* (Ridl.) Carr, from Peninsular Malaysia and Sumatra, by the lip which has distinct oblong side lobes, a trilobulate mid-lobe and a narrower spur.

DERIVATION OF NAME: The generic name is derived from *Ascochilus*, a genus of orchids and the Greek *opsis*, appearance, referring to its *Ascochilus*-like appearance. The specific epithet is derived from the Latin *lobatus*, lobed, in reference to the lobing of the lip.

Figure 10. Brachypeza zamboangensis (Ames) Garay. - A: habit. - B: flower, oblique view. - C: dorsal sepal. - D: lateral sepal. - E: petal. - F: lip and column, side view. - G: lip and column, longitudinal section. - H: anther-cap, back view. - I: pollinarium. - J: ovary, transverse section. All drawn from a plant cultivated at Tenom Orchid Centre, TOC 765 by C.L. Chan and Chin Wan Wai.

10. BRACHYPEZA ZAMBOANGENSIS (Ames) Garay

Brachypeza zamboangensis (*Ames*) *Garay* in Bot. Mus. Leafl. Harv. Univ., 23: 164 (1972). Type: Philippines, Mindanao, Zamboanga Province, Flecha Point, 10 m, 30 September 1922, *Merrill* 11640 (holotype PNH, isotype AMES).

Sarcochilus zamboangensis Ames, Sched. Orch., 5: 39 (1923).

Pendulous ***epiphyte***. ***Stems*** 3–4 cm long, stout, concealed by leaf bases. ***Leaves*** 6–7, 20–30 × 4.5–7 cm, distichous, broadly oblong-ligulate, coriaceous, obtuse, unequally bilobed, sheaths 1 cm long. ***Inflorescence*** many-flowered, 1 or 2 flowers opening at a time in succession; peduncle 10–15 cm long, 1 mm wide, slender, wiry, with several triangular-ovate sterile bracts 1–1.5 mm long; rachis 8 cm long, 2–2.5 mm wide, olive-green to purple-red; floral bracts 1 mm long, triangular-ovate, obtuse, thickened, rigid, quaquaversal, olive-green to purple-red. ***Flowers*** ephemeral, lasting for one day only; sepals and petals yellow, mottled and flushed dark red; lip white, the side lobes flushed pale yellow, the mid-lobe with pink stripes above and a purple flush on the underside; column pale yellow; anther-cap and pollinia white. ***Pedicel*** with ***ovary*** 1 cm long, narrow. ***Dorsal sepal*** 7–8 × 4.5 mm, oblong-elliptic, subacute, strongly concave. ***Lateral sepals*** 8–9 × 4–5.5 mm, obliquely ovate-elliptic or obovate, obtuse, concave. ***Petals*** 7–8 × 3 mm, oblong or oblong-spathulate, obtuse. ***Lip*** 3-lobed, spurred, aseptate, slightly moveably articulated to column-foot; side lobes 2 × 1–1.5 mm, oblong, obtuse to truncate, erect; mid-lobe 1.5–2 × 2–2.5 mm when flattened, transversely oblong-elliptic, obtuse, conduplicate, with a short basal thickened area; spur 8–9 mm long, 2 mm wide at apex, strongly curved, dilated and infundibuliform at the throat, tubular below with an inflated, dorsi-ventrally compressed apical portion. ***Column*** 3–4 mm long, gently curved; foot 2–2.8 mm long; anther-cap ovate, apiculate; pollinia two, deeply divided to the middle. ***Fruit*** 3.5 × 0.4 cm, cylindrical. Plate 2C–E.

HABITAT AND ECOLOGY: Epiphyte low down and near ground level on trees and large lianas in hill forest on limestone. Alt. 500 m. Flowering observed in February.

DISTRIBUTION IN BORNEO: SABAH: Batu Urun.

GENERAL DISTRIBUTION: Borneo and the Philippines.

NOTES: Four species of *Brachypeza* are recorded from Borneo. *B. zamboangensis* is at once distinguished by its curious incurved spur which is shaped rather like a crumhorn, a wind instrument popular in Renaissance Europe.

DERIVATION OF NAME: The generic name is derived from the Greek *brachys*, short and *peza*, foot, in reference to the characteristically short column-foot. The specific epithet refers to the type locality in the Philippines.

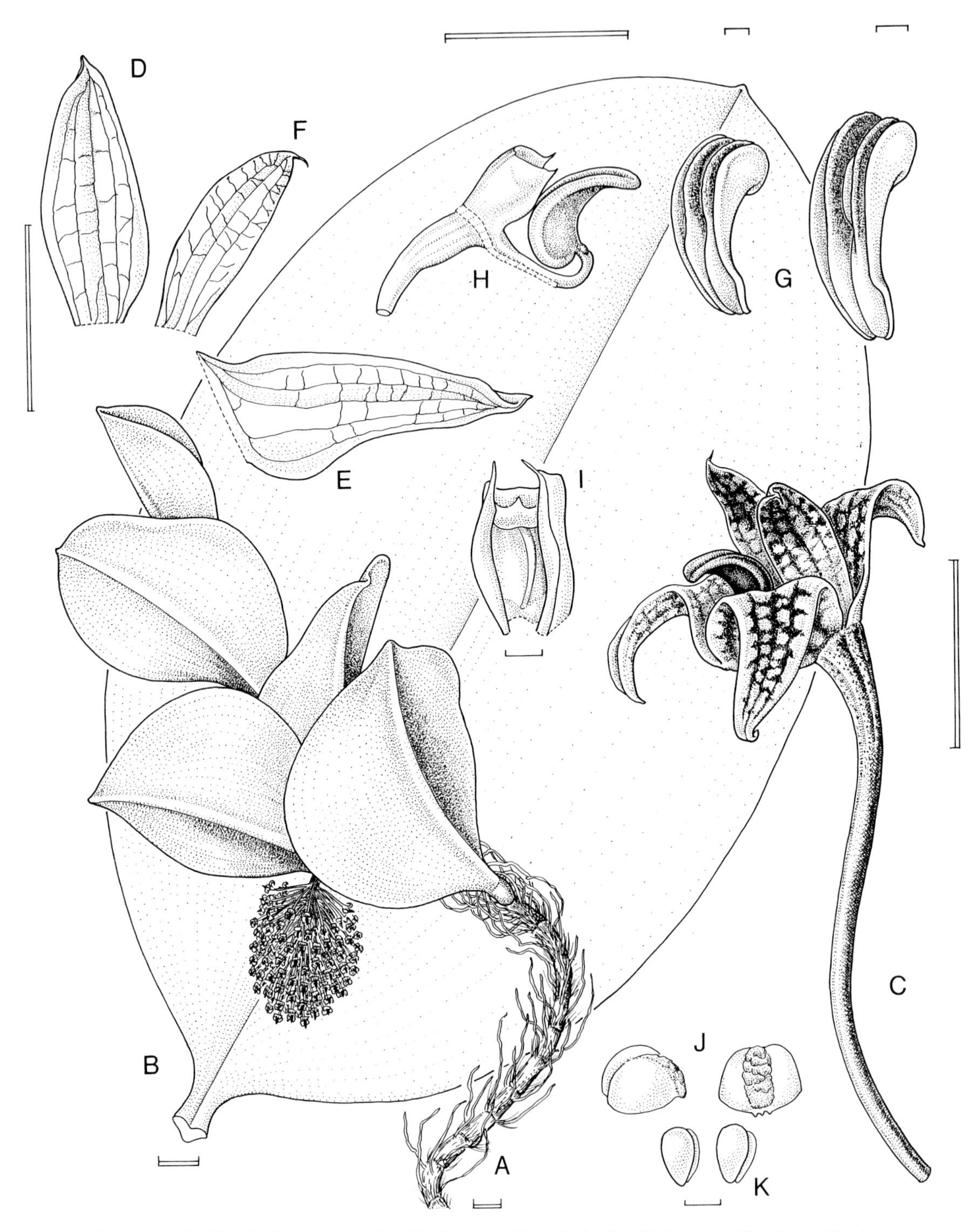

Figure 11. Bulbophyllum beccarii Rchb.f. - A: habit. - B: leaf. - C: flower, side view. - D: dorsal sepal. - E: lateral sepal. - F: petal. - G: lip, oblique views. - H: pedicel with ovary, lip and column. - I: column, front view. - J: anther-cap, side and back views. - K: pollinia. A (habit) drawn from slides taken by *Collenette*, B from *Anderson* 13258 and *Lamb* in SAN 88514, and C–K from cult. *Henderson and Sons* by Susanna Stuart-Smith. Scale: A & B: double bar = 2 cm; remainder: single bar = 1 mm; double bar = 1 cm.

11. BULBOPHYLLUM BECCARII Rchb. f.

Bulbophyllum beccarii *Rchb. f.* in Gard. Chron., ser. 2, 11: 41 (1879). Type: Borneo, Sarawak, Kuching, March 1867, *Beccari* 3515 (holotype FI).

Robust ***climber***. ***Rhizome*** up to 2.5 cm in diameter, fleshy, climbing spirally around tree trunks for several metres, sheaths decaying to leave coarse, persistent fibres. ***Roots*** numerous and grouped together along rhizome, 1–3 mm in diameter. ***Pseudobulbs*** (15–)18–19 cm apart, 1.8–3.5 × 2–3.5 cm, ovoid, pale yellow-green. ***Leaves*** erect, forming a cup to gather dead leaves and other debris; blade 26–50 × 18.5–38 cm, oblong to broadly elliptic, obtuse to acute, thick and leathery in texture, with numerous transverse veins forming a reticulate pattern; petiole 1–4 cm long, sulcate. ***Inflorescence*** 12–43 × 7–14 cm, pendulous, densely many-flowered, ovoid to cylindrical; peduncle 9 cm long, 1 cm in diameter, rosy with violet lines, usually clothed in five imbricate ovate to ovate-elliptic, acute to acuminate, dorsally carinate sheaths, the longest 4.5 cm long; rachis around 34 cm long; floral bracts 3–3.5 × 0.6 cm, lanceolate, acuminate, flesh-coloured with darker pink or purplish flecks and spots. ***Flowers*** 2 × 1.5 cm, smelling of rotten carrion or bad fish; pedicel with ovary cream, finely speckled with purple; sepals and petals with a whitish cream to ochreous yellow ground colour, with dark blackish purple or maroon spots and reticulations, apex of petals often deep maroon-purple; lip deep maroon-purple with a whitish or yellow apex; column yellow; anther-cap cream. ***Pedicel*** with ***ovary*** 2–2.5(–3) cm long, narrow. ***Dorsal sepal*** 1.2 × 0.4–0.45 cm, oblong, obtuse to subacute, recurved. ***Lateral sepals*** 1.5–1.6 × 0.5–0.7 cm, triangular-ovate, acute, adnate to column-foot to form an obtuse mentum, becoming recurved. ***Petals*** 0.9–1 × 0.25–0.3 cm, oblong, obtuse, apiculate. ***Lip*** 0.5–0.6 cm long, 0.3 cm wide at base, cordate, oblong, obtuse, somewhat retuse, strongly curved, with two distinct ridges inside. ***Column*** 3–3.5 mm long, unistriate on either side; stelidia short, subulate; anther-cap ovate; pollinia four. Plate 3A–C.

HABITAT AND ECOLOGY: Podsol forest with *Dacrydium*, *Rhododendron*, *Tristania*, etc.; peat swamp forest with *Shorea albida*, etc.; lowland dipterocarp forest. Alt. sea level to 600 m. Flowering observed in January, February, March, April and June.

DISTRIBUTION IN BORNEO: BRUNEI: Locality unknown. KALIMANTAN: Locality unknown. SABAH: Nabawan area; upper Kinabatangan River. SARAWAK: Bako National Park; Betong District, Saribas Forest Reserve.

GENERAL DISTRIBUTION: Endemic to Borneo.

NOTES: This bizarre species has probably the broadest leaves of any Bornean orchid, some measuring as much as 38 cm across. The plant climbs in a spiral fashion around the trunks of trees such as *Shorea albida*, the thick leathery leaves lying erect against the trunk and forming a cup into which falling debris is collected from the canopy above. The captured debris rapidly rots and is invaded by a thick tangle of roots which tap the rich source of nutrients provided.

Reichenbach (1879) commented that it "excites the highest interest by its grand leaves." In the Gardeners' Chronicle for September 1880 he relates that "a great stir has been caused by the unexpected flowering of this remarkable plant; we had not expected to see it so soon in such a state. When I obtained the case containing the inflorescence at hand I smelt the most hideous stench of old fish that can be imagined."

DERIVATION OF NAME: The generic name is derived from the Greek *bulbos*, bulb, and *phyllon*, leaf, referring to the leafy pseudobulbs of most species. The specific epithet honours Odoardo Beccari (1843–1920), the Italian naturalist whose book "Wanderings in the great forests of Borneo" (1904) gives a thrilling picture of Bornean forests before the widespread deforestation of modern times. He was described by Reichenbach (1879) as an "energetic and most successful traveller" who would have been "agreeably surprised to see his Bornean friend (*B. beccarii*) in flower in Europe."

12. BULBOPHYLLUM KEMULENSE J.J. Sm.

Bulbophyllum kemulense *J.J. Sm.* in Bull. Jard. Bot. Buitenzorg, ser. 3, 11: 144 (1931). Type: Borneo, Kalimantan Timur, West Koetai, Mt. Kemal (Kemoel, Kemul), 1800 m, 17 October 1925, *Endert* 4278 (holotype BO).

Medium-sized **epiphyte**. **Rhizome** short, creeping, up to 6 cm long. **Roots** numerous, fleshy. **Pseudobulbs** 0.8–1.5 cm long, 1–1.1 cm in diameter, ovoid, truncate, closely spaced, covered, when young, by fibrous sheaths *c.* 3.5–4 cm long. **Leaves** erect; blade 19–21 × 3.2–3.7 cm, oblong-elliptic or narrowly elliptic, apex acute, shortly conduplicate, fleshy, coriaceous; petiole canaliculate, 4.5–7 cm long. **Inflorescence** erect, densely many-flowered, globose-capituliform; peduncle 3 cm long, broadest above, *c.* 3–4 mm wide, white with pink streaks, emerging from a white basal bract 0.5 cm long; rachis 1.1 cm long, 0.43 cm in diameter, gently curving, especially at bud stage, fleshy, angular; floral bracts 3–3.5 × 2–3.2 mm, broadly ovate-triangular, acute to acuminate, concave, adpressed, white spotted with pinkish purple. **Flowers** dorsally compressed, *c.* 9 mm long, 7–9 mm wide; pedicel with ovary white and purple; sepals translucent white to pink, spotted and stained with dark purple-violet, blackish purple at apex; petals white, dark purple at apex; lip off-white to cream with dark reddish violet or dark purple papillae above, with pinkish spots on undersurface; column white with dark purple blotches. **Pedicel** with **ovary** 4–5 mm long, gently curved, the ovary obconical. **Dorsal sepal** 7.5–8 × 5 mm, laterally adpressed, oblong-ovate, obtuse, apex minutely ciliolate, veins 5–6, prominent and verruculose on outer surface. **Lateral sepals** 7–9 × 4–5.5 mm, obliquely oblong, obtuse, minutely obtusely apiculate, concave at base, convex at middle, apex slightly concave, fleshy. **Petals** 4 × 2–2.2 mm, obliquely oblong, apex slightly dilated, obtuse and minutely ciliolate to papillose, concave, porrect, veins 3. **Lip** 5–5.3 mm long, 3 mm wide at base, 3.5–4.5 mm wide above, entire, shortly stipitate to column-foot, oblong to oblong-ovate, obtuse, papillose, the lower margins erect, sulcate-canaliculate below, bent and decurved at middle, convex above, with an obscure, slightly thickened ridge either side below edge of lower margins. **Column** 2.5 mm long; foot 3–3.3 mm long, incurved, apex bidentate; stelidia 1 mm long, subulate, parallel, porrect; anther-cap cucullate. Plate 3D.

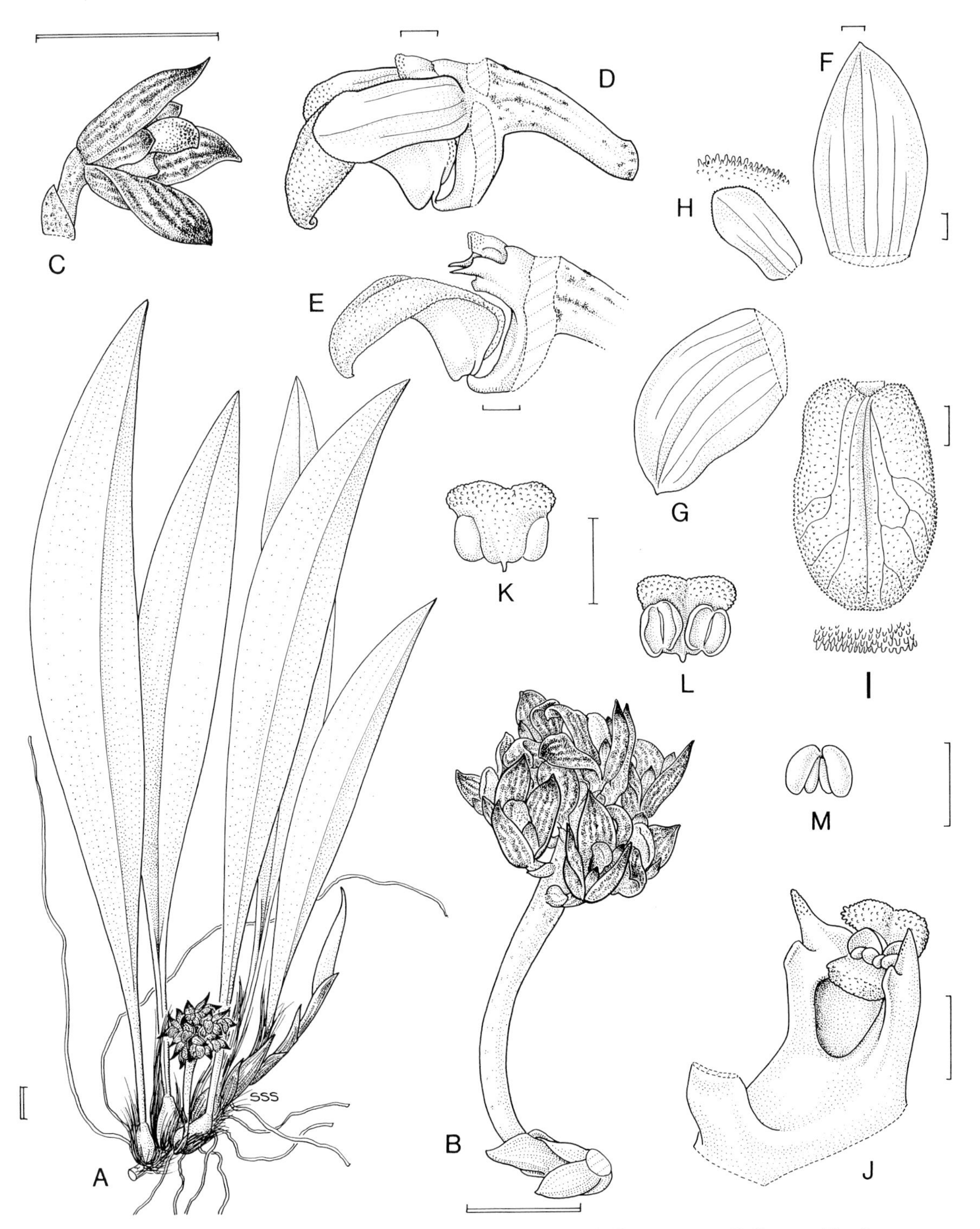

Figure 12. Bulbophyllum kemulense J.J. Sm. - A: habit. - B: inflorescence. - C: flower, side view. - D: flower, side view, sepals removed. - E: pedicel with ovary, lip and column, side view. - F: dorsal sepal. - G: lateral sepal. - H: petal with close-up of papillae. - I: lip, flattened, with close-up of papillae. - J: column with anther-cap, oblique view. - K: anther-cap, back view. - L: anther-cap, interior. - M: pollinia. All drawn from *Lamb* AL 1324/91 by Susanna Stuart-Smith. Scale: single bar = 1 mm; double bar = 1 cm.

HABITAT AND ECOLOGY: Mixed *Lithocarpus/Castanopsis* ridge forest on sandstone. Alt. 1300 to 1800 m. Flowering observed in February and October.

DISTRIBUTION IN BORNEO: KALIMANTAN TIMUR: Apokayan; Mt. Kemal. SABAH: Mt. Alab.

GENERAL DISTRIBUTION: Endemic to Borneo.

NOTES: *B. kemulense* was the first member of section *Globiceps* to be recorded from Borneo. A further nine species have been discovered since it was described in 1931.

DERIVATION OF NAME: The specific epithet refers to Mt. Kemal (2053 m), the type locality.

13. BULBOPHYLLUM POLYGALIFLORUM J.J. Wood

Bulbophyllum polygaliflorum *J.J. Wood* in Kew Bull. 39(1): 96, fig. 15 (1984). Type: Borneo, Sarawak, Mt. Mulu National Park, Camp 4, along path to camp, 18 March 1978, *Hansen* 499 (holotype C, isotype K).

Epiphyte. ***Rhizome*** to 10 cm or more long, creeping, producing many wiry roots. ***Pseudobulbs*** to 2 × 0.8 cm, borne 2–3.5 cm apart, cylindrical to narrowly ovoid, narrowed above, 1-leaved. ***Leaf blade*** 9 × 3.5 cm, oblong-elliptic, retuse and rounded at apex, coriaceous; petiole 1.8 cm long, sulcate. ***Inflorescence*** a rather dense raceme with 30 or more flowers; peduncle to 67 cm long, erect, wiry, emerging from 2 small purplish-brown sheaths near base of pseudobulb, purplish-green, bearing 5 small remote purplish-green, ovate, acute, amplexicaul sterile bracts; rachis 10–11 cm long; floral bracts 2–3 mm long, ovate-elliptic, acuminate. ***Flowers*** not opening widely, purple with white hairs. ***Pedicel*** with ***ovary*** 5–6 mm long, narrowly clavate. ***Dorsal sepal*** 6 × 1.5 mm, triangular-acuminate, margins revolute and ciliate, 3-veined. ***Lateral sepals*** 8.5–9 × 5 mm, connate to *c.* 1.5 mm from base, ovate, obtuse, fleshy, margins somewhat thickened and shortly ciliate, 5-veined. ***Petals*** 3 × 2 mm, broadly spathulate, rounded, obtuse, pilose except at base, with long white ciliate hairs along margins, 1–2-veined. ***Lip*** 6 × 2 mm, oblong-elliptic with two rounded basal auricles, fleshy, with four rows of irregular, fleshy, pubescent coral-like calli, margins and lower surface densely ciliate. ***Column*** 1 mm long, hirsute below stigma; stelidia narrow, irregularly and acutely tridentate; foot 2–2.5 mm long; anther-cap 0.8 mm long, cucullate, retuse, dorsally papillose, ciliate below. Plate 3E.

HABITAT AND ECOLOGY: Upper montane mossy forest. Alt. 1700 m. Flowering observed in March.

DISTRIBUTION IN BORNEO: SABAH: Maliau Basin. SARAWAK: Mt. Mulu National Park.

GENERAL DISTRIBUTION: Endemic to Borneo.

NOTE: *B. polygaliflorum* belongs to section *Hirtula* which is represented by thirteen species in Borneo, five of which remain undescribed.

DERIVATION OF NAME: The specific epithet refers to the flowers which bear a superficial resemblance to the genus *Polygala* (Polygalaceae).

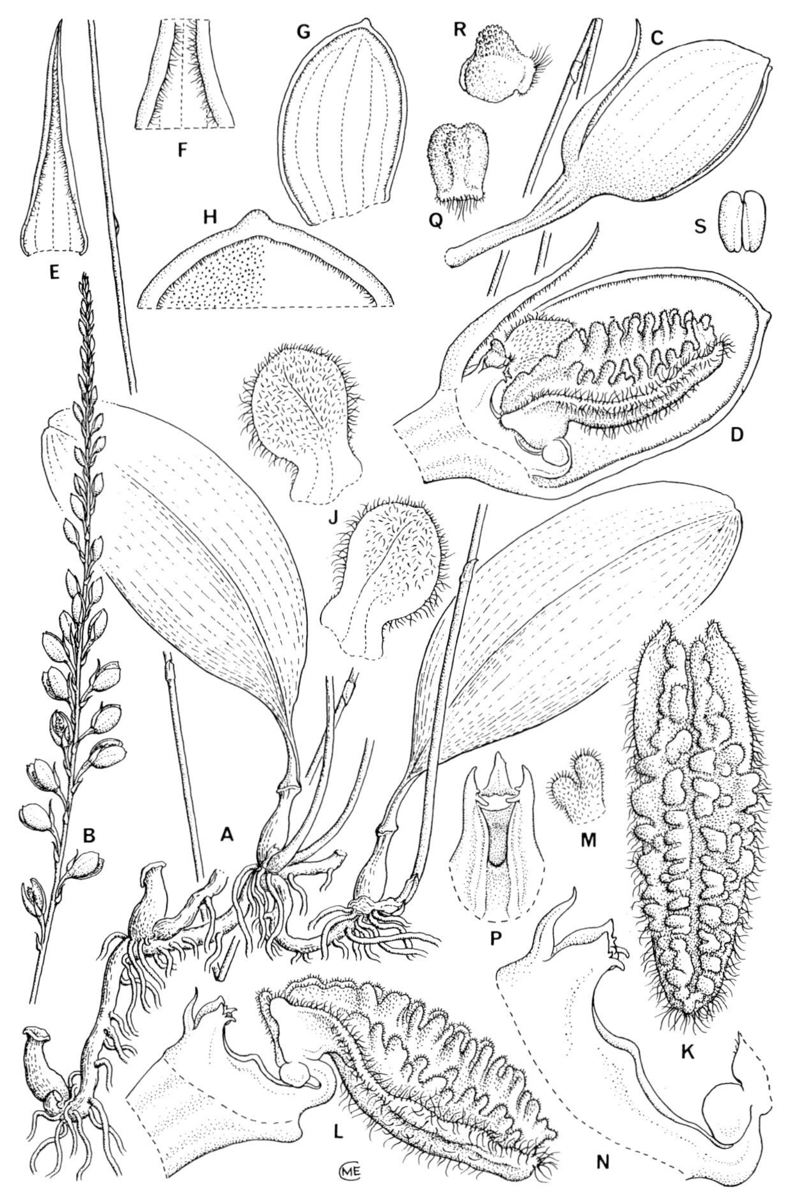

Figure 13. Bulbophyllum polygaliflorum J.J. Wood. - A: habit ×2/3. - B: inflorescence ×2/3. - C: flower, side view ×4. - D: flower with lateral sepal and petal removed, side view ×6. - E: dorsal sepal ×6. - F: portion of dorsal sepal, reverse ×14. - G: lateral sepal, inner surface ×4. - H: apex of lateral sepal, inner surface ×14. - J: petals ×8. - K: lip, front view ×10. - L: lip and column, side view ×8. - M: detail of lip callosity ×20. - N: column with anther-cap removed and lip attached, side view ×14. - P: column apex, front view ×16. - Q: anther-cap, front view ×16. - R: anther-cap, side view ×16. - S: pollinia ×16. All drawn from *Hansen* 499 (holotype) by Maureen Church.

Figure 14. Calanthe undulata J.J. Sm. - A: habit. - B: flower, side view. - C: dorsal sepal. - D: lateral sepal. - E: apex of lateral sepal, back view. - F: petal. - G: apex of pedicel with ovary, lip and column, front view. - H: column and lip, back view. - I: column apex, oblique view. - J: apex of pedicel with ovary, column and base of lip, longitudinal section. - K: anther-cap, back view and front view showing pollinia. - L: pollinia. - M: fruit. A & M drawn after a sketch by *Lamb* of *Lamb* AL 632/86, and B–L from *de Vogel & Cribb* 9140 by Susanna Stuart-Smith. Scale: single bar = 1 mm; double bar = 1 cm.

14. CALANTHE UNDULATA J.J. Sm

Calanthe undulata *J.J. Sm.* in Icones Bog., 2: 67, t. 112, fig. B (1903). Type: Borneo, Kalimantan, *Nieuwenhuis* s.n. (holotype BO).

Erect ***terrestrial*** 60–70 cm tall. ***Rhizome*** *c.* 1 cm thick, short, vertical. ***Pseudobulbs*** 8 × 1 cm, concealed by cataphylls and leaf bases. ***Cataphylls*** 9–13 × 2 cm, narrowly elliptic, acute. ***Leaves*** 4–6; blade (27–)36–55 × 6–8.5 cm, narrowly elliptic to elliptic, acuminate, plicate, erect to spreading, dark green; petioles 5–15 cm long, 5-ribbed, base sheathing. ***Inflorescence*** racemose, up to 20-flowered; peduncle 15–34 cm long, dark green, furfuraceous with dark brown stellate hairs, enclosed below by several imbricate sheaths, with 3 or 4 elliptic, acute, pale green sheathing bracts 3–3.5 cm long above; rachis 30–34 cm long, angular, sulcate, green, covered with brown and white stellate hairs; floral bracts deciduous, lowermost 2.6 × 5 mm, uppermost smaller, narrowly elliptic, acute, concave, pale green. ***Flowers*** 1.8–2 cm across; pedicel with ovary green, furfuraceous with dark brown stellate hairs; sepals and petals brownish orange to cinnamon-orange; lip white with dark yellow keels and median line; column white turning dark yellow; anther-cap white. ***Pedicel*** with ***ovary*** 1.7–2 cm long, narrowly clavate. ***Sepals*** recurved, rather fleshy and convex, dark brown-furfuraceous on reverse. ***Dorsal sepal*** 1.3–1.4 × 0.5–0.6 cm, oblong-elliptic, acuminate, 3-veined. ***Lateral sepals*** 1.2–1.3 × 0.4–0.5 cm, oblong or narrowly elliptic, acuminate, 3- to 4-veined. ***Petals*** 1.2 × 0.4–0.5 cm, narrowly elliptic to spathulate-elliptic, acuminate, recurved, convex, glabrous, 3-veined. ***Lip*** entire, spurless, adnate along almost the entire length of the column, sparsely hairy beneath, particularly on the adnate portion, free portion 0.5–0.6 × 0.6 cm, ovate, subacute, the margin undulate, somewhat erose, disc with two prominent rounded, verrucose keels, a shorter, less prominent keel on the outer side of each and a slightly raised median line, throat with white woolly hairs extending to the base of the keels. ***Column*** 0.8 cm long, truncate, fleshy, white-hairy; anther-cap 2 mm long, ovate, acute, sparsely hairy; pollinia eight, clavate. Plate 4A.

HABITAT AND ECOLOGY: Podsolic dipterocarp/*Dacrydium* forest, with understorey of *Rhododendron malayanum*, on very wet sandy soil, in shade. Alt. 400 to 1000 m. Flowering observed in October.

DISTRIBUTION IN BORNEO: KALIMANTAN TIMUR: Apokayan area. SABAH: Nabawan area.

GENERAL DISTRIBUTION: Endemic to Borneo.

DERIVATION OF NAME: The generic name is derived from the Greek *kalos*, beautiful, and *anthe*, bloom, in allusion to the beautiful flowers of many species. The specific epithet is derived from the Latin *undulatus*, wavy, referring to the wavy margins of the free portion of the lip.

Figure 15. Cleisocentron merrillianum (Ames) Christenson. - A & B: habits. - C: close-up of portion of leaf. - D: leaf apex variation. - E: leaf showing two cross-sections. - F: flower, oblique view. - G: pedicel with ovary, lip and column with dorsal sepal, longitudinal section. - H: upper portion of ovary, lip and column with anther-cap, side view. - I: dorsal sepal. - J: lateral sepal. - K: petal. - L: column, front view, with anther-cap (left), without anther-cap (right). - M: anther-cap, back view. - N: anther-cap, interior, flattened. - O: pollinarium. - P: stipes and viscidium, side view. A & C drawn from *Lamb* in SAN 89677, B from *Lamb* in SAN 89678, D from *Collenette* 756 (left), *Lamb* in SAN 89678 (centre) and *Lamb* in SAN 89677 (right), and E–P from *Collenette* 1, cult. R.B.G. Kew EN. 657-60 by Susanna Stuart-Smith. Scale: single bar = 1 mm; double bar = 1 cm.

15. CLEISOCENTRON MERRILLIANUM (Ames) Christenson

Cleisocentron merrillianum (*Ames*) *Christenson* in Amer. Orchid. Soc. Bull. 61, 3: 246 (1992). Type: Borneo, Sabah, Mt. Kinabalu, Marai Parai Spur, 23 November 1915, *J. Clemens* s.n. (holotype AMES, isotype K).

Sarcanthus merrillianus Ames, Orchidaceae 6: 230, plate 97 (1920).

Robiquetia merrilliana (Ames) Lueckel, M. Wolff & J.J. Wood in Die Orchidee 40, 3: 109 (1989).

Robust, erect ***epiphyte. Roots*** grouped at base of stem, stout, rigid, fibrous, flexuous, longitudinally sulcate, branched, 2–3 mm in diameter. ***Stems*** 20–90 cm long, 0.5 cm in diameter, terete, entirely concealed by leaf sheaths, internodes 1.3–4 cm long. ***Juvenile leaves*** 12–13 × 0.8–1.1 cm, linear-ligulate, conduplicate at base, obtusely unequally bilobed or acute, thickly coriaceous, grading on older shoots into ***mature leaves*** 12–20 × 0.2–0.4 cm, terete, conduplicate at base, acute, falcate or straight; leaf sheaths 1.3–4 cm long, tubular, thickly coriaceous, rigid, striate-veined and finely rugose with transverse reticulations between the longitudinal veins. ***Inflorescence*** 8- to 15-flowered; peduncle 0.5–1.5 cm long, stout, emerging through leaf sheaths opposite leaf bases with 2–3 obtuse bracts; rachis abbreviated, 6–7 mm long; floral bracts 1–2 mm long, ovate, acute, scale-like, papery, mauvish-brown, persistent. ***Flowers*** *c.* 1.5 cm long, opening together, long lasting, unscented, variously described as aquamarine-blue, translucent Cambridge-blue, or translucent lavender-blue, with deep indigo-blue markings on lip, the intensity of blue varying; column white or pale yellow. ***Pedicel*** with ***ovary*** 1.5 cm long, very slender, spreading, with a slight twist. ***Dorsal sepal*** 4–6 mm long, 2.8 mm wide when flattened, ovate, strongly concave, broadly rounded and sometimes obliquely bilobed at apex, minutely denticulate above middle. ***Lateral sepals*** 4.5–6 × 3 mm, obovate-oblong, oblique, obtuse, slightly cucullate and dorsally cornute at apex, prominently 1-veined. ***Petals*** 3.5–5 × 2 mm, oblong-spathulate, apex broadly rounded, minutely denticulate above middle. ***Lip*** 3-lobed, prominently spurred; side lobes less than 1 mm long, 2–2.5 mm across apex, transversely oblong, obscure, truncate, erect; mid-lobe 1.5–2 mm long, triangular, acute or obtuse, upcurved, fleshy and thickened within below apex; spur 1.1–1.2 mm long, cylindric, curved, obtuse, provided on anterior wall below base of mid-lobe with a small callus and on posterior wall about 4 mm from apex with an upcurved ligulate, blunt callus that is longitudinally sulcate beneath. ***Column*** 3–4 mm long, erect, cylindric, fleshy; foot decurrent to lip; anther-cap triangular, cucullate, rostrate in front; pollinia four, appearing as two pollen masses, each completely divided into free halves; stipes ligulate, membranaceous; viscidium elliptic. Plate 4B & C.

HABITAT AND ECOLOGY: Lower and upper montane forest; lower montane riverine forest; crowns of large trees on river banks; recorded as epiphytic on branches of *Agathis*, in light shade; often on ultramafic substrate. Alt. 1100 to 3000 m. Flowering observed in July, August, October and November.

DISTRIBUTION IN BORNEO: SABAH: Mt. Kinabalu.

GENERAL DISTRIBUTION: Endemic to Borneo.

NOTES: *Cleisocentron* contains five species distributed in the Himalayan region, Myanmar (Burma), Vietnam and Borneo. *C. merrillianum* exhibits leaf polymorphy as well as having unusually coloured flowers. The holotype at AMES of *C. merrillianum* consists of flowering stems displaying two leaf types, the lower terete and acute, the upper broader, strap-shaped and acute. The isotype at K consists only of two detached leaves, one strap-shaped, one terete, and some loose flowers. Ames clearly overlooked or ignored these odd leaf types when drawing up his description of *Sarcanthus merrillianus*. Christenson (1992) comments that the "young shoots produce strap-shaped leaves very similar to *C. trichromum*, but which grade in older shoots to terete leaves."

A second species from Mt. Kinabalu (see Wood *et al.,* plate 21C, 1993), with a longer inflorescence of purple flowers, has been recorded by Lamb. No preserved material has been available for examination.

DERIVATION OF NAME: The generic name is derived from the Greek *kleistos*, closed, and *kentron*, spur, relating to the spur of the lip which is almost completely closed by a thickening of its inner wall. The specific epithet honours Elmer Drew Merrill (1876–1956) who collaborated with Oakes Ames to produce a bibliographic enumeration of Bornean plants published in 1921.

16. CORYBAS MULUENSIS J. Dransf.

Corybas muluensis *J. Dransf.* in Kew Bull. 41(3): 588, fig. 4, plate 1C (1986). Type: Borneo, Sarawak, Mt. Mulu National Park, summit ridge of Mt. Mulu, *G. Lewis* 352 (holotype K).

Terrestrial. Tuber not seen. ***Stem*** *c.* 8 mm long, subangular with a basal sheath up to 4 × 1 mm. ***Leaf*** to 9.5 mm from origin to apex, 12 mm long from basal auricles to apex, 9.5 mm wide at widest point, sessile, heart-shaped, acuminate, flat with smooth margins, auricles shallow, shining mid green, with few inconspicuous veins. ***Bract*** to 4 × 1 mm, pale green. ***Flower*** sessile, to 13.5 mm high, held somewhat backwards; dorsal sepal white marked with 5 dark purple lines within; lateral sepals whitish flushed with crimson; lip with a conspicuous white vertically held patch *c.* 4 mm in diameter, lateral wings whitish and translucent, conspicuously marked with 4 deep purple-black lines, bright purple in the middle, with scattered dark purple spots and dense papillae; spurs deep purple. ***Dorsal sepal*** 12 mm long, 1.5 mm wide at base, gradually widening to 3.75 mm wide *c.* 10 mm from base, then narrowed to apex, ± acute, gently curved, 5-veined. ***Lateral sepals*** *c.* 9.5 mm long, thread-like, acute. ***Petals*** *c.* 11 mm long, thread-like, acute, laterally adnate to lateral sepals at base. ***Lip*** erect in basal half, bent through *c.* 100° at middle, margins in basal half forming a tube with sepal claw, partly expanded in upper half, ± cup-like, ± orbicular, *c.* 10.5 mm wide when flattened out, throat ± convex, distal lip margin coarsely toothed; spurs to 4 × 1 mm, slender, conical. ***Column*** *c.* 1.7 × 0.7 mm; stigma *c.* 0.5 mm wide; pollinia two, each bilobed. Plate 4D & E.

HABITAT AND ECOLOGY: Mossy banks in mossy montane forest. Alt. 1350 to 2000 m. Flowering observed in May and October.

DISTRIBUTION IN BORNEO: SABAH: Crocker Range, Kimanis road. SARAWAK: Mt. Mulu National Park.

GENERAL DISTRIBUTION: Endemic to Borneo.

DERIVATION OF NAME: The generic name is derived from the Greek *korybas*, drunken man or, priest of Cybele, which describes the dorsal sepal which simulates a veiled drooping head, alluding to a priest's head-dress or to the nodding of a drunken man. The specific epithet refers to the type locality.

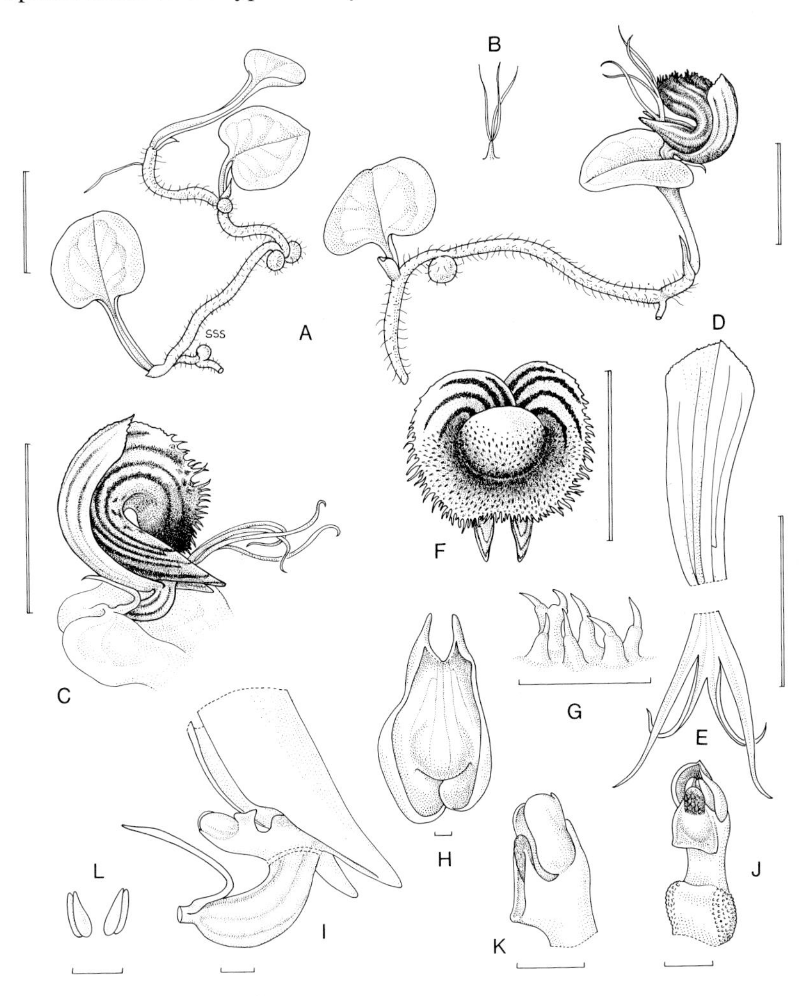

Figure 16. Corybas muluensis J. Dransf. - A: habit. - B: rhizome hair, close-up. - C: flower and portion of leaf, side view. - D: dorsal sepal. - E: adnate lateral sepals and petals. - F: lip, front view. - G: close-up of lip papillae. - H: lower portion of lip showing spurs. - I: pedicel with ovary, column and lip spurs, side view. - J: column, oblique view. - K: upper portion of column, side view. - L: two pollinia, each bilobed. All drawn from *Wood* 823 by Susanna Stuart-Smith. Scale: single bar = 1 mm; double bar = 1 cm.

Figure 17. Cymbidium borneense J.J. Wood. - A: habit. - B: leaf, distal portion. - C: flower, oblique view. - D: dorsal sepal. - E: lateral sepal. - F: petal. - G: lip, front and side views. - H: pedicel with ovary and column, side view. - I: column, front view. - J: lip, front and side views. - K: anther-cap, back and front views. - L: pollinia. A–C drawn from *Lamb* C18, D–I from *G. Lewis* 314 (holotype) and J-L from *Lamb* 7 by Susanna Stuart-Smith. Scale: single bar = 1 mm; double bar = 1 cm.

17. CYMBIDIUM BORNEENSE J.J. Wood

Cymbidium borneense *J.J. Wood* in Kew Bull. 38(1): 69, t. 1 (1983). Type: Borneo, Sarawak, Mt. Mulu National Park, Melinau Gorge area, *G. Lewis* 314 (holotype K, isotype SAR).

A medium-sized ***terrestrial***. ***Pseudobulbs*** 8 × 1.5 cm, fusiform, with 6–13 distichous leaves and covered by the sheathing leaf bases with a 2 mm broad membranous margin, and occasionally 2–3 scarious or fibrous cataphylls on young specimens. ***Leaves*** (12–)40–79 × 0.5–2.1 cm, linear-ligulate, acute, arching, somewhat coriaceous, articulated 8–12 cm from the base, not constricted into a petiole. ***Inflorescence*** 16–18 cm long, suberect, arising from within the lower leaf bases, with 3–5 flowers produced in the apical third of the scape and held below the leaves; peduncle arching upwards, with 3–5 distant, amplexicaul sheaths up to 1.5 cm long; floral bracts 0.5–1.5 cm long, narrowly ovate, acute. ***Flowers*** small, *c.* 4 cm across, coconut-scented; pedicel with ovary pale olive-green, stained reddish brown; sepals and petals cream with a narrow white margin, strongly stained and blotched with maroon-purple, especially in the centre; lip white, the side-lobes speckled with maroon, the mid-lobe with some maroon blotches and a pale yellow patch at the base, the callus ridges pale yellow; column cream above, stained maroon-purple below; anther-cap pale yellow. ***Pedicel*** with ***ovary*** 2–3.5 cm long. ***Dorsal sepal*** 2.2–2.8 × 0.5–0.6 cm, narrowly oblong-elliptic, apiculate, erect, margins somewhat revolute. ***Lateral sepals*** similar, subacute, slightly curved, spreading. ***Petals*** 2–2.4 × 0.6–0.8 cm, narrowly ovate, oblique, slightly broader than the sepals, porrect, but not forming a hood over the column. ***Lip*** about 1.5 cm long when flattened; side lobes erect, rounded, subacute at the apex, minutely papillose; mid-lobe 7–8 × 7 mm, ovate to oblong, obtuse to subacute, weakly recurved, minutely papillose, margin entire; callus very reduced, composed of two small swellings at the base of the mid-lobe. ***Column*** 1.2–1.3 cm long, arching, narrowly winged, minutely papillose; pollinia four, ovate-elliptic, in two unequal pairs; viscidium triangular, with two short processes at the lower corners. Plate 5A & B.

HABITAT AND ECOLOGY: Lowland forest on limestone; lower montane ridge-top forest with *Gymnostoma sumatrana* on ultramafic substrate; in humus-rich soils over limestone or ultramafic rocks, including serpentine boulders, often near streams. Alt. 100 to 1300 m. Flowering observed in March and April in Sabah, October in Sarawak.

DISTRIBUTION IN BORNEO: SABAH: Mt. Kinabalu; Sepilok. SARAWAK: Mt. Mulu National Park.

GENERAL DISTRIBUTION: Endemic to Borneo.

NOTE: *C. borneense* is the sole representative of section *Borneense* established by Du Puy and Cribb in 1988.

DERIVATION OF NAME: The generic name is derived from the Greek *kymbes*, a boat-shaped cup, alluding to the boat-shaped lip. The specific epithet refers to the island of Borneo.

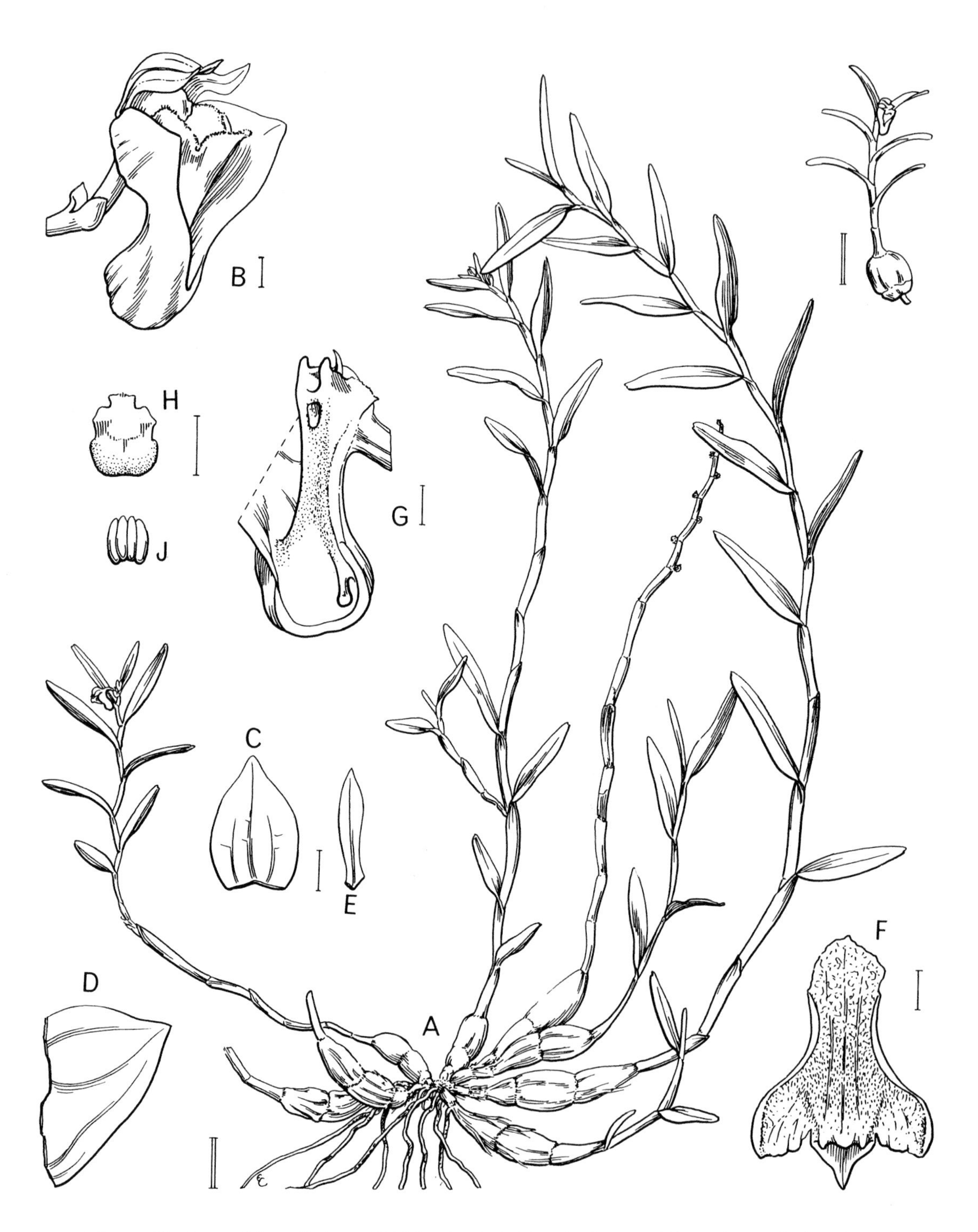

Figure 18. Dendrobium alabense J.J. Wood. - A: habit. - B: flower, oblique view. - C: dorsal sepal. - D: lateral sepal. - E: petal. - F: lip, flattened. - G: column and longitudinal section through mentum. - H: anther-cap, back view. - J: pollinia. All drawn from *Wood* 777 (holotype) by Eleanor Catherine. Scale: single bar = 1 mm; double bar = 1 cm.

18. DENDROBIUM ALABENSE J.J. Wood

Dendrobium alabense *J.J. Wood* in Lindleyana 5(2): 90, fig. 6 (1990). Type: Borneo, Sabah, Mt. Alab, *Wood* 777 (holotype K).

Clump-forming ***epiphyte***. ***Stems*** (4–)10–30 cm long, 2–2.5 mm wide, flattened, concealed by leaf sheaths, leafy to apex, branching above, young growth stained purple, two internodes near base swollen and ovoid or flattened to form elliptical, shiny straw-yellow ***pseudobulbs*** (1–)1.6–3 × 1 cm. ***Leaves*** 1.8–3 × 0.3–0.5 cm, linear-ligulate to ligulate-elliptic, obtusely unequally bilobed, sheaths 0.8–1.5 cm long. ***Inflorescence*** borne on uppermost nodes, flowers opening in succession from a tuft of grey-brown, chaffy bracts 1–2 mm long. ***Flowers*** ephemeral; sepals and petals creamy white with several dark purple spots; lip creamy white, the disc purple, the base of the mid-lobe yellow; column and anther-cap creamy white. ***Pedicel*** with ***ovary*** 4 mm long, narrowly clavate. ***Dorsal sepal*** 4.5 × 3 mm, ovate, acute, curving over column. ***Lateral sepals*** 4.5 × 6mm, obliquely triangular-ovate, acute, slightly concave, lower margin adnate to column-foot 6 mm long. ***Mentum*** 4–5.5 mm long, 2 mm wide at apex, obtuse, somewhat swollen at apex. ***Petals*** 4.5 × 0.9–1 mm, linear-ligulate, acute, curving over column. ***Lip*** 3-lobed, 6 mm long, 6 mm wide across side lobes, 2 mm wide at base; side lobes 2 × *c.* 1.8 mm, erect, oblong, rounded, margin slightly irregular, minutely papillose-hairy, with long hairs at base; mid-lobe *c.* 1 × 2 mm, triangular-ovate, acute, glabrous, horizontal; disc densely pilose, with a fleshy, flat, pilose to papillose callus, shallowly 3-keeled at apex. ***Column*** shortly toothed, 1 mm long; foot 5 mm long, swollen at base; anther-cap *c.* 1.2 × 1.2 mm, oblong, concave above, minutely papillose; pollinia four. Plate 5C & D.

HABITAT AND ECOLOGY: Lower montane ridge-top forest with *Dacrydium*, *Leptospermum*, *Phyllocladus*, *Podocarpus*, *Rhododendron durionifolium* and *R. fallacinum*, bamboo, ferns, rattans, *Gahnia*, etc.; low mossy and xerophyllous scrub forest on extreme ultramafic substrate; montane oak forest; *Dacrydium*, *Drimys*, *Leptospermum* and *Phyllocladus* scrub on steep rocky slopes. Alt. 1500 to 2400 m. Flowering observed in May and June.

DISTRIBUTION IN BORNEO: SABAH: Crocker Range; Mt. Kinabalu.

GENERAL DISTRIBUTION: Endemic to Borneo.

DERIVATION OF NAME: The generic name is derived from the Greek *dendron*, tree, and *bios*, life, referring to the epiphytic habit. The specific epithet refers to the type locality Mt. Alab (1932 m), in the Crocker Range of Sabah.

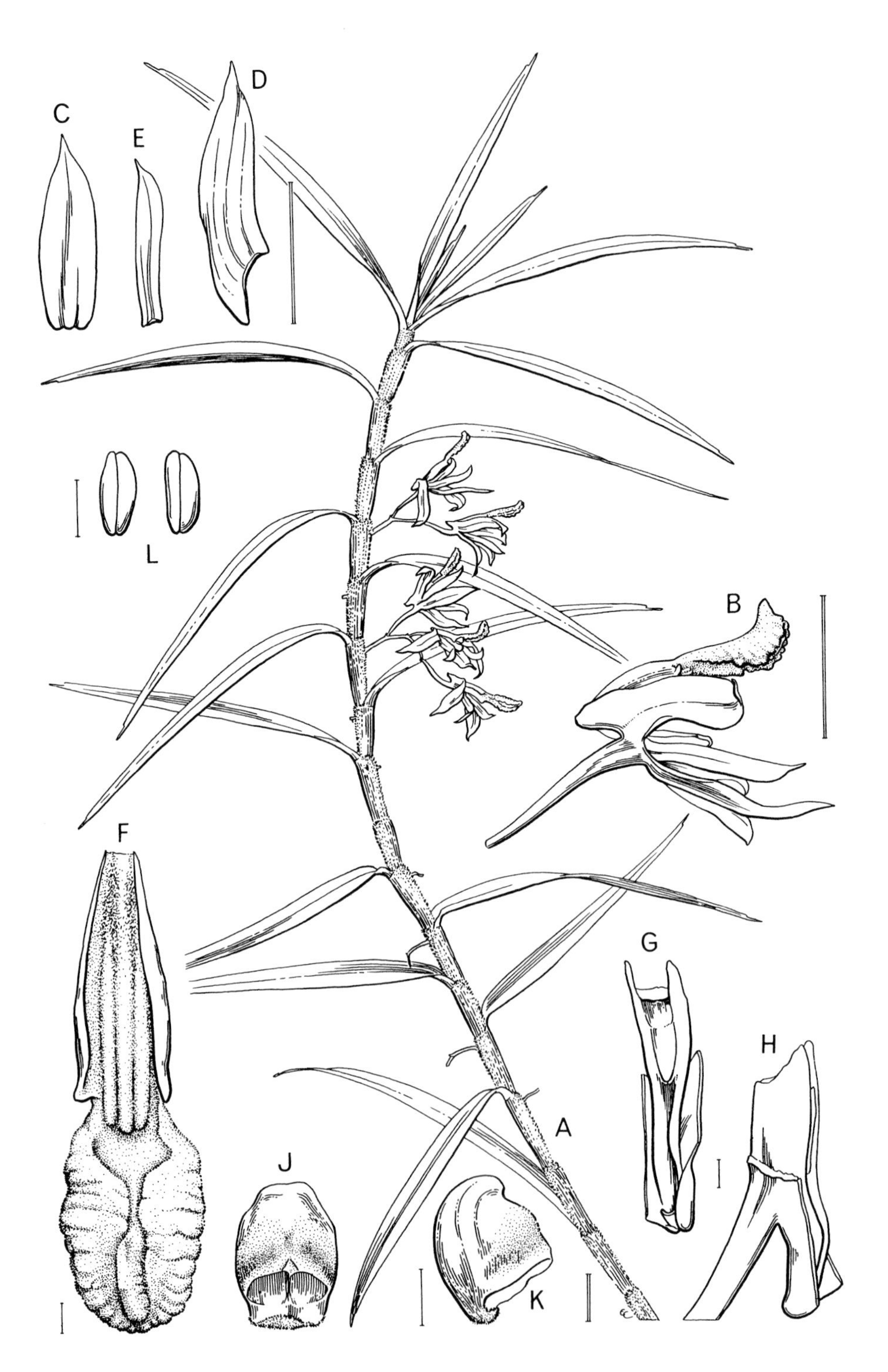

Figure 19. Dendrobium beamanianum J.J. Wood. - A: habit. - B: flower, side view. - C: dorsal sepal. - D: lateral sepal. - E: petal. - F: lip, front view. - G: column, front view. - H: pedicel with ovary and column, side view. - J: anther-cap, front view. - K: anther-cap, side view. - L: pollinia. A and F-L drawn from *Beaman* 10722 (holotype) and B-E from *Lamb* AL 52/83 by Eleanor Catherine. Scale: single bar = 1 mm; double bar = 1 cm.

19. DENDROBIUM BEAMANIANUM J.J. Wood

Dendrobium beamanianum *J.J. Wood* in Wood, Beaman & Beaman, Plants of Mt. Kinabalu 2, Orchids: 163, fig. 18 (1993). Type: Borneo, Sabah, Mt. Kinabalu, Pinosuk Plateau, *c.* 8 km E.S.E. of Desa Dairy, *Beaman* 10722 (holotype K).

Epiphyte. Stems 30–60 × 0.3–0.5 cm, erect to spreading. ***Leaves*** linear-ligulate, acutely unequally bilobed; blade 3.5–8(–9) × 0.3–0.6(–0.8) cm; sheaths 1.5–2 cm long, densely black-hirsute, the lowermost becoming glabrous. ***Inflorescences*** 1– to 2(–3)-flowered, usually only 1 open at a time, emerging from below top of leaf sheath opposite blade; peduncle and rachis (0.5–)1–1.5 cm long, glabrous; floral bracts 1 mm long, triangular-ovate, acute, yellowish. ***Flowers*** 2.5–2.8 cm across, non-resupinate; pedicel with ovary orange and greenish; sepals pale brown or pale creamy yellow, with pale orange or orange-brown veins; petals pale brown or pale creamy yellow; lip pale yellow or pale brown with a white keel and brownish yellow wrinkled epichile; column pale green; anther-cap white. ***Pedicel*** with ***ovary*** 1–1.2 cm long, straight. ***Sepals*** spreading, acute to acuminate, stiff. ***Dorsal sepal*** 1.1–1.4 × 0.45–0.5 cm, oblong-elliptic. ***Lateral sepals*** 1.7–1.8 × 0.4 cm, obliquely narrowly elliptic. ***Mentum*** 4–5 mm long, conical. ***Petals*** 1–1.5 × 0.2–0.25 cm, linear-spathulate, acute, very minutely erose. ***Lip*** 1.5–1.6 cm long, 3-lobed, fleshy, slightly to densely white farinose, divided into a hypochile and epichile; hypochile 8–9 × 3 mm, lobes erect, rounded, 1.5–2 mm high, disc with two fleshy keels extending from base and terminating on base of epichile, between which is a shorter apical median keel, outer keels farinose, particularly near the base; epichile 6–8 × 3–5 mm, oblong, obtuse, margin at first conduplicate, then parting, very fleshy and rugose, with a variably sized but usually prominent erect fleshy apical keel in between. ***Column*** 2 mm long, oblong, with obtuse triangular-oblong stelidia *c.* 1 mm long; foot 4–5 mm long, sulcate; anther-cap 2 × 1 mm, cucullate.

HABITAT AND ECOLOGY: Lower montane forest on ultramafic substrate. Alt. 1200 to 1700 m. Flowering observed in July.

DISTRIBUTION IN BORNEO: SABAH: Mt. Kinabalu.

GENERAL DISTRIBUTION: Endemic to Borneo.

NOTES: *D. beamanianum* has so far only been recorded from lower montane forest on ultramafic substrate. It is closely related to the Javan *D. corrugatilobum* J.J. Sm. (section *Conostalix*), but is distinguished by its longer leaves, larger flowers with longer sepals and a lip epichile with a distinct fleshy apical keel.

DERIVATION OF NAME: The specific epithet honours Professor John H. Beaman, formerly of Michigan State University, USA, who, together with his wife Dr Teofila E. Beaman and son Reed S. Beaman, has contributed significantly to our knowledge of the flora of Mount Kinabalu.

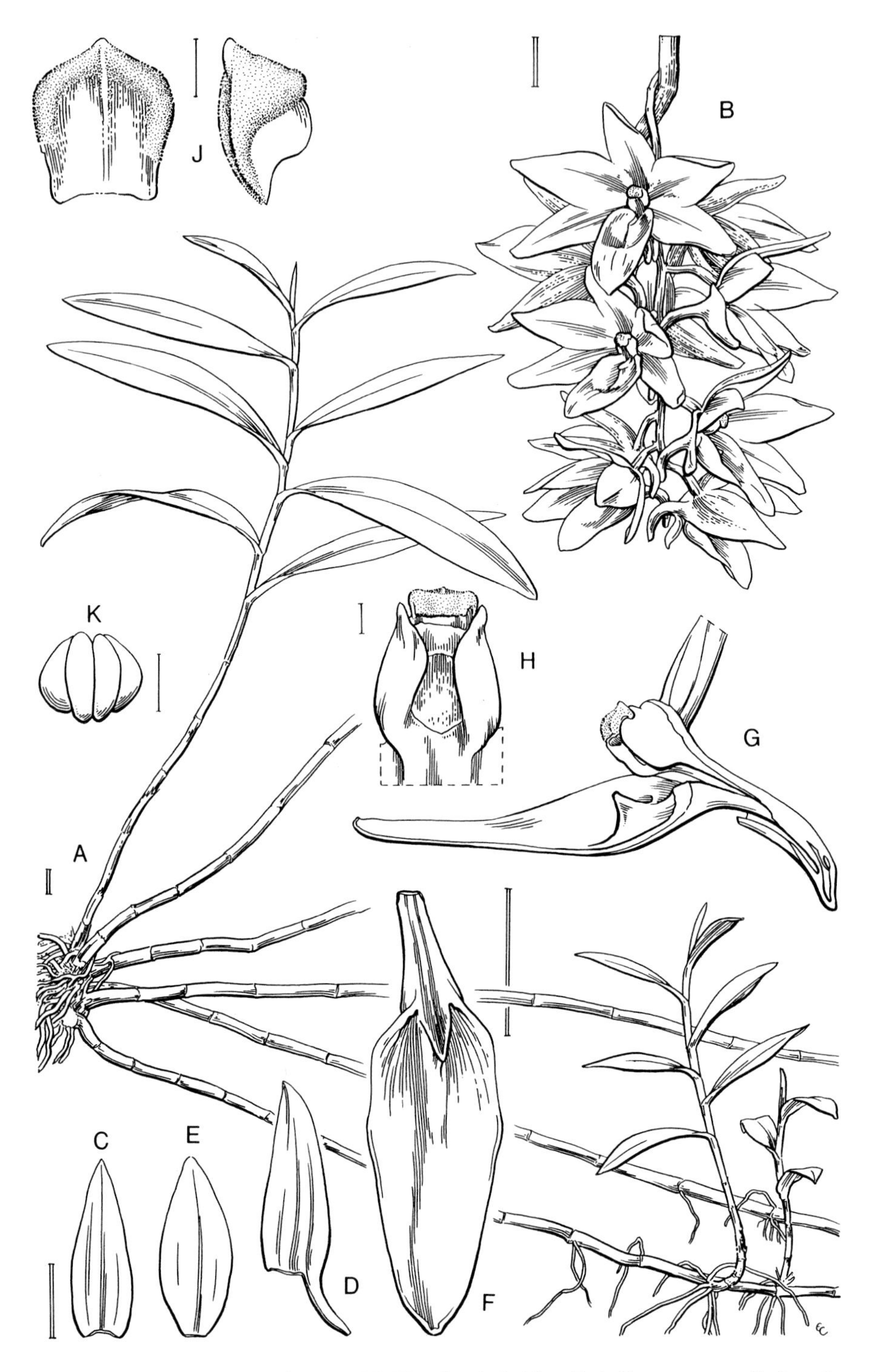

Figure 20. Dendrobium cymboglossum J.J. Wood. - A: habit. - B: inflorescence. - C: dorsal sepal. - D: lateral sepal. - E: petal. - F: lip, front view. - G: pedicel with ovary, column with anther-cap, and longitudinal section through lip and mentum, side view. - H: column and anther-cap, front view. - J: anther-cap, back and side views. - K: pollinia. A (habit) drawn from *Creagh* s.n. (holotype), B (inflorescence) from a photograph by *Lamb* and C–K from *Lim Weng Hee* 6.18 by Eleanor Catherine. Scale: single bar = 1 mm; double bar = 1 cm.

20. DENDROBIUM CYMBOGLOSSUM J.J. Wood & A. Lamb

Dendrobium cymboglossum *J.J. Wood & A. Lamb* in Wood & Cribb, Checklist Orch. Borneo: 247, fig. 29, plate 8C (1994). Type: Borneo, Sabah, Sandakan, *Governor Creagh* s.n. (holotype K).

Epiphyte. ***Stems*** 30–100 cm long, to 0.8 cm wide, producing keikis from several nodes. ***Leaves*** 11–12 × 1.8–2 cm, narrowly elliptic, acute to acuminate. ***Inflorescence*** up to 8-flowered, pendulous; peduncle 1.5–2 cm long; rachis 3.5–5.5 cm long; floral bracts 3–4 mm long, acute to acuminate. ***Flowers*** *c.* 3.5 cm across; pedicel with ovary pale green, streaked purple-red; sepals cream to yellowish, densely covered in purple-red streaks on exterior, mentum streaked purple-red; petals cream to whitish yellow flushed pink; lip yellow to off-white flushed orange on the sides towards the base, with pinkish apical spots, the callus yellow with dark orange banding on disc in front; column cream. ***Pedicel*** with ***ovary*** 1.5–1.7 cm long, narrowly clavate. ***Dorsal sepal*** 2.4–2.9 × 0.9–1 cm, narrowly elliptic, apex acute and slightly cucullate. ***Lateral sepals*** fused to column-foot, free portion 2.6–3 × 0.9–1 cm, elliptic, asymmetrical, apex acute and slightly cucullate. ***Mentum*** 1.3–2 cm long, 0.3 cm wide at base, narrow, strongly hamate, acute. ***Petals*** 2.5–3 × 1–1.2 cm, elliptic, acute. ***Lip*** 2.5 cm long, 0.8–1 cm wide unflattened, blade entire, oblong-elliptic, obtuse, deeply concave, cymbiform; claw 2 mm wide; callus 4 mm long, *c.* 2.5 mm wide at base, upcurved, fleshy, triangular-acute, thorn-like. ***Column*** 5 × 4 mm, with short oblong, obtuse wings; foot 2 cm long, hamate; anther-cap 3 × 2 mm, oblong-ovate, apiculate. Plate 6A & B.

HABITAT AND ECOLOGY: Lowland and hill forest. Alt. not recorded. Flowering observed in March, July and November.

DISTRIBUTION IN BORNEO: SABAH: Sandakan area; Tawau Hills National Park.

GENERAL DISTRIBUTION: Endemic to Borneo.

NOTES: This member of section *Calcarifera* is distinguished by the narrow, strongly incurved hook-shaped mentum and the unusual entire boat-shaped lip with an upturned thorn-like basal callus. The flowers superficially resemble *D. sarawakense* Ames (syn. *D. multiflorum* Ridl.) which also has a strongly hooked, though broader, mentum. *D. cymboglossum* can be distinguished at once by the curious boat-shaped lip.

DERIVATION OF NAME: The specific epithet is derived from the Latin *cymba*, a boat, and the Greek *glosso*, a tongue, in reference to the lip.

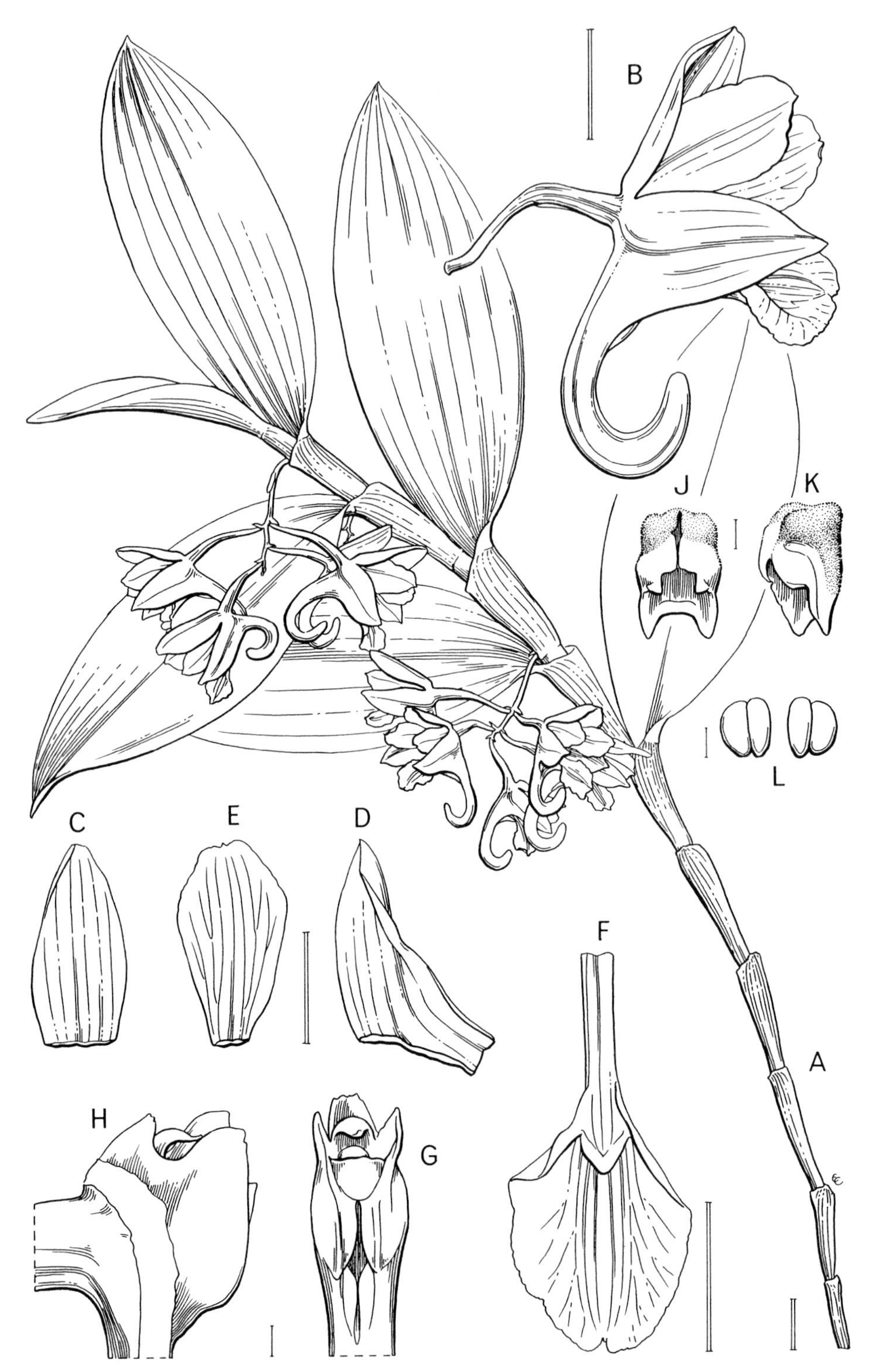

Figure 21. Dendrobium hamaticalcar J.J. Wood & Dauncey. - A: habit. - B: flower, side view. - C: dorsal sepal. - D: lateral sepal. - E: petal. - F: lip, front view. - G: column, front view. - H: column, side view. - J: anther-cap, front view. - K: anther-cap, side view. - L: pollinia. All drawn from *Bacon* 142 (holotype) by Eleanor Catherine. Scale: single bar = 1 mm; double bar = 1 cm.

21. DENDROBIUM HAMATICALCAR J.J. Wood & Dauncey

Dendrobium hamaticalcar *J.J. Wood & Dauncey* in Wood, Beaman & Beaman, Plants of Mt. Kinabalu 2, Orchids: 168, fig. 20, plate 35C (1993). Type: Borneo, Sabah, foothills of Mt. Trus Madi, near Kaingaran, *Bacon* 142, cult. Edinburgh Botanic Garden no. C6614 (holotype E, isotype K).

Epiphyte. ***Stems*** 60–100 cm long, internodes *c*. 2 cm long, *c*. 1 cm thick, arching, somewhat thicker toward apex, sheaths flushed violet. ***Leaves*** 6–10 × 2.5–3.5 cm, thin-textured. ***Inflorescence*** pendulous, more or less perpendicular to stem, from upper nodes of leafy and leafless stems, 4- to 9-flowered; peduncle *c*. 1.2–2.1 cm long, *c*. 1 mm thick; rachis *c*. 0.8–1.6 cm long, pink; floral bracts 2–3 mm long. ***Flowers*** medium-sized, 2.4–3.6 cm long, pale dull yellow, the reverse of sepals and spur darker coloured, veins magenta especially on lip; lip callus yellow; column with a dark magenta line on each side. ***Pedicel*** with ***ovary*** 1.1–1.5 cm long, 1.8–2.6 mm thick, straight or gently curving, perpendicular to mentum. ***Dorsal sepal*** 10–14 × 5.5–8 mm, ovate-elliptic, obtusely rounded. ***Lateral sepals*** obliquely ovate, apex obtuse, apiculate, free distal portion 10.5–15 × 7–9 mm, lower portion extended into a narrow ***mentum*** 24–30 mm long, 2–3.5 mm thick, curving through 180–300°, fused at base for 2.5–9.5 mm. ***Petals*** 10–14.5 × 6–8.5 mm, oblong-spathulate, apex obtusely rounded, somewhat retuse. ***Lip*** 18–23.5 × 2–2.5 mm, spathulate; claw narrow, linear, with callus 5.5–10 mm from base, abruptly expanding into an oblong lamina, apex obtusely rounded, somewhat prominently retuse, apical margin coarsely dentate; callus thick, U-shaped, at base of lamina, 10.5–14.5 × 6–10.5 mm. ***Column*** short, with prominent rhombic stelidia; anther-filament broadly triangular, with a somewhat triangular dorsal protrusion; foot 26–30 × 2–2.5 mm, curved, with a nectary *c*. 2 mm from apex; rostellum absent; stigma narrow; anther-cap 3.5–4.2 × 2.5–3.2 mm, somewhat dorsi-ventrally flattened, lower edge concave; pollinia 2 mm long. Plate 6C.

HABITAT AND ECOLOGY: Lowland and hill forest. Alt. 400 to 900 m. Flowering observed in March, October and November.

DISTRIBUTION IN BORNEO: SABAH: Mt. Kinabalu; Mt. Trus Madi.

GENERAL DISTRIBUTION: Endemic to Borneo.

NOTES: A striking species belonging to section *Calcarifera* with a long mentum that forms a semicircular hook or may even be curved into an almost complete circle. It is closely related to *D. sarawakense* Ames which has a broader lip with two or three longitudinal keels and a shorter, less hooked mentum.

DERIVATION OF NAME: The specific epithet is derived from the Latin *hamatus*, hooked at the tip, and *calcar*, spur, in reference to the hooked, spur-like mentum.

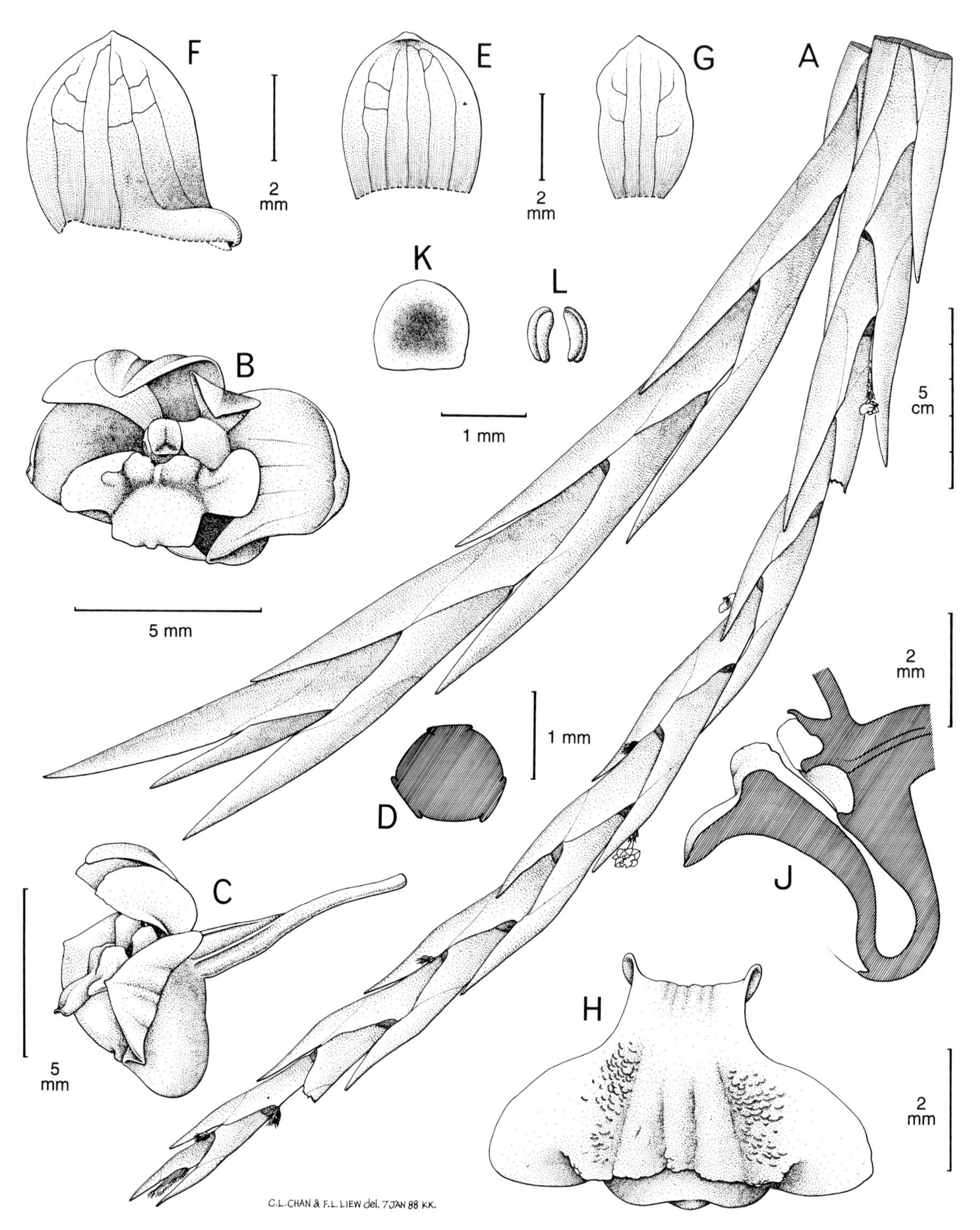

Figure 22. Dendrobium kiauense Ames & C. Schweinf. - A: habit. - B: flower, front view. - C: flower, side view. - D: ovary, transverse section. - E: dorsal sepal. - F: lateral sepal. - G: petal. - H: lip, front view. - J: pedicel with ovary, column and lip, longitudinal section. - K: anther-cap, back view. - L: pollinia. All drawn from a plant cultivated at Tenom Orchid Centre, ex Lohan River, Mt. Kinabalu by C.L. Chan and Lucy F.L. Liew.

22. DENDROBIUM KIAUENSE Ames & C. Schweinf.

Dendrobium kiauense *Ames & C. Schweinf.*, Orchidaceae 6: 103 (1920). Type: Borneo, Sabah, Mt. Kinabalu, Kiau, 900 m, November 1915, *J. Clemens* 176 (holotype AMES, isotype BO).

Dendrobium rajanum J.J. Sm. in Mitt. Inst. Allg. Bot. Hamburg 7: 56, t. 9, fig. 46 (1927). Type: Borneo, Kalimantan Barat/Tengah border, Bukit Raja, *c.* 1250 m, 18 December 1924, *Winkler* 940 (holotype HBG, isotype BO).

Epiphyte. ***Roots*** numerous, fibrous, flexuous, branching, glabrous, up to 1 mm in diameter. ***Stems*** up to 20 per plant, suberect to pendulous, 18–50 cm long, strongly complanate from a suberect slender base, almost or entirely concealed by sheathing leaf bases, yellow, shiny, internodes 1.5–2 cm long. ***Leaves*** 3.5–8.5 cm long (posterior margin), 2.7–6 cm long (anterior margin), 0.6–1.5 cm wide at base, much narrower above, distichous, equitant, scalpelliform, ensiform, acute, falcately incurved and erect or scarcely falcate and ascending, rigidly veined, coriaceous, dark olive green. ***Flowers*** produced successively from few-flowered glomerules of scarious fibrous bracts occurring at the bases of the upper leaves and at stem apex; sepals and petals translucent pale yellow or cream, with pale purple or red veins; lip cream or yellow, the disc with 2 red lines; callus pale to dark yellow; anther-cap white, unscented. ***Pedicel*** with ***ovary*** 5–6 mm long, clavate. ***Dorsal sepal*** 3–3.3 × 2.5 mm, broadly ovate, acute or obtuse. ***Lateral sepals*** 3.7–4 mm long (posterior margin), 7 mm long (anterior margin), 3.7–4 mm wide, broadly triangular-ovate, oblique, obtuse or acute. ***Mentum*** deeply saccate, obtuse. ***Petals*** 3.2–4 × 1.8–2 mm, narrowly ovate to elliptic, acute. ***Lip*** connate with column-foot, parallel to column in natural position, sharply decurved near base, 3-lobed, 4–4.5 mm long, 4–6 mm wide across side lobes (when flattened), spathulate-flabellate in outline; claw broad, concave-involute, abruptly spreading into blade; side lobes 2 × 1 mm, suberect to spreading, oblong, rounded; mid-lobe *c.* 0.7–0.9 × 2–2.8(–3) mm, reflexed, subequal to side lobes, subquadrate, broadly truncate, shallowly retuse or bilobed; disc with a transverse, fleshy, shallowly bilobed callus extending between side lobes at junction with mid-lobe. ***Column*** 3–4 mm long, hood fleshy, broadly truncate, foot broad, saccate, unciform; anther-cap oblong-ovate, cucullate. Plate 6D & E.

HABITAT AND ECOLOGY: Hill forest; lower montane oak-chestnut forest; sometimes on ultramafic substrate. Alt. 800 to 1500 m. Flowering observed in February, April, October and November.

DISTRIBUTION IN BORNEO: KALIMANTAN BARAT/TENGAH border: Bukit Raja. SABAH: Crocker Range; Mt. Kinabalu; Long Pa Sia area.

GENERAL DISTRIBUTION: Endemic to Borneo.

NOTE: *D. kiauense* belongs in section *Aporum* which is represented by 21 named species in Borneo.

DERIVATION OF NAME: The specific epithet refers to the village of Kiau in the western foothills of Mt. Kinabalu, the type locality.

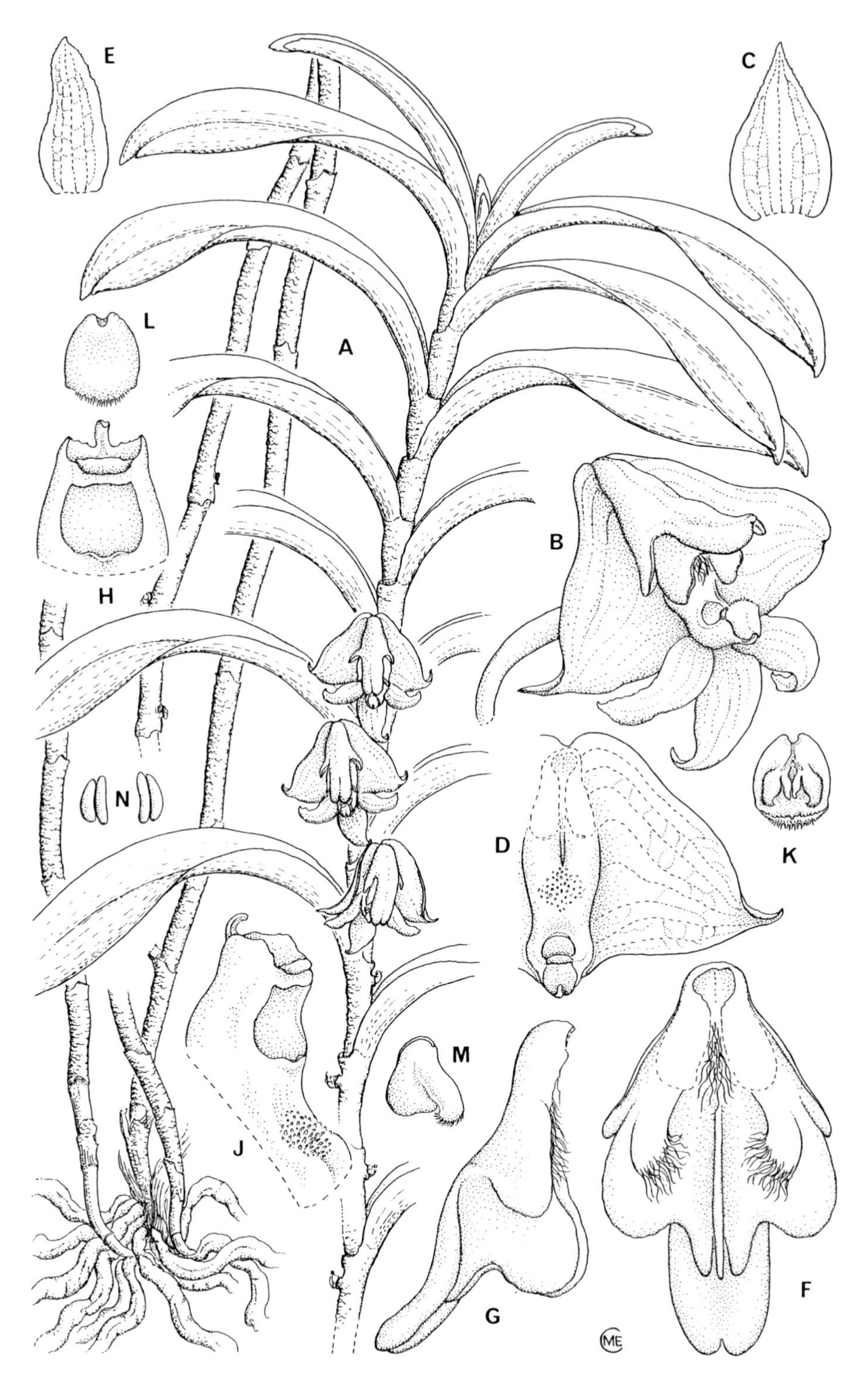

Figure 23. Dendrobium lambii J.J. Wood. - A: habit ×1. - B: flower, front view ×3. - C: dorsal sepal ×3. - D: lateral sepal and column, front view ×3. - E: petal ×3. - F: lip, flattened ×4.5. - G: lip, side view ×4.5. - H: column apex, front view ×6. - J: column, oblique view ×6. - K: anther-cap, front view ×6. - L: anther-cap, back view ×6. - M: anther-cap, side view ×6. - N: pollinia ×6. All drawn from *Lamb* s.n. (holotype) by Maureen Church.

23. DENDROBIUM LAMBII J.J. Wood

Dendrobium lambii *J.J. Wood* in Kew Bull. 38(1): 79, fig. 1 (1983). Type: Borneo, Sabah, Mt. Alab, above Sinsuron, *Lamb* s.n. (holotype K).

Erect ***epiphyte*** to 65 cm high. ***Stems*** leafy, the oldest remaining entirely covered by persistent greyish-brown, distinctly rugose leaf sheaths reminiscent of the genera *Glomera* and *Glossorhyncha*. ***Leaves*** distichous; blade 6–9 × 0.9–1.6 cm, oblong-ligulate, curved, only slightly narrowed at base, apex unequally acutely bilobed and minutely erose, coriaceous; sheaths 1.5–3.5 cm long. ***Inflorescence*** solitary, sessile, emerging opposite leaves; floral bracts *c.* 2–4 mm long, minute, ovate, acute, membranous. ***Flowers*** non-resupinate, white shading to cinnamon-brown and olive-green at first, ageing to orange, apricot or salmon-pink, long-lasting, scented like overipe fruit. ***Pedicel*** with ***ovary*** 1.7–2.3 cm long, clavate, gently curving, obscurely ridged. ***Sepals*** and ***petals*** stiff, reflexed, acute to acuminate. ***Dorsal sepal*** 1.3–1.4 × 0.6–0.7 cm, ovate. ***Lateral sepals*** 1.2–1.5 × 1.2–1.5 cm, broadly triangular-ovate. ***Mentum*** 1–1.5 cm long, obtuse. ***Petals*** 0.9–1.2 × 0.3–0.5 cm, oblong-ovate or oblong-elliptic. ***Lip*** 1.2–1.8 cm long, fleshy, 3-lobed; side lobes 7–9 × 2–3 mm, small, thin, oblong, obtuse, spreading; mid-lobe larger, 8–10 × 2.5–5 mm, fleshy, oblong, usually retuse or emarginate, sometimes mucronate, calli 5, the 4 outer large, thick and fleshy, the lowermost 2 of which are flat and plate-like, the uppermost 2 erect, rounded, 4–7 mm high, each abruptly tapering to 2 low keels which terminate half-way along the mid-lobe and are provided on the inner surfaces with a 3–4 mm long free, strap-like appendage terminating in a fringe of silky hairs; median callus a very narrow, fleshy ridge emerging from a dense tuft of silky hairs and terminating half-way along the mid-lobe. ***Column*** 2–3 × 3.5–5 mm, oblong; foot 1–1.5 cm long, papillose, sulcate below; anther-cap 2 × 2–3 × 3 mm, ovate-cucullate, retuse, hirsute below. Plate 7A.

HABITAT AND ECOLOGY: Lower montane forest; somewhat mossy forest; recorded as an epiphyte on *Dacrydium*. Alt. 1600 to 1800 m. Flowering observed in February and March.

DISTRIBUTION IN BORNEO: SABAH: Mt. Alab.

GENERAL DISTRIBUTION: Endemic to Borneo.

NOTE: *D. lambii*, a member of section *Distichophyllum* Hook.f., differs from the closely related *D. revolutum* Lindl., which is widespread from Myanmar (Burma) and Peninsular Malaysia to Borneo and Sulawesi, in having rugose leaf sheaths, non-resupinate flowers and a lip with spreading but never deflexed, side lobes, a much narrower mid-lobe and two prominent erect calli each provided with an extraordinary fimbriate strap-like appendage.

DERIVATION OF NAME: The specific epithet honours Anthony Lamb, an agricultural officer based in Tenom, Sabah and contributor to this series of volumes.

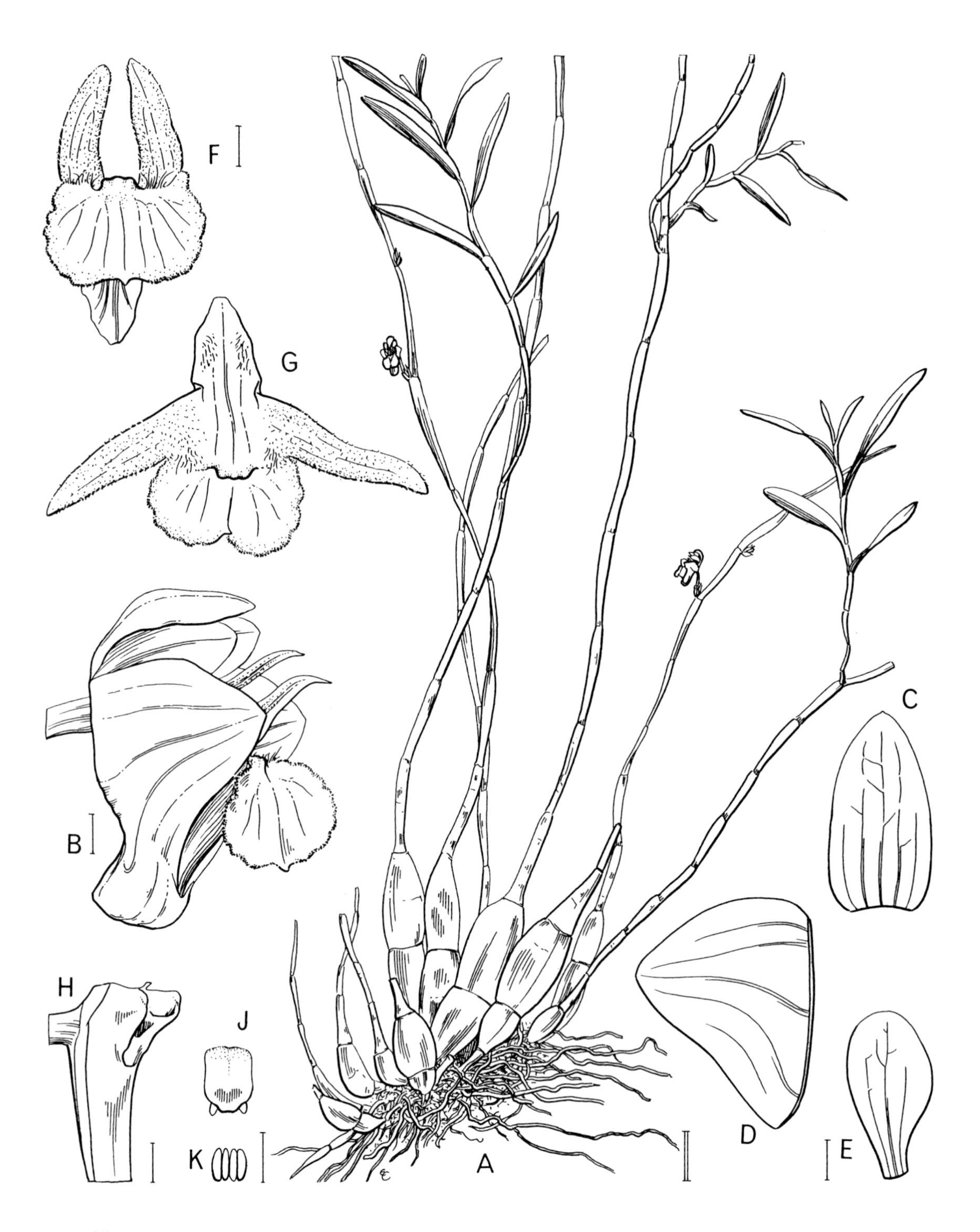

Figure 24. Dendrobium lancilobum J.J. Wood. - A: habit. - B: flower, side view. - C: dorsal sepal. - D: lateral sepal. - E: petal. - F: lip, front view. - G: lip, flattened. - H: column, oblique view. - J: anther-cap, back view. - K: pollinia. All drawn from *Vermeulen & Duistermaat* 1131 (holotype) by Eleanor Catherine. Scale: single bar = 1 mm; double bar = 1 cm.

24. DENDROBIUM LANCILOBUM J.J. Wood

Dendrobium lancilobum *J.J. Wood* in Lindleyana 5(2): 90, fig. 7 (1990). Type: Borneo, Sabah, Sipitang District, east of Long Pa Sia to Long Miau trail, *Vermeulen & Duistermaat* 1131 (holotype K).

Clump-forming ***epiphyte***. ***Stems*** 20–35 cm long, 2–2.5 mm wide, flattened, covered in leaf sheaths, branching and producing roots from nodes, two internodes near base expanded to form flattened, elliptical, shiny straw-yellow pseudobulbs 2–3.5 × 1–1.2 cm. ***Leaves*** 1.8–4.5 × 0.3–0.5 cm, linear-ligulate, attenuate towards the slightly obliquely-obtuse apex, thin-textured; sheaths 1.5–2 cm long. ***Inflorescences*** borne at the nodes of the upper branches, flowers appearing in succession from a tuft of grey-brown, chaffy bracts 2–3 mm long. ***Flowers*** ephemeral, sepals delicate pale apricot or pinkish violet; lip pale yellow to lemon-apricot; column cream-coloured. ***Pedicel*** with ***ovary*** 5 mm long, narrowly clavate. ***Dorsal sepal*** 6 × 3–3.5 mm, oblong, subacute. ***Lateral sepals*** 6 × 7–8 mm, obliquely triangular-ovate, obtuse. ***Mentum*** 6–6.5 × 2.5 mm, with an obtuse, swollen apex. ***Petals*** 5.5 × 3 mm, spathulate, obtuse. ***Lip*** deeply 3-lobed; hypochile 4.5 mm long, with 2 groups of hairs at the base, side lobes 5 × *c.* 1.5 mm, lanceolate, acute, erect to spreading, minutely pubescent; mid-lobe 2.8–3 × 5.8–6 mm, sharply deflexed, transversely oblong, truncate, irregularly and shallowly 3-lobed, the margin irregular; disc with a flat, glabrous, central callus and a group of hairs on either side. ***Column*** 1 mm long; foot 6 mm long, concave at base; anther-cap *c.* 1.5 × 1.1 mm, oblong, truncate. Plate 7B.

HABITAT AND ECOLOGY: Very open "kerangas" forest consisting of an open field layer of Araceae, *Bromheadia* and *Cyperus*, low, dense shrubs and areas of open, taller-canopied forest of *Eugenia*, *Ficus* and *Tristania*, with climbing bamboo; lower montane forest. Alt. 900 to 1700 m. Flowering observed in October and December.

DISTRIBUTION IN BORNEO: KALIMANTAN: Locality unknown. SABAH: Sipitang District.

GENERAL DISTRIBUTION: Endemic to Borneo.

NOTES: *D. lancilobum* belongs to section *Rhopalanthe* and is allied to the Javanese *D. tricuspe* Lindl. from which it differs by its shorter, lower pseudobulbs of two internodes, narrower leaves, shorter column-foot, broader mentum, lip with much longer, acute side lobes, strongly deflexed mid-lobe, and a disc with a flattened fleshy callus. *D. incurvociliatum* J.J. Sm., also from Borneo, is distinguished from *D. lancilobum* by the triangular, acute mid-lobe of the lip with inflexed ciliate margins.

DERIVATION OF NAME: The specific epithet is derived from the Latin *lancea*, a lance or light spear, and *lobus*, lobe, referring to the lanceolate lip side lobes.

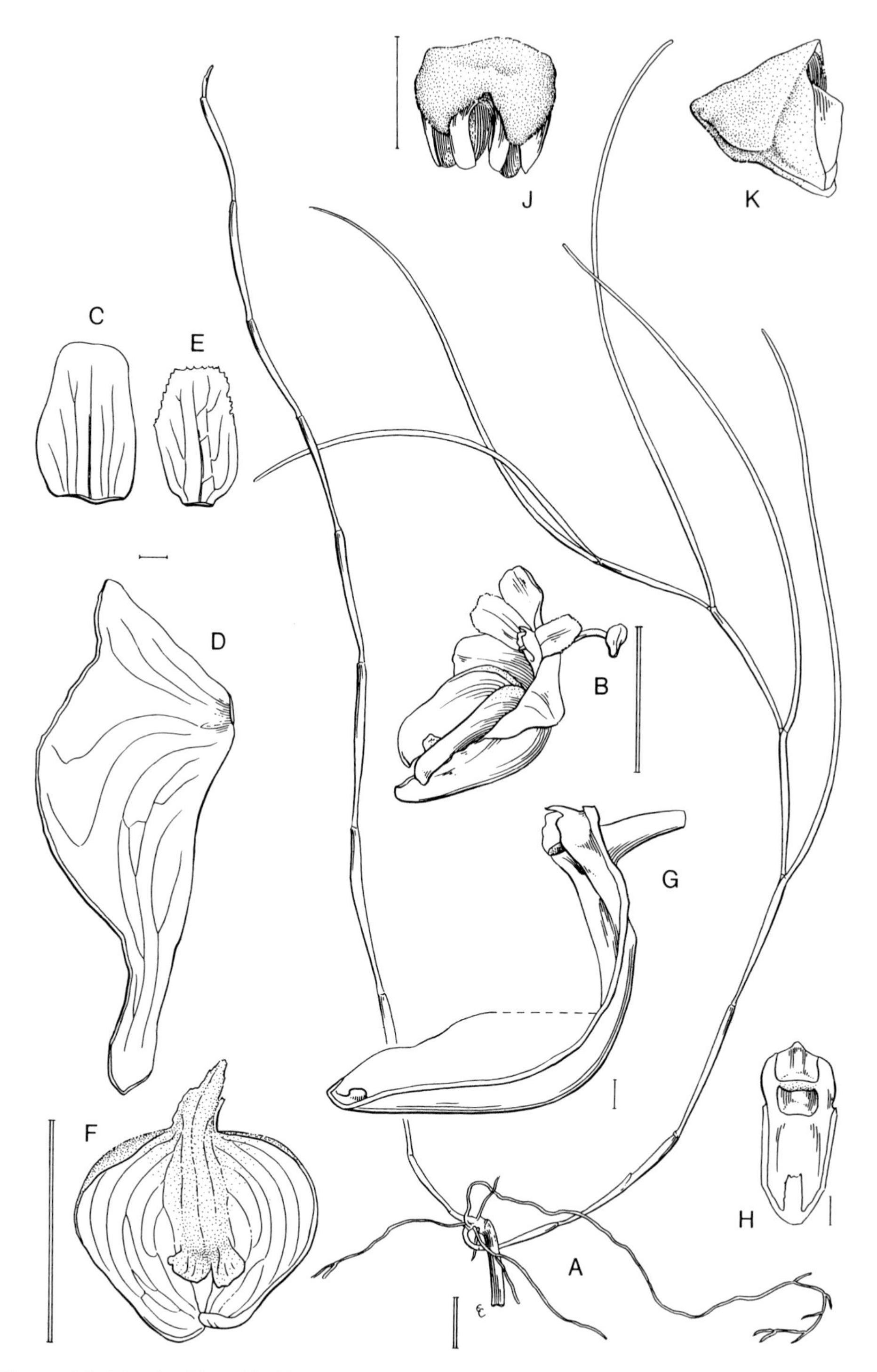

Figure 25. Dendrobium limii J.J. Wood. - A: habit. - B: flower, side view. - C: dorsal sepal. - D: lateral sepal. - E: petal. - F: lip, flattened. - G: pedicel with ovary, column and part of mentum, side view. - H: column, front view. - J: anther-cap with pollinia, front view. - K: anther-cap, side view. All drawn from *Lim Weng Hee* TT5.4 (holotype) by Eleanor Catherine. Scale: single bar = 1 mm; double bar = 1 cm.

25. DENDROBIUM LIMII J.J. Wood

Dendrobium limii *J.J. Wood* in Wood & Cribb, Checklist Orchids Borneo: 254, fig. 31 (1994). Type: Borneo, Sabah, Tawau Hills National Park, *Lim Weng Hee* TT 5.4 (holotype K).

Epiphyte. **Stems** 20–60(–90) cm long, slender, 1.8 mm wide, lowermost node swollen to 3 mm wide, internodes 3–5 cm long. ***Leaves*** 4.5–12 cm long, to 1.7 mm wide, terete, not sulcate. ***Flowers*** 1.5 cm long, 0.5 cm wide, borne in succession from small groups of chaffy bracts at the nodes of the leafless apical part of the stem, ephemeral; sepals translucent white, tip of lateral sepals suffused pink; petals white, suffused pink at base; lip white with 2 purple basal spots. ***Floral bracts*** 2.5 mm long, oblong-elliptic, obtuse, hyaline. ***Pedicel*** with ***ovary*** 6 mm long. ***Dorsal sepal*** 4.5–5 × 3 mm, oblong, truncate, apical margins somewhat cucullate. ***Lateral sepals*** 7 mm wide, triangular-ovate, obtuse, anterior margin 1.2 cm long, fused to column-foot. ***Mentum*** 1.5 cm long, conical, curved. ***Petals*** 4 × 2.5 mm, oblong, truncate, margins erose above. ***Lip*** 1 × 1.3 cm; side lobes large, 1 cm long, *c.* 4.8 mm wide at the middle, 2 mm wide near apex, falcate, obtuse; mid-lobe small, bilobulate, lobules 2 × 2 mm, oblong-spathulate, margins irregularly crenulate and overlapping side lobes; disc without ridges. ***Column*** 1.5 mm long; foot 1.5 mm long, with a rounded apical gland; anther-cap 1 × 1 mm, ovate, obtuse, minutely papillose.

HABITAT AND ECOLOGY: Not recorded. Flowering observed in Decmber.

DISTRIBUTION IN BORNEO: SABAH: Tawau Hills National Park.

GENERAL DISTRIBUTION: Endemic to Borneo.

NOTES: *D. limii* is distinguished from other members of section *Rhopalanthe*, to which it belongs, by the character combination of a distinctive broad lip with large falcate side lobes, a small bilobulate mid-lobe, oblong, truncate, erose petals and terete leaves.

DERIVATION OF NAME: The specific epithet honours Dr W.H. Lim of the Malaysian Agricultural Research and Development Institute (MARDI), who collected the type.

26. DENDROBIUM LOWII Lindl.

Dendrobium lowii *Lindl.* in Gard. Chron. 1861: 1046 (1861). Type: Borneo, cult. *Messrs. H. Low & Co., Low* s.n. (holotype K).

Dendrobium lowii Lindl. var. *pleiotrichum* Rchb.f. in Gard. Chron., n.s., vol. 24: 424 (1885). Type: Borneo, ex cult. *Edward Low* (holotype W).

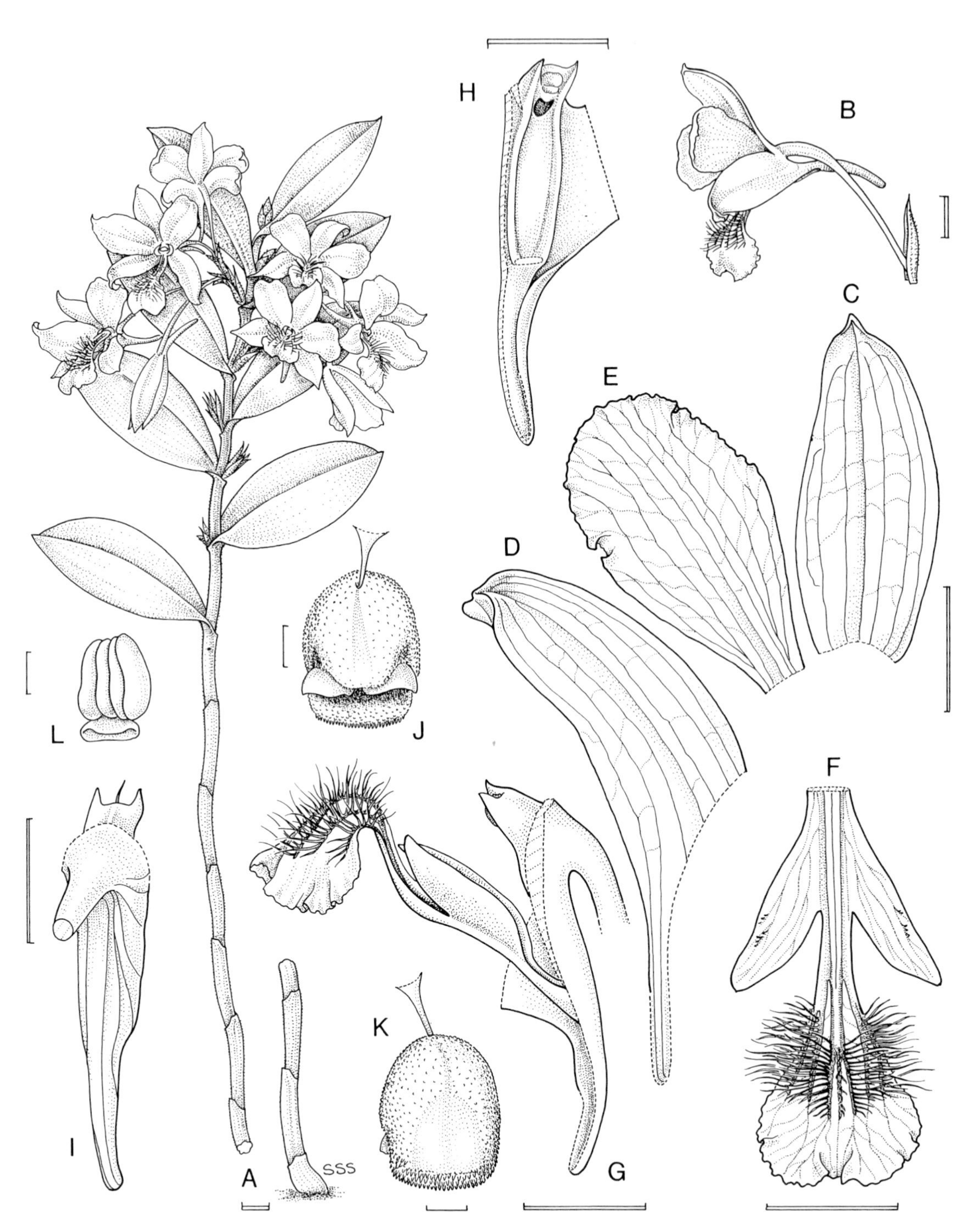

Figure 26. Dendrobium lowii Lindl. - A: habit. - B: flower, side view. - C: dorsal sepal. - D: lateral sepal. - E: petal. - F: lip, flattened. - G: apex of pedicel with ovary, mentum, lip and column, side view. - H: column and mentum, oblique view. - I: apex of pedicel with ovary, mentum and column, back view. - J: anther-cap, front view. - K: anther-cap, back view. - L: pollinarium. A (habit) drawn after a specimen in Herbarium Lehmannianum at Kew and Curtis's Botanical Magazine plate 5303, and B–L from *Lamb & Tan* s.n. by Susanna Stuart-Smith. Scale: single bar = 1 mm; double bar = 1 cm.

Epiphyte. Stems to 40 cm long, erect, internodes 2–3.5 cm long. ***Leaves*** 6.5–6.7 × 2.4–2.8 cm, oblong-elliptic or oblong-ovate, apex obtusely unequally bilobed, nigrohirsute on reverse, sheaths 2–3.5 cm long, nigrohirsute, entirely enclosing stem. ***Inflorescence*** abbreviated, 2- to 7-flowered; peduncle and rachis 0.8–1 cm long; floral bracts 1–2 × 0.5–0.6 cm, ovate-elliptic, acute to acuminate, nigrohirsute. ***Flowers*** up to 5 cm across, yellow, usually with 6–8 orange-red to crimson keels on lip mid-lobe, each bearing orange-red to crimson laciniae, isthmus between side and mid-lobes with orange veins, the side lobes with obscure orange veins (orange and crimson markings lacking in var. *pleiotrichum*). ***Pedicel*** with ***ovary*** 3–4.5 cm long, pedicel slender, ovary gently curved. ***Sepals*** with prominent veins, especially median, on reverse. ***Dorsal sepal*** 2.3–2.5 × 1.1–1.2 cm, oblong-elliptic, prominently obtusely or acutely dorsally carinate at apex, erect. ***Lateral sepals*** posterior margin 2.5–2.7 cm long, anterior margin 4 cm long, 1.2 cm wide at base, 0.8 cm wide at apex, obliquely triangular-ovate, prominently obtusely or acutely dorsally carinate at apex, spreading. ***Petals*** 2.5–3.2 × 1.4–1.6 cm, oblong-elliptic, obtuse, margin undulate. ***Mentum*** 2.3–3 cm long, narrowly conical, infundibuliform, narrowly obtuse, gently decurved. ***Lip*** distinctly 3-lobed, 3-lobed blade 2.2–3.2 cm long; claw 1.3–1.5 cm long; side lobes 1.1–1.2 × 0.4 cm, linear-lanceolate or oblong, obtuse, incurved, central portion of veins on inner surface with or without numerous hairs; distal part of mid-lobe spathulate-flabellate to subrotund, emarginate, decurved and abruptly expanded from a long narrow claw, 1.7–2.2 × 1.2–1.6 cm, margin undulate, disc with six to eight low keels bearing numerous finely branched laciniae, producing a 'bearded' effect. ***Column*** 0.4–0.6 × 0.4–0.5 cm; foot 2.3–3 cm long; stelidia short, triangular, acute; anther-cap 3.5 × 3–3.1 mm, ovate, cucullate, papillose. Plate 7C & D.

HABITAT AND ECOLOGY: Lower montane forest. Alt. 900 m. Flowering observed (in cultivation) from May to September, November.

DISTRIBUTION IN BORNEO: BRUNEI: Locality unknown. KALIMANTAN SELATAN: Banjarmasin area (probably Muratus Mountains to the east). SABAH: Tawau District.

GENERAL DISTRIBUTION: Endemic to Borneo.

NOTES: A beautiful species belonging to section *Formosae* (syn. *Nigrohirsutae*) to which the Bornean *D. parthenium* Rchb.f. and *D. spectatissimum* Rchb.f. also belong. It is unusual among species of the section in having yellow rather than white flowers and consequently caused quite a stir in Victorian England soon after its discovery in 1861. Sir W.J. Hooker, writing in Curtis's Botanical Magazine for 1862 (t. 5303), describes it as a "splendid and remarkable new species, deservedly dedicated to Hugh Low, Esq. of Labuan, who found it on a mountain on the north west of Borneo, at an altitude of 3000 feet above the sea-level, growing on trees in exposed situations, whence it was introduced to his father's nursery at Clapton. It flowered in November, 1861."

A variant with a pale lemon-yellow lip bearing yellow keels with short yellow laciniae and hairs has been described as var. *pleiotrichum* Rchb.f. (see plate 7D).

Another Bornean species, belonging to section *Formosae* and having yellow sepals, is *D. erythropogon* Rchb.f. This is very similar to *D. lowii*, but has rather longer leaves and white rather than yellow petals and lip. It also differs in lip details, the side lobes being much broader and the mid-lobe lacking the narrow claw of *D. lowii* and having short hairs on the keels.

DERIVATION OF NAME: The specific epithet honours Sir Hugh Low, the first European to ascend Mt. Kinabalu, who collected the type.

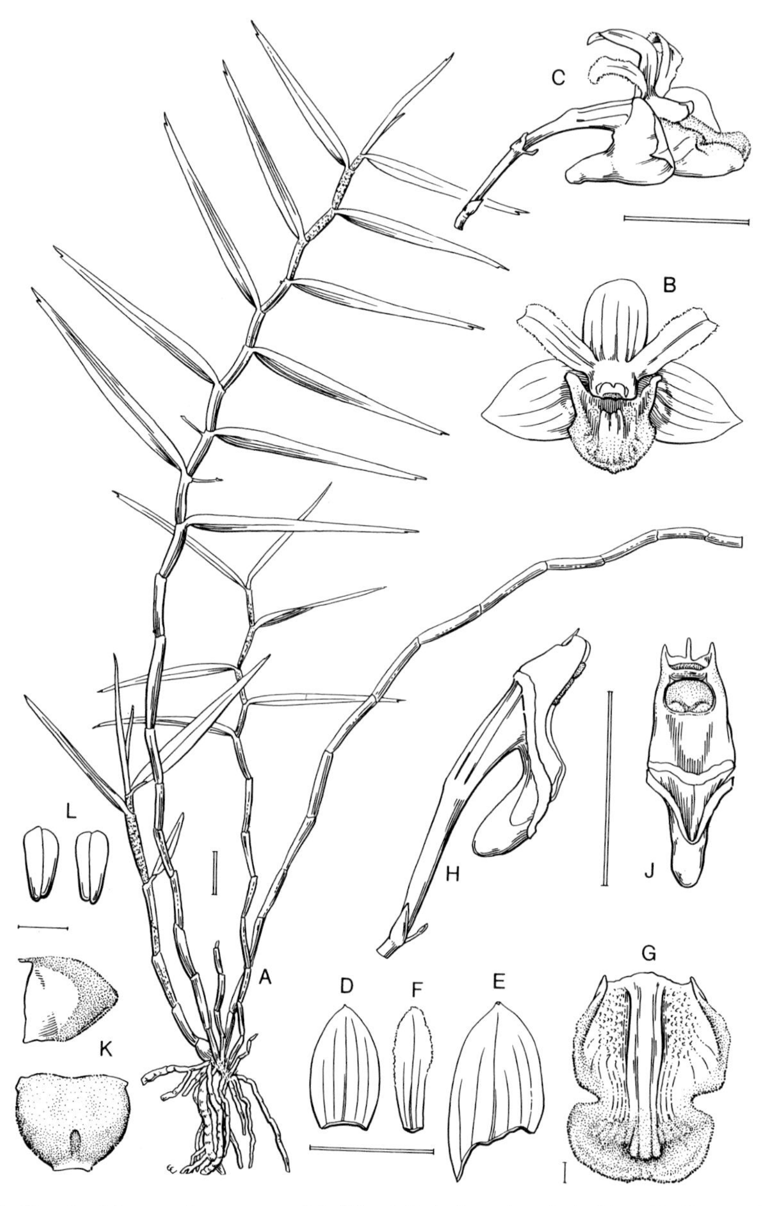

Figure 27. Dendrobium nabawanense J.J. Wood & A. Lamb. - A: habit. - B: flower, front view. - C: flower, side view. - D: dorsal sepal. - E: lateral sepal. - F: petal. - G: lip, flattened. - H: pedicel with ovary, column and mentum, side view. - J: column and mentum, front view. - K: anther-cap, back and side views. - L: pollinia. A (habit) and C-K drawn from *Collenette* 2292 and B (flower) from *Wood* 750 (holotype) by Eleanor Catherine. Scale: single bar = 1 mm; double bar = 1 cm.

27. DENDROBIUM NABAWANENSE J.J. Wood & A. Lamb

Dendrobium nabawanense *J.J. Wood & A. Lamb* in Wood & Cribb, Checklist Orchids Borneo: 258, fig. 32 (1994). Type: Borneo, Sabah, Keningau District, near Nabawan, S.E. of Keningau, *Wood* 750 (holotype K).

Terrestrial or ***epiphyte. Stems*** 24–45(–60) cm high, erect. ***Leaves*** linear-lanceolate, acutely narrowly unequally bilobed, blade (3.5–)6–8 × 0.4–0.6 cm; sheaths 1.2–2.5 cm long, black-ramentaceous, older sheaths becoming less so. ***Inflorescence*** 1-flowered, emerging from base of sheath opposite leaf blade, horizontal to porrect; peduncle and rachis 0.8–1 cm long, filiform, stiff; floral bracts 3, 1 cm long, ovate, acute, minutely ramentaceous. ***Flower*** 2 cm across, unscented, resupinate; sepals and petals white; mentum white with pale salmon-pink veins; lip white, the hypochile veined ochre-brown inside, the epichile flushed apricot-yellow, the disc with a shiny pale ochre-brown ridge; column white, the foot brown, the margins of the stigmatic cavity green; anther-cap white. ***Pedicel*** with ***ovary*** 1.3 cm long, narrowly clavate. ***Sepals*** and ***petals*** stiff, spreading. ***Dorsal sepal*** 9 × 5 mm, ovate-elliptic, apiculate, reflexed. ***Lateral sepals*** obliquely triangular-ovate, acute, lower margin 1.5–1.6 cm long, upper margin 0.9–1 cm long, 0.7 cm wide near base. ***Mentum*** 4–5 mm long, straight, obtuse. ***Petals*** 8–9 × 2.5–3 mm, ligulate to spathulate, subacute, margin minutely erose above, somewhat reflexed. ***Lip*** 1.4–1.6 cm long, pandurate, minutely papillose, fleshy, waxy in appearance; hypochile 6–7 mm long, (6–)7–8 mm wide unflattened, with erect rounded margins, and scattered papillose hairs at base each side of central ridge; epichile 4–5 × 6–9 mm, reniform, very shallowly bilobulate, spreading; disc with a central fleshy, raised, three-ribbed, rather flat, glabrous ridge extending from near base of hypochile to just below apex of epichile, the three ribs more distinct and minutely papillose at the apex, the three veins each side sometimes thickened. ***Column*** 3 mm long, oblong, with short, obtuse apical wings and an acicular anther connective; foot 4–5 mm long; anther-cap 2 × 2 mm, ovate, cucullate, minutely papillose. Plate 7E.

HABITAT AND ECOLOGY: Podsol forest on sandstone soils, composed of *Dacrydium*, *Eugenia* and *Garcinia*, with an understorey of *Rhododendron longiflorum* and *R. malayanum*; common on the lower branches of trees, but seems to avoid *Dacrydium*. Alt. 500 to 700 m. Flowering observed in February, May and July.

DISTRIBUTION IN BORNEO: SABAH: Nabawan area; Lahad Datu District, Mt. Nicola.

GENERAL DISTRIBUTION: Endemic to Borneo.

NOTE: *D. nabawanense* is closely related to *D. hosei* Ridl. (section *Distichophyllum*), from Peninsular Malaysia and Borneo, but differs in having gently rounded lip hypochile lobes, a disc with a central fleshy, raised, rather flat, 3-ribbed elongate ridge and a straight mentum.

DERIVATION OF NAME: The specific epithet refers to the type locality.

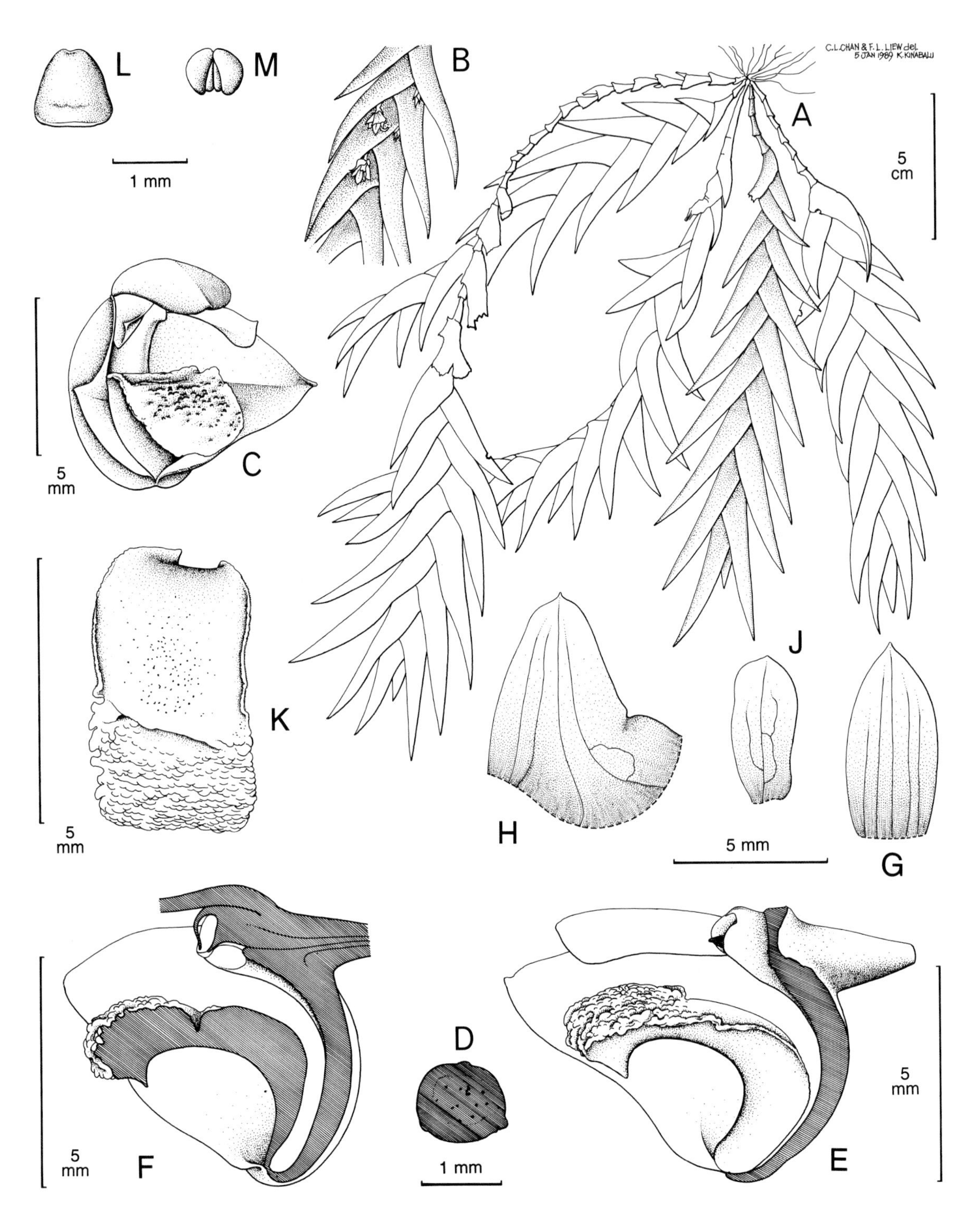

Figure 28. Dendrobium oblongum Ames & C. Schweinf. - A: habit. - B: portion of flowering stem. - C: flower, oblique view. - D: ovary, transverse section. - E. flower with dorsal sepal, lateral sepal and petal removed, side view. - F: flower with lateral sepal and petals removed, longitudinal section. - G: dorsal sepal. - H: lateral sepal. - J: petal. - K: lip, front view. - L: anther-cap, back view. - M: pollinia. All drawn from *Chan & Gunsalam* s.n. by C.L. Chan and Lucy F.L. Liew.

28. DENDROBIUM OBLONGUM Ames & C. Schweinf.

Dendrobium oblongum *Ames & C. Schweinf.*, Orchidaceae 6: 108 (1920). Type: Borneo, Sabah, Mt. Kinabalu, July–August 1916, *Haslam* s.n. (holotype AMES).

Pendulous ***epiphyte. Roots*** numerous, very slender, fibrous, flexuous, *c.* 0.5 mm in diameter, forming a tangled mat. ***Stem*** simple or 2- to several-branched, 6–45 cm long, entirely concealed by imbricate leaves, when exposed slender, complanate, internodes 1 cm long, clavate. ***Leaves*** 3–7 cm long, 0.8–1 cm wide at base, distichous, equitant, closely imbricate, laterally compressed, ensiform, linear-scalpelliform, somewhat spreading to falcate, acute to somewhat mucronate. ***Inflorescences*** abbreviate, peduncle and rachis lateral, to 1 cm long, arising from the bases of the middle and upper leaves, mostly concealed by the imbricate leaves, simple or bifurcate, clothed with distichous imbricate deltoid, acute floral bracts 0.4–1.2 cm long. ***Flowers*** solitary, rather fleshy, pale green, greenish yellow or creamy green; petals often pale yellow with pink edging; lip dull red, purple-red, sometimes dark red proximally, yellowish green with red warts distally. ***Pedicel*** with ***ovary*** 1.8–2 mm long, clavate. ***Dorsal sepal*** 6 × 2.7–3 mm, narrowly elliptic, acute, apiculate, 3-veined. ***Lateral sepals*** 6.5–8 × 6 mm, obliquely, broadly triangular-ovate, apex acuminate, complicate, prominently 4-veined. ***Mentum*** 5 mm long, deeply saccate, obtuse. ***Petals*** 4.2–4.8 × 1.5–2 mm, linear-oblong, the apex broadly rounded, 1-veined. ***Lip*** 6–7 mm long, 3 mm wide at the slightly broader base and apex, ligulate-oblong, shallowly retuse, decurved, anterior and lateral portions thickened, anterior margins erose-crenate, the lateral margins ciliolate, rugose distally, with a small tooth below apex abaxially. ***Column*** 1 mm long, stout, broadly winged; foot 5 mm long, incurved; anther-cap cucullate, papillose.

HABITAT AND ECOLOGY: Hill and lower montane forest; oak-laurel forest with *Agathis*; sometimes on ultramafic substrate. Alt. 900 to 1700 m. Flowering observed in February, June to September, November and December.

DISTRIBUTION IN BORNEO: KALIMANTAN TIMUR: Kutai, Mt. Beratus. SABAH: Mt. Alab; Mt. Kinabalu; Mt. Trus Madi. SARAWAK: Mt. Penrissen.

GENERAL DISTRIBUTION: Endemic to Borneo.

NOTES: *D. oblongum* belongs to section *Oxystophyllum* which is represented by numerous poorly-known species, mostly from Indonesia and New Guinea. *Oxystophyllum* is distinguished from the closely related section *Aporum* by the presence of a small conical wart beneath the apex of the lip.

DERIVATION OF NAME: The specific epithet refers to the oblong-shaped lip.

29. DENDROBIUM PANDURIFERUM Hook. f.

Dendrobium panduriferum *Hook. f.*, Fl. Brit. Ind. 6: 186 (1890). Type: Myanmar (Burma), Pegu, Rangoon, *Gilbert* s.n. (holotype K).

Dendrobium panduriferum Hook.f. var. *serpens* Hook. f., Fl. Brit. Ind. 6: 186 (1890). Type: Peninsular Malaysia, Perak, *Kunstler* s.n. (holotype drawing at CAL).

D. serpens (Hook. f.) Hook. f. in Ann. Bot. Gard. Calc. 5: 10, t. 16 (1895).

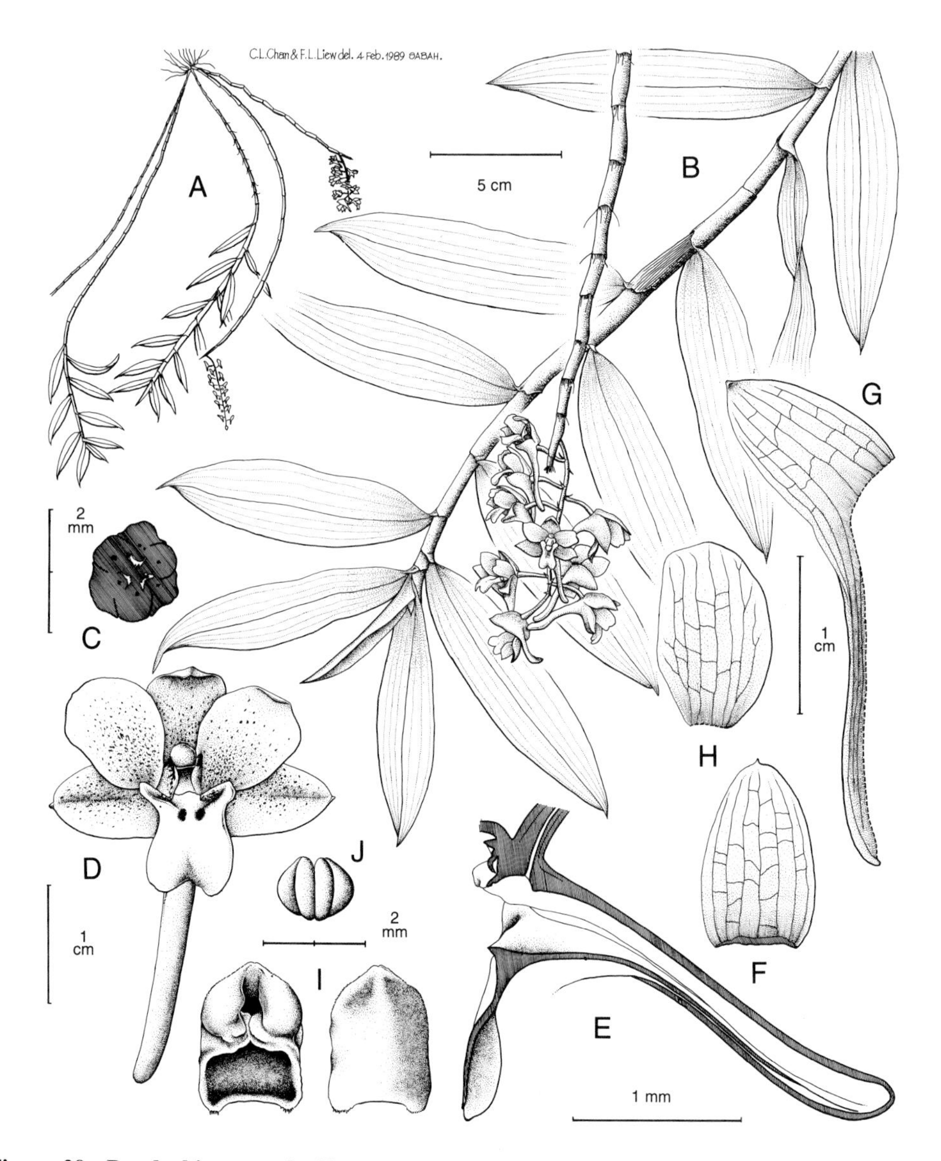

Figure 29. Dendrobium panduriferum Hook. f. - A: habit. - B: leafy stem and flowering stem. - C: ovary, transverse section. - D: flower, front view. - E: column and lip, longitudinal section. - F: dorsal sepal. - G: lateral sepal. - H: petal. - I: anther-cap, front and back views. - J: pollinia. All drawn from *Lamb* TOC 1276 by C.L. Chan and Lucy F.L. Liew.

Callista pandurifera (Hook. f.) Kuntze, Rev. Gen. Pl. 2: 655 (1891), as "*panduriformis*" sphalm.

Dendrobium virescens Ridl. in J. Linn. Soc., Bot. 32: 259 (1896). Type: Thailand, Legeh at Tomoh, *Macado* s.n. (holotype SING).

D. ionopus auct. non Rchb. f.: Kraenzlin 1910: 122 p.p.

Pedilonum panduriferum (Hook. f.) Brieger in Schlechter, Orchideen ed. 3, 1: 681 (1981).

P. serpens (Hook. f.) Brieger, loc. cit. (1981).

Pendulous ***epiphyte. Stems*** 60–100 cm long, 0.6–1 cm wide at middle, narrowest at base, sulcate, internodes 2–4.5 cm long, leafless when old. ***Leaves*** (2.5–)8–12.5 × 1.8–2.2 cm, oblong-elliptic to narrowly elliptic, acuminate, thin-textured, sheaths 2–4.5 cm long, entirely enclosing stem, purplish when exposed to sun. ***Inflorescences*** mostly produced on older, leafless stems, subdensely 6- to 13-flowered, pendulous; flowers borne 2–5 mm apart; peduncle 1.2–1.4 cm long; rachis 3.5–6.5 cm long; floral bracts 2–4 mm long, triangular, acute, pink. ***Flowers*** greenish yellow to yellow speckled red; pedicel with ovary olive-green; mentum yellowish olive-green, flecked purple; lip yellow, spotted red on side lobes and with two red spots or a red patch at base of mid-lobe. ***Pedicel*** with ***ovary*** 1.2–1.5 cm long. ***Dorsal sepal*** 1.3–1.5 × 0.7–0.8 cm, oblong-elliptic, obtuse, shallowly retuse, minutely mucronate, spreading. ***Lateral sepals*** 1.4 cm long, obliquely triangular-ovate, apex dorsally apiculate, spreading. ***Mentum*** 2–2.5 cm long, spur-like, slender, obtuse. ***Petals*** 1.3 × 0.8 cm, oblong to oblong-elliptic, truncate, obtuse, spreading. ***Lip*** 2–2.5 cm long, canaliculate above; limb pandurate; side lobes 0.4–0.5 × 0.4–0.5 cm, auriculate, subacute, porrect; mid-lobe 0.8–0.9 cm long, 0.6 cm wide at base, 0.8–0.9 cm wide at apex, oblong-spathulate, obcordate, retuse, decurved; disc with a raised broadly V-shaped callus between side lobes. ***Column*** 0.4 × 0.3 cm; foot 2–2.5 cm long; anther-cap 0.2–0.3 × 0.2 cm, oblong. Plate 8A.

HABITAT AND ECOLOGY: Hill forest on ultramafic substrate. Alt. 500 to 600 m. Flowering observed in January and May.

DISTRIBUTION IN BORNEO: SABAH: Mt. Kinabalu.

GENERAL DISTRIBUTION: Myanmar (Burma), Peninsular Malaysia, Thailand and Borneo.

NOTES: *D. panduriferum* belongs to section *Calcarifera* which contains many attractive though often little known species, the largest concentration of which is found in Malaysia, Indonesia and the Philippines. Some 21 named species are recorded from Borneo.

DERIVATION OF NAME: The specific epithet is derived from the Latin *panduratus*, fiddle-shaped, i.e. broadest near the top, curving inwards in the lower part, then curving outwards again above the base, in reference to the shape of the lip.

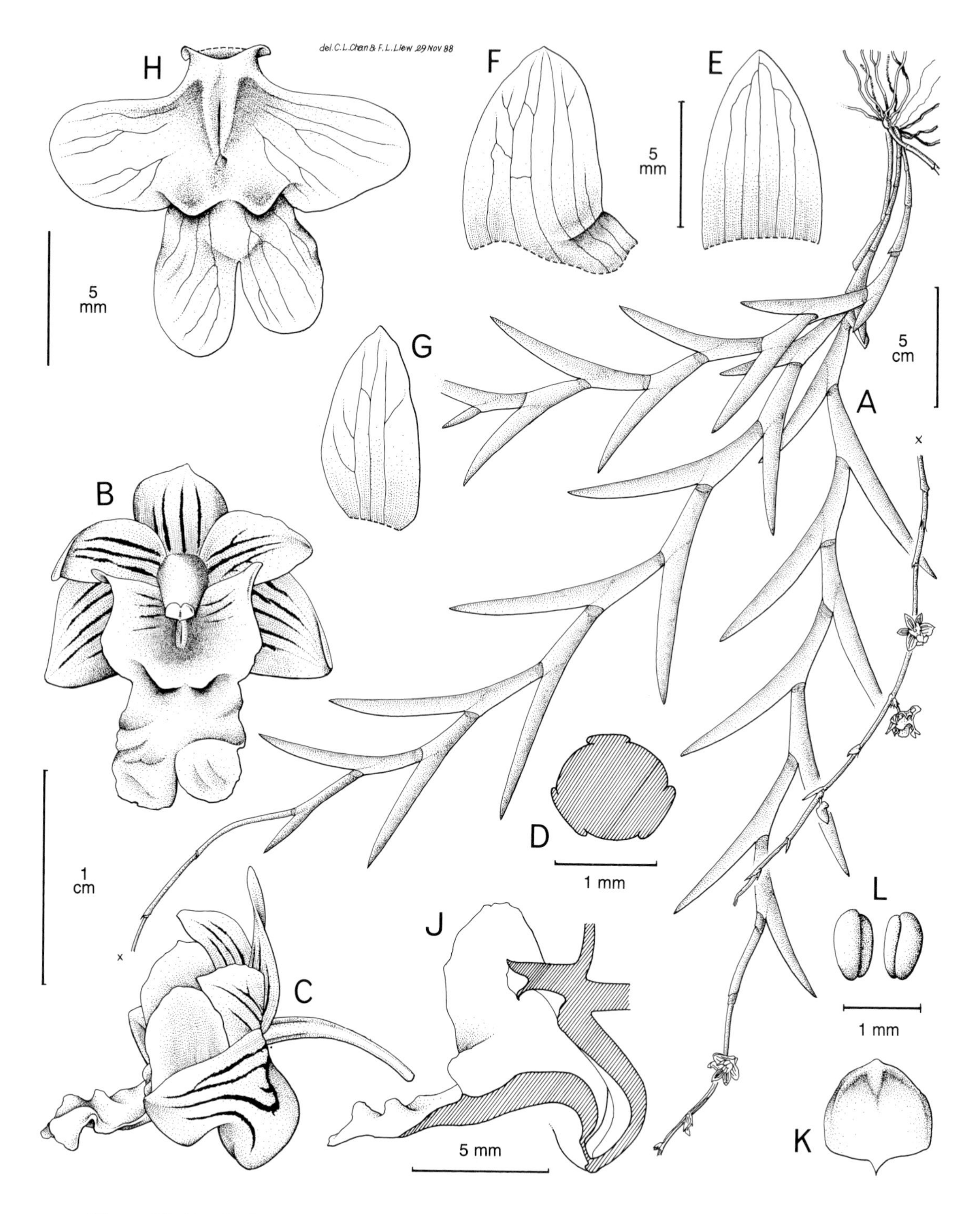

Figure 30. Dendrobium patentilobum Ames & C. Schweinf. - A: habit. - B: flower, front view. - C: flower, side view. - D: ovary, transverse section. - E: dorsal sepal. - F: lateral sepal. - G: petal. - H: lip, flattened. - J: pedicel with ovary, column and lip, longitudinal section. - K: anther-cap, back view. - L: pollinia. All drawn from *Lamb* AL 1117/89 by C.L. Chan and Lucy F.L. Liew.

30. DENDROBIUM PATENTILOBUM Ames & C. Schweinf.

Dendrobium patentilobum *Ames & C. Schweinf.*, Orchidaceae 6: 110 (1920). Type: Borneo, Sabah, Mt. Kinabalu, Marai Parai, *J. Clemens* 366 (holotype AMES, isotypes BO, K, SING).

Erect to spreading ***epiphyte. Rhizome*** abbreviated, to 2 cm long. ***Roots*** numerous, slender, to 1 mm in diameter, wiry, glabrous. ***Stems*** several from rhizome, 28–50 cm long, sinuous, strongly complanate from a slender subterete base, young internodes almost entirely covered by leaf sheaths, upper portion, i.e. $^1/_3$ to $^1/_4$ of the entire length, leafless and producing inflorescences only, internodes 1.2–3.3 cm long, yellow and shiny when exposed, terminated by a dark band. ***Leaves*** 2.5–9.5 cm long, 0.3–0.5 cm wide at base, lowermost and uppermost blades the smallest, distichous, equitant, laterally compressed, narrowly linear, falcate, decurved, obtuse to acute, spreading, coriaceous, rigid; sheaths 1.2–3.3 × 0.3–0.6 cm. ***Inflorescences*** borne at the nodes on the leafless portions of the stems, flowers borne successively one at a time in axils of fasciculate, scarious, becoming fibrous, bracts. ***Flowers*** *c.* 1.6 cm across; pedicel with ovary pink; sepals and petals cream, white or pale lemon-yellow, with purple-red veins, particularly on reverse; lip translucent white or pale yellow, the disc and proximal portion dark purple-red, the side lobes with or without purple veins; calli pale to dark yellow; column pale yellow, flecked purple-red on foot. ***Pedicel*** with ***ovary*** 5–7 mm long. ***Dorsal sepal*** 6.5–7 × 4–4.2 mm, ovate-elliptic, acute or subacute, concave, 3-veined. ***Lateral sepals*** 4–8 mm long, 6.8–7 mm wide at base, *c.* 4.5 mm wide at middle, broadly and obliquely triangular-ovate, obtuse or subacute, falcate, dorsal margin very convex, forming a deeply saccate mentum with column-foot. ***Mentum*** 3–5 mm long, obtuse, retuse, 3-veined. ***Petals*** 6.5–7 × 2.5–3.8 mm, oblong to elliptic-oblong, obtuse or acute, 1- to 3-veined, very variable in size and shape. ***Lip*** 1–1.3 cm long, clawed, semi-orbicular in outline, 3-lobed; claw *c.* 2.2 mm long, ligulate, canaliculate, thickened above; limb sharply reflexed, usually *c.* 7 mm long, *c.* 1.3 cm wide across side lobes; side lobes 5–7 mm long, 4–5 mm wide at base, 3 mm wide above, oval to flabellate or broadly oblong, rounded, margin somewhat undulate, erect; mid-lobe 4–5 × 5–6 mm, suborbicular in outline, deeply bilobed to below the middle, lobules broadly oblong, obtuse; disc with a small, median, narrow V-shaped basal callus near the claw and a transverse, shallowly bilobed fleshy keel extending from centre to centre of each side lobe. ***Column*** 2.7–3 × 1.5 mm, without wings; foot 3–5 mm long, broad, saccate, incurved; anther-cap ovate, cucullate. Plate 8B & C.

HABITAT AND ECOLOGY: Hill and lower montane forest; low stature forest on ultramafic substrate; dipterocarp, Fagaceae, *Gymnostoma*, *Agathis* forest on sandstone and ultramafic substrate; recorded as epiphytic on *Gymnostoma sumatrana*. Alt. 800 to 1500 m. Flowering recorded in January, February, March, June, November and December.

DISTRIBUTION IN BORNEO: SABAH: Danum Valley, Mt. Nicola; Maliau Basin; Mt. Kinabalu.

GENERAL DISTRIBUTION: Endemic to Borneo.

NOTE: *D. patentilobum* belongs to section *Aporum*.

DERIVATION OF NAME: The specific epithet is derived from the Latin *patens*, spreading, and *lobus*, a lobe, in reference to the side lobes of the lip when spread apart.

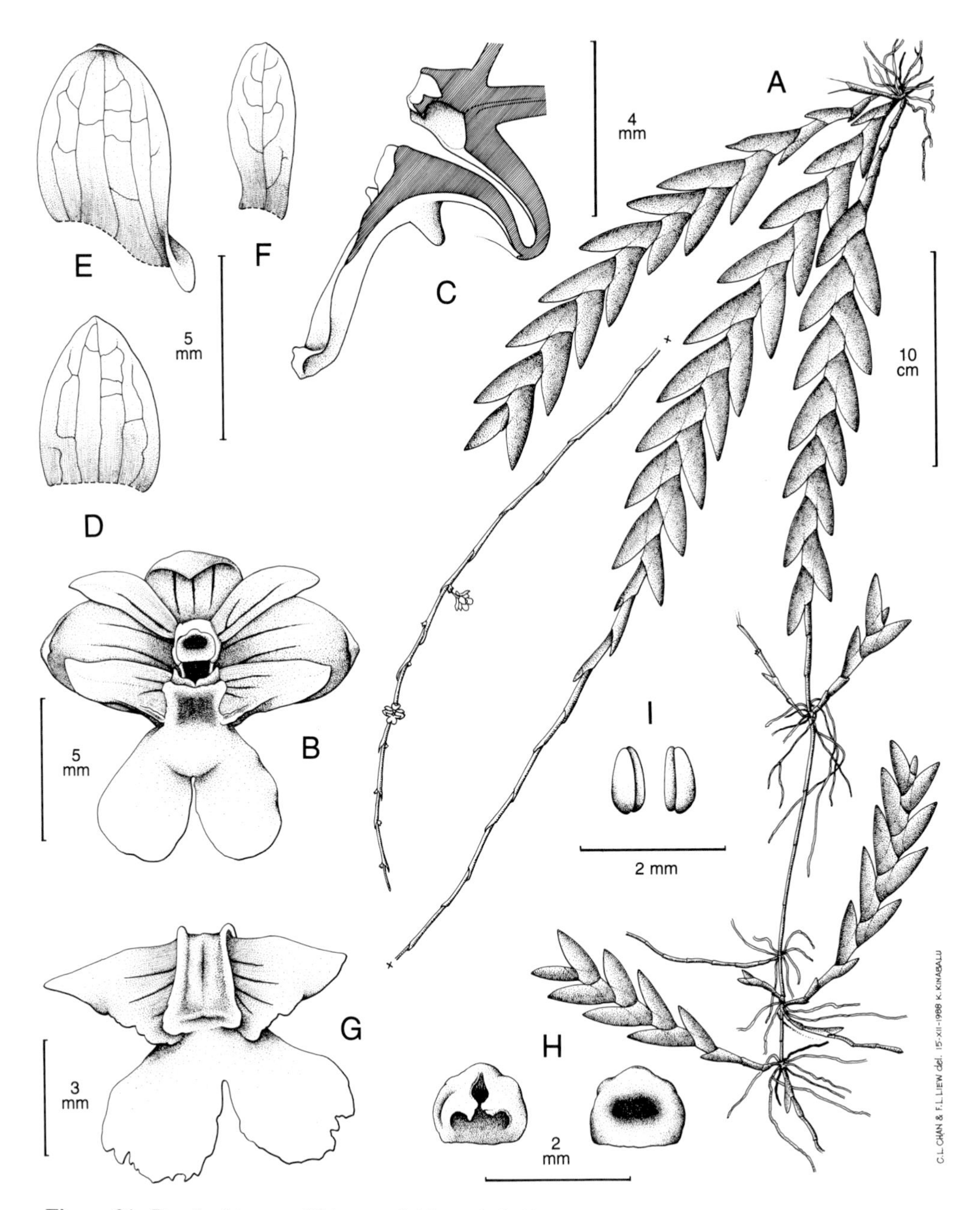

Figure 31. Dendrobium smithianum Schltr. - A: habit. - B: flower, front view. - C: column and lip, longitudinal section. - D: dorsal sepal. - E: lateral sepal. - F: petal. - G: lip, front view. - H: anther-cap, front and back views. - I. pollinia. All drawn from *Lamb* AL 1115/89 by C.L. Chan and Lucy F.L. Liew.

31. DENDROBIUM SMITHIANUM Schltr.

Dendrobium smithianum *Schltr.* in Repert. Spec. Nov. Regni Veg. 10: 74 (1911). Types: Sulawesi, Minahassa, Ayermadidi, *Schlechter* 20522 (syntype B, destroyed, isosyntype K); Toli Toli District, Lampasioe, *Schlechter* 20681 (syntype B, destroyed, isosyntype K).

Dendrobium smithianum Schltr. var. *nebularum* Schltr., loc. cit.: 75 (1911). Type: Sulawesi, Minahassa, Tomohon, *Schlechter* 20433 (holotype B, destroyed, isotype K).

Erect to pendent ***epiphyte. Rhizome*** very abbreviated. ***Roots*** produced freely, filiform, flexuose, glabrous, elongate. ***Stems*** several from rhizome, 15–63 cm long, straight or flexuose, simple or producing keikis from the distal leafless portion, strongly complanate from a narrow base, young internodes almost entirely covered by leaf sheaths, internodes 1–1.5(–2) cm long, yellow and shiny when exposed. ***Leaves*** 2.5–4.5 cm long, 0.5–1.1 cm wide at base, narrower above, lowermost and uppermost blades the smallest, distichous, equitant, laterally compressed, narrowly elliptic, oblong-elliptic, ensiform, acute to acuminate, straight; sheaths 1–1.5(–2) × 0.5–0.9 cm. ***Inflorescences*** borne at the nodes on the leafless portions of the stems, flowers borne successively one at a time in axils of fasciculate, scarious, becoming fibrous, bracts. ***Flowers*** 1.1–1.3 cm across; sepals and petals translucent whitish cream, usually with pink or red veins; lip whitish cream or pale yellow, the callus bright yellow; column purple, the foot purple-red. ***Pedicel*** with ***ovary*** 5–7 mm long, narrowly clavate. ***Dorsal sepal*** 5 × 3.5 mm, ovate, obtuse, main veins 3. ***Lateral sepals*** 6–7 mm long along anterior margin, 5 mm long along posterior margin, 4.2 mm wide at base, 4 mm wide above, obliquely oblong-ovate, obtuse. ***Mentum*** 3–4.5 mm long, obtuse. ***Petals*** 4.5–4.7 × 2 mm, oblong-elliptic, obtuse. ***Lip*** 7–8 mm long, 9 mm wide across side lobes, shortly clawed, decurved at junction of claw and side lobes; claw 2 × 1.5–1.7 mm; side lobes 3–3.2 × 2 mm wide at base, auriculate, oblong-triangular, obtuse, spreading; mid-lobe flabellate in outline, deeply bilobed to below middle, lobules 3–4 × 3 mm, oblong, rounded, margins sometimes erose; disc with a median callus extending from claw to junction of side and mid-lobes, oblong, shallowly canaliculate, apex raised, fleshy, bilobed, with a small swelling in the sinus; also an undulate fleshy crest-like transverse keel extending along anterior edge of side lobes. ***Column*** 1–2 mm long, 1.4–1.5 mm wide at base; foot 3–4.5 mm long, canaliculate, gently incurved at apex; anther-cap broadly ovate, cucullate.

HABITAT AND ECOLOGY: Hill forest on limestone; podsol forest. Alt. 300 to 1300 m. Flowering observed in June and October.

DISTRIBUTION IN BORNEO: SABAH: Batu Ponggol; Long Pa Sia area.

GENERAL DISTRIBUTION: Sulawesi and Borneo.

NOTES: This Bornean member of section *Aporum* has, for the time being, been assigned to *D. smithianum* Schltr., described from Sulawesi. The shorter lip side lobes and narrower mid-lobe more closely resemble var. *nebularum* Schltr. Differences exist in the callus structure, however, and further study of a wider range of material than currently at hand may show it to warrant recognition at specific level.

DERIVATION OF NAME: The specific epithet honours the eminent Dutch orchidologist, Johannes Jacobus Smith (1867–1947).

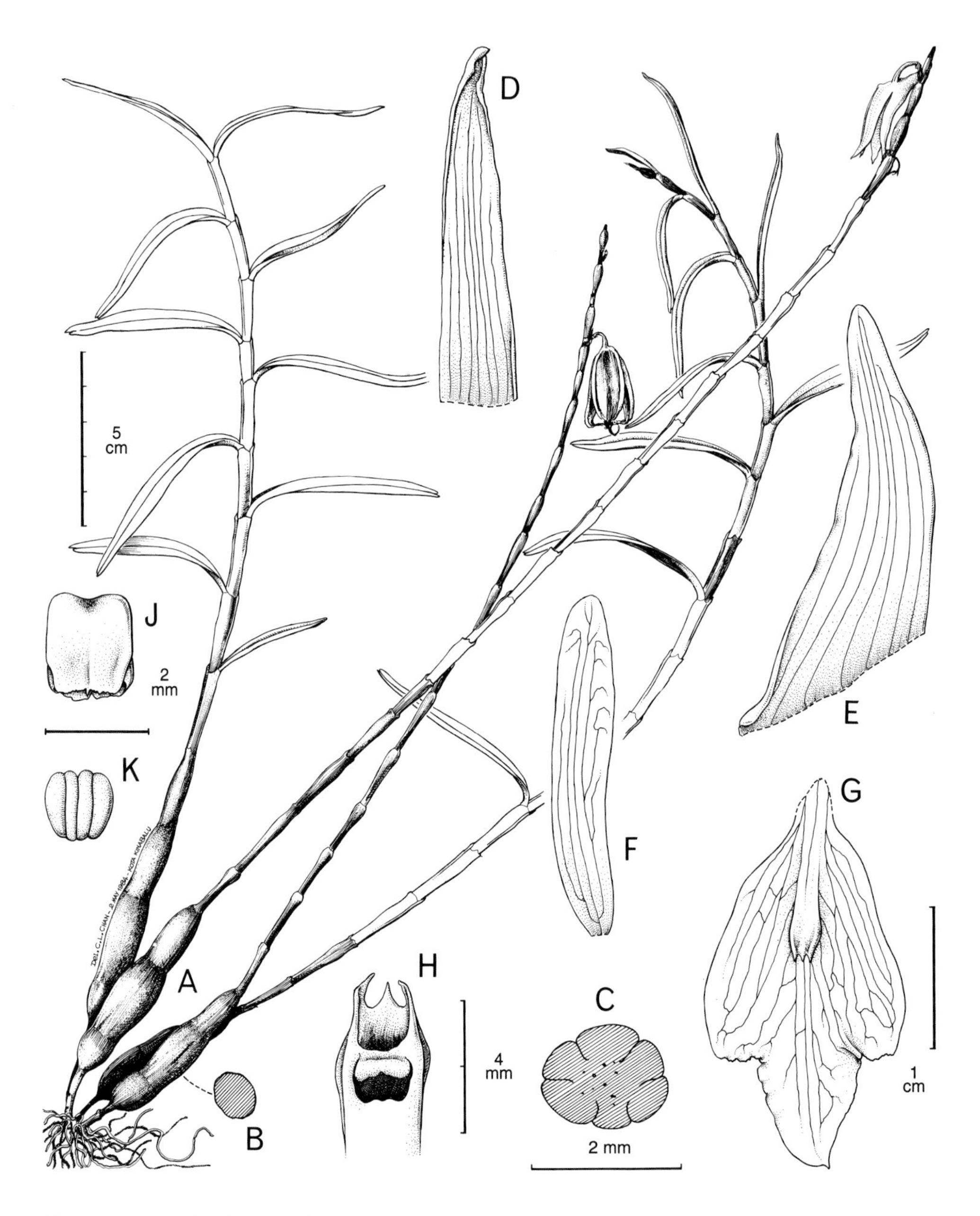

Figure 32. Dendrobium tridentatum Ames & C. Schweinf. - A: habit. - B: pseudobulb, transverse section. - C: ovary, transverse section. - D: dorsal sepal. - E. lateral sepal. - F: petal. - G: lip, flattened. - H: column, front view. - J: anther-cap, back view. - K: pollinia. All drawn from *Beaman* 9574 by C.L. Chan.

32. DENDROBIUM TRIDENTATUM Ames & C. Schweinf.

Dendrobium tridentatum *Ames & C. Schweinf.*, Orchidaceae 6: 115 (1921). Type: Borneo, Sabah, Mt. Kinabalu, Marai Parai Spur, *J. Clemens* 257 (holotype AMES).

Tufted ***epiphyte*** 18–60 cm high. ***Stems*** covered in tubular leaf sheaths, slightly fractiflex above, internodes 1.6–3 cm long, thickened near the base into an oblong, one- to three-jointed, rugose pseudobulbous thickening 1.5–6 cm long, yellow to greenish-brown. ***Leaves*** 5–6.5 × 0.3–0.5 cm, narrowly linear to linear-ligulate, unequally lobed and usually sharply mucronate at apex, sessile, rigid, often very thick and fleshy. ***Inflorescence*** one-flowered, pendulous from the upper nodes. ***Flowers*** translucent white; petals often flushed pink at apex; lip pink at the apex and orange at the base; column white with an orange spot at the base; lip also described as having thin lilac streaks at centre of mid-lobe, yellow 'throat' and central scarlet blotch. ***Pedicel*** with ***ovary*** 8 mm long. ***Sepals*** cucullate at apex. ***Dorsal sepal*** 2 cm long, *c.* 4.6 mm wide near base, linear-oblong, acute to acuminate. ***Lateral sepals*** 2–2.8 cm long, 0.8–1 cm wide at base, 0.3–0.4 cm wide above, narrowly triangular-elliptic, very oblique, somewhat falcate, obtuse to acuminate. ***Mentum*** *c.* 1 cm long, slightly saccate. ***Petals*** *c.* 1.9 cm long, 3.4–5 mm wide above the middle, linear-ligulate, obtuse to acute. ***Lip*** *c.* 2.25 cm long, 1.3 cm wide across side lobes; side lobes shallowly triangular to oblong-triangular, margin erose; mid-lobe *c.* 7.5 mm long, 6.5 mm wide across base, triangular, acuminate; disc hirsute, with a 3-toothed fleshy callus just below the middle. ***Column*** 2 mm long; foot gently curved, to 1 cm long.

HABITAT AND ECOLOGY: Lower and upper montane forest; oak-laurel forest, on ultramafic substrate. Alt. 1200 to 2200 m. Flowering observed in June, August, September and October.

DISTRIBUTION IN BORNEO: SABAH: Mt. Alab; Mt. Kinabalu.

GENERAL DISTRIBUTION: Endemic to Borneo.

NOTES: *D. tridentatum* belongs to section *Rhopalanthe*. Two forms occur, one with narrow, fleshy leaves with a mucronate apex, described by Ames and Schweinfurth, the other, figured here, having broader, less fleshy leaves with an unequally bilobed apex. It is not known whether intermediates occur.

DERIVATION OF NAME: The specific epithet is derived from the Latin *tri-*, three, and *dentatus*, toothed, in reference to the three-toothed fleshy callus on the lip.

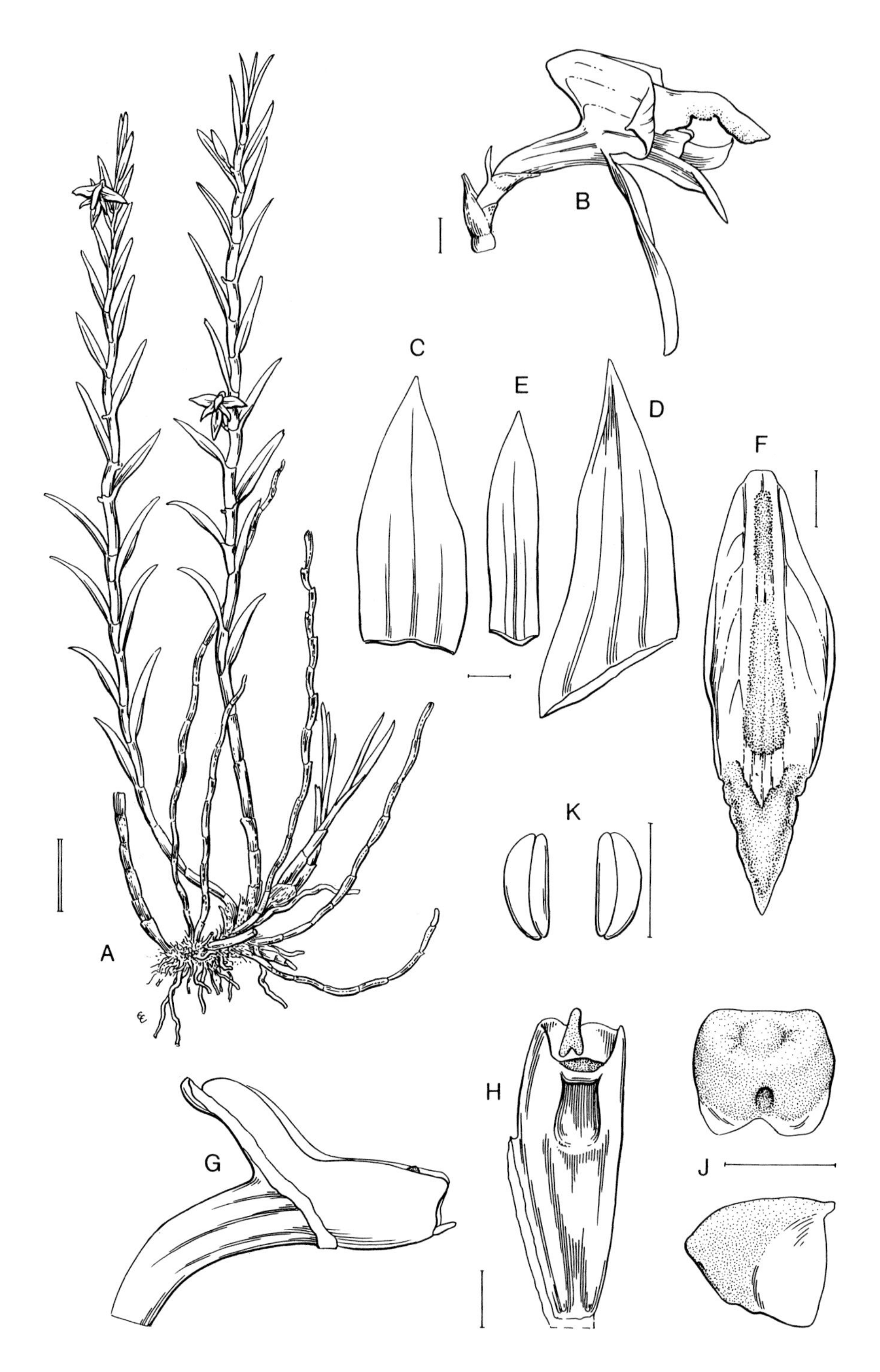

Figure 33. Dendrobium trullatum J.J. Wood & A. Lamb. - A: habit. - B: flower, side view. - C: dorsal sepal. - D: lateral sepal. - E: petal. - F: lip, front view. - G: pedicel with ovary and column, side view. - H: column, front view. - J: anther-cap, back and side views. - K: pollinia. A (habit) drawn from *Lamb* AL 841/87 and B–K from *Wood* 598 (holotype) by Eleanor Catherine. Scale: single bar = 1 mm; double bar = 1 cm.

33. DENDROBIUM TRULLATUM J.J. Wood & A. Lamb

Dendrobium trullatum *J.J. Wood & A. Lamb* in Wood & Cribb, Checklist Orchids Borneo: 269, fig. 33 (1994). Type: Borneo, Sabah, Keningau District, near Nabawan, S.E. of Keningau, *Wood* 598 (holotype K).

Epiphyte. Stems 12–14 cm long, 0.3 cm wide, caespitose, slightly sinuate-fractiflex. ***Leaves*** linear-ligulate, slightly acutely unequally bilobed, stiff; blade 1.2–2 × 0.25–0.4 cm; sheaths 0.5–0.7 cm long, glabrous. ***Inflorescences*** 1–(–2)-flowered, emerging from sheath base opposite leaf blade, sessile; floral bracts 3–4, 1 mm long, ovate, acute. ***Flowers*** *c.* 1.3 cm across, non-resupinate, scented; sepals and petals pale green or bright yellow; lip cream. ***Pedicel*** with ***ovary*** 3 mm long, clavate. ***Sepals*** stiff, spreading. ***Dorsal sepal*** 6 × 2–2.1 mm, triangular-ovate, acute. ***Lateral sepals*** 7.5–8 × 2.5 mm, obliquely triangular-ovate, acute. ***Mentum*** 2 mm long, conical, subacute. ***Petals*** 5.5–6 × 1 mm, linear, acute. ***Lip*** stiff, fleshy, 3-lobed, 6.5–7 mm long, 3 mm wide across side lobes when flattened, trullate in outline; disc ecallose, farinose; mid-lobe 2–2.5 mm long, narrowly triangular-ovate, acute, papillose, margin strongly undulate-convolute and meeting in the middle; side lobes 4.5 mm long, 0.5 mm wide at middle, erect, broadly triangular, rounded. ***Column*** 2 mm long, oblong; foot 2 mm long, straight; anther-cap 1 × 1 mm, ovate, cucullate. Plate 8D.

HABITAT AND ECOLOGY: Podsol forest on sandstone soils, composed on *Dacrydium*, *Eugenia* and *Garcinia*, with an understorey of *Rhododendron longiflorum* and *R. malayanum*. Alt. 400 to 600 m. Flowering observed in February, September and October.

DISTRIBUTION IN BORNEO: SABAH: Nabawan area.

GENERAL DISTRIBUTION: Endemic to Borneo.

NOTES: *D. trullatum* belongs to section *Conostalix* and is related to *D. pachyglossum* C.S.P. Parish & Rchb.f. from Myanmar (Burma), Laos, Vietnam, Thailand and Peninsular Malaysia, and *D. pinifolium* Ridl. from Borneo. It differs from both in having shorter, unequally bilobed leaves with glabrous sheaths and a trullate ecallose lip with a narrowly triangular, acute mid-lobe having an undulate-convolute, papillose margin.

DERIVATION OF NAME: The specific epithet is derived from the Latin *trullatus*, angular-ovate or trowel-shaped, in reference to the lip shape.

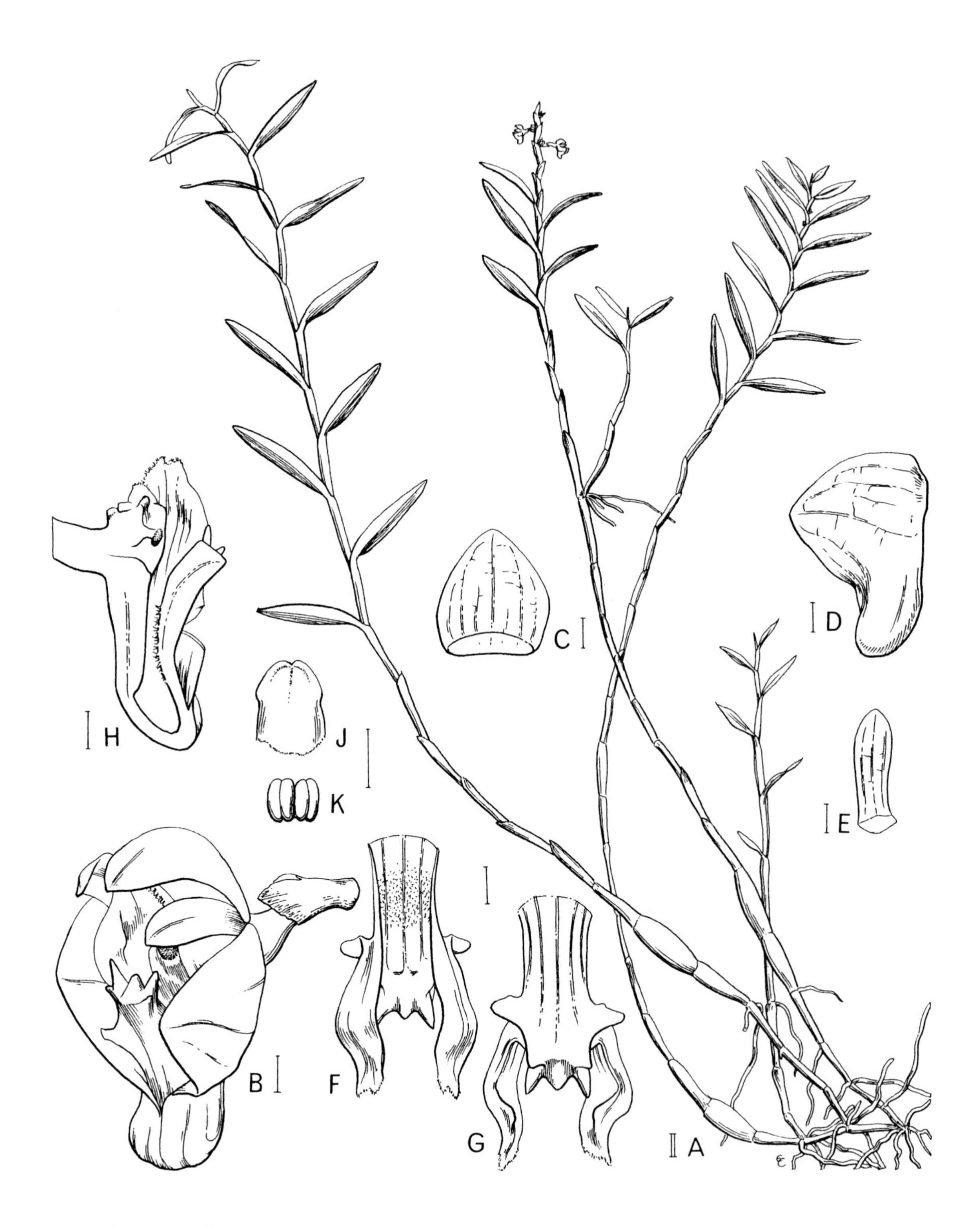

Figure 34. Dendrobium ventripes Carr. - A: habit. - B: flower, oblique view. - C: dorsal sepal. - D: lateral sepal. - E: petal. - F: lip, front view. - G: lip, back view. - H: pedicel with ovary, column and lip, longitudinal section. - J: anther-cap, back view. - K: pollinia. A (habit) drawn *Synge* S. 461 and B-K from *Wood* 879 by Eleanor Catherine. Scale: single bar = 1 mm; double bar = 1 cm.

34. DENDROBIUM VENTRIPES Carr

Dendrobium ventripes *Carr* in Gard. Bull. Straits Settlem. 8: 103 (1935). Type: Borneo, Sarawak, Mt. Dulit, Dulit Ridge, 1200 m, 6 September 1932, *P.M. Synge* S.422 (holotype SING, isotype K).

Epiphyte or ***terrestrial. Stems*** up to 65 cm long, cylindrical at base, 2–3 internodes above the base thickened into a narrowly fusiform many-grooved pseudobulb constricted at the nodes and up to 9.5 cm long, above the pseudobulb much flattened, narrowly elliptic in section, leafy to apex, lowermost internodes to 4.5 cm long, to 0.5 cm wide, uppermost 5 mm long. ***Leaves*** up to 5.5 × 1 cm below, upper leaves 1–2 cm long, narrowly elliptic or oblong-elliptic, usually narrowed below the shortly unequally, obtusely bilobed apex, thinly coriaceous, grooved above, keeled beneath, base twisted, sheaths tubular, winged. ***Inflorescences*** borne from tufts of dry scales from the uppermost nodes among the leaves, very short, one-flowered. ***Flowers*** dark orange, splashed purple-red or greenish yellow with crimson markings; column whitish. ***Pedicel*** with ***ovary*** 6 mm long, clavate. ***Sepals*** *and* ***petals*** rather fleshy. ***Dorsal sepal*** 5–6 × 4 mm, ovate, subacute or obtuse. ***Lateral sepals*** 5.5 × 5.5 mm, ovate, obtuse, subfalcate. ***Mentum*** *c.* 4 mm long, oblong, obtuse. ***Petals*** 4.5 × 1.5 mm, oblong-spathulate, obtuse. ***Lip*** *c.* 1.3 cm long, *c.* 0.5 cm wide across side lobes when flattened; side lobes *c.* 3.5 × *c.* 1.5 mm, twisted, erect, much longer than mid-lobe, oblong, apex incurved, obtuse, denticulate, shortly hairy towards apex, margins ciliate; mid-lobe *c.* 1.8 mm long, thickly fleshy, quadrilaterally cylindric, widened towards base and apex, the upper half recurved, abruptly triangularly acuminate, keeled with a short tubercular keel on each side below the apex. ***Column*** *c.* 1.5 mm long, straight; stelidia triangular, obtuse; foot *c.* 6.7 mm long, roundly spathulately dilate in the upper half and ventricose with a minute tubercle in the middle of the depression. Plate 8E.

HABITAT AND ECOLOGY: Oak/Ericaceae mossy forest; upper montane forest; ridge-top forest. Alt. 1200 to 2100 m. Flowering reported in March, June and September.

DISTRIBUTION IN BORNEO: SABAH: Mt. Kinabalu; Mt. Lumaku; Mt. Trus Madi. SARAWAK: Mt. Dulit.

GENERAL DISTRIBUTION: Endemic to Borneo.

NOTE: *D. ventripes* belongs to section *Rhopalanthe*.

DERIVATION OF NAME: The specific epithet is derived from the Latin *ventricosus*, swollen, especially on one side, and refers to the column-foot which is dilated and swollen in the upper half.

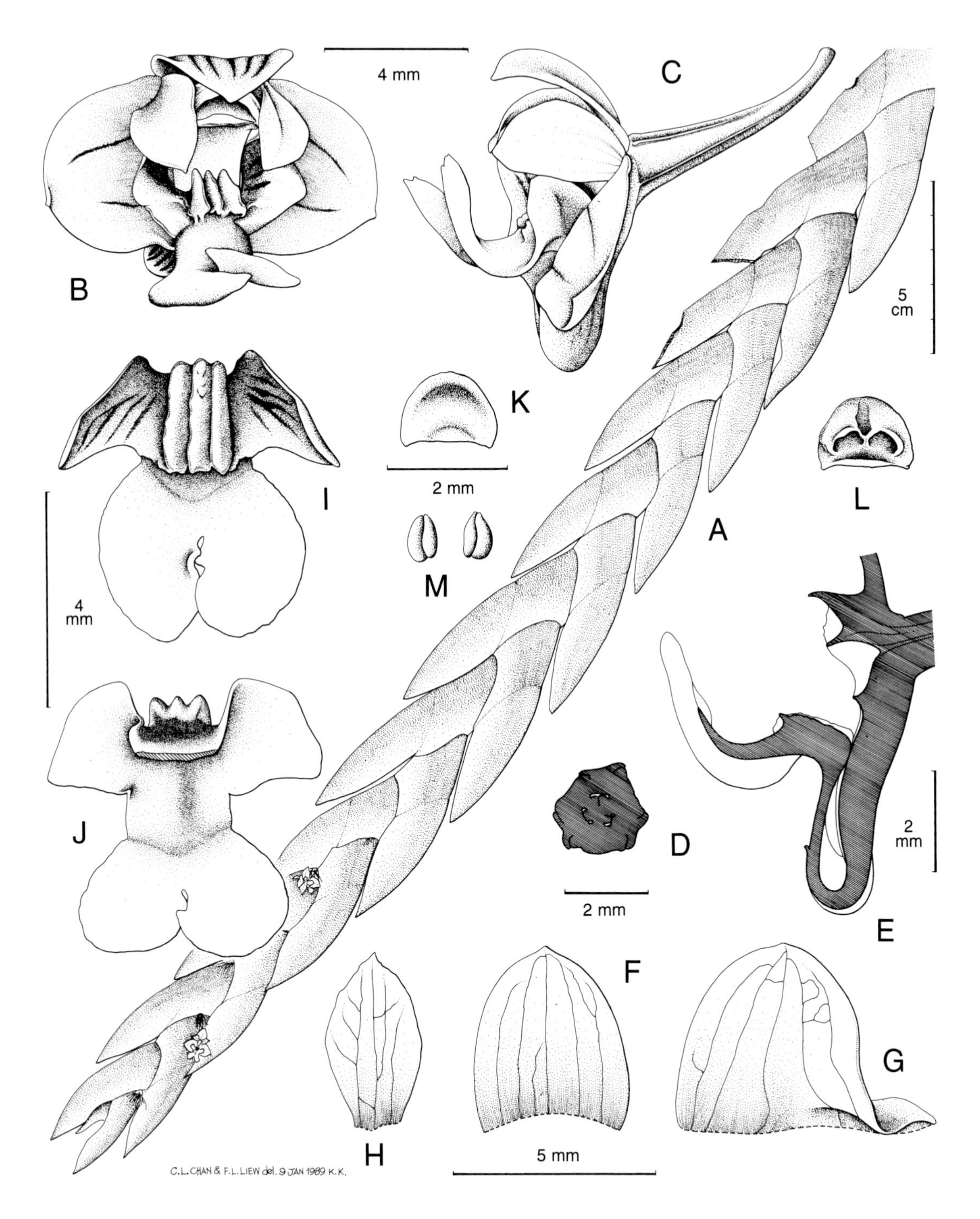

Figure 35. Dendrobium xiphophyllum Schltr. - A: habit. - B: flower, front view. - C: flower, side view. - D: ovary, transverse section. - E: column and lip, longitudinal section. - F: dorsal sepal. - G: lateral sepal. - H: petal. - I: lip, front view. - J: lip, back view. - K: anther-cap, back view. - L: anther-cap, front view. - M: pollinia. All drawn from a plant cultivated at Tenom Orchid Centre (TOC) by C.L. Chan and Lucy F.L. Liew.

35. DENDROBIUM XIPHOPHYLLUM Schltr.

Dendrobium xiphophyllum *Schltr.* in Repert. Spec. Nov. Regni Veg. 9: 291 (1911). Type: Borneo, Sarawak, Mt. Matang, *Beccari* 2049 (syntype B, destroyed, isosyntype ?FI); 'Pennisen' (= Mt. Penrissen), *Brooks* s.n. (syntype B, destroyed).

Erect to pendent ***epiphyte*** with the habit of *D. grande* Hook.f. ***Rhizome*** very abbreviated. ***Roots*** filiform, elongate, flexuose, glabrous. ***Stems*** 30–48 cm long, leafy to apex, straight, strongly complanate from a narrow base, young internodes entirely covered by leaf sheaths, internodes 2–2.5 cm long, yellow and shiny when exposed. ***Leaves*** (1.8–)5–8 cm long, 0.5–1 cm wide at base, distichous, equitant, ensiform, acute, fleshy, straight to slightly falcate; sheaths (0.8–)2–2.5 × 0.8–1 cm. ***Inflorescences*** borne at the upper nodes, emerging from base of leaf sheaths, flowers borne successively one at a time in axils of fasciculate fibrous hyaline bracts. ***Flowers*** *c.* 7 mm across; sepals yellow stained pale red; petals yellow; mentum yellow, stained dark purple-red; side lobes of lip dark purple-red, yellow along anterior margin; mid-lobe yellow; callus yellow, flushed red. ***Pedicel*** with ***ovary*** 7–8 mm long, narrowly clavate. ***Dorsal sepal*** 4 × 4 mm, ovate, obtuse, erect or slightly curving forward. ***Lateral sepals*** 4–5 × 5.1 mm, obliquely triangular-ovate, obtuse, spreading to reflexed. ***Mentum*** 3.5–5 mm long, obtuse, truncate, straight. ***Petals*** 4–4.1 × 2–2.1 mm, narrowly elliptic or obliquely spathulate-ligulate, obtuse to acute, horizontal. ***Lip*** 6–7 mm long, 5 mm wide across side lobes, shortly clawed, sharply decurved at junction of claw and side lobes; claw 1.8–2 × 2 mm; side lobes 3 × 2 mm, auriculate, obliquely oblong, obtuse, spreading or porrect; mid-lobe 3.5–4 × 3.5–4.1 mm, subquadrate to orbicular in outline, obtusely bilobed from at or below the middle, lobules oblong, usually slightly overlapping; disc with a callus composed of three or four fleshy ridges extending from base of side lobes, terminating at junction of side lobes and mid-lobe. ***Column*** 1 × 1.1 mm; foot 3.5–5 mm long, canaliculate; anther-cap broadly ovate, obtuse. Plate 9A.

HABITAT AND ECOLOGY: Lower montane forest. Alt. 700 to 800 m. Flowering observed in January, May and July.

DISTRIBUTION IN BORNEO: SABAH: Locality unknown. SARAWAK: Mt. Matang; Mt. Penrissen.

GENERAL DISTRIBUTION: Endemic to Borneo.

NOTE: *D. xiphophyllum* belongs to section *Aporum*.

DERIVATION OF NAME: The specific epithet is derived from the Greek *xiphos*, a sword, and *phyllum*, a leaf, in reference to the shape of the leaves.

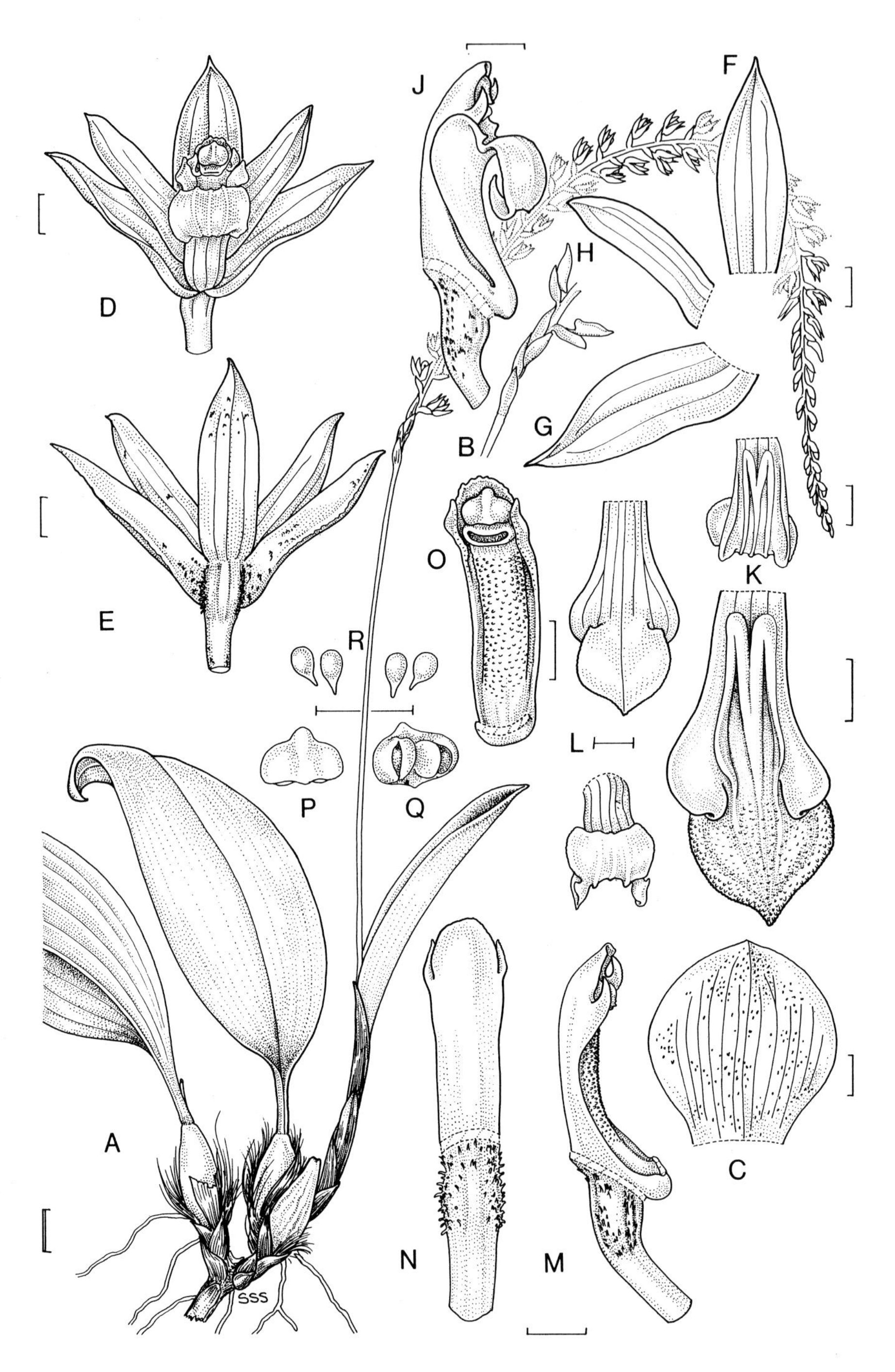

Figure 36. Dendrochilum anomalum Carr. - A: habit. - B: base of inflorescence. - C: floral bract, flattened. - D: flower, front view. - E: flower, back view. - F: dorsal sepal. - G: lateral sepal. - H: petal. - J: pedicel with ovary, lip and column, side view. - K: lip, natural position and flattened, front view. - L: lip, natural position and flattened, back view. - M: pedicel with ovary and column, side view. - N: pedicel with ovary and column, back view. - O: column, front view. - P: anther-cap, back view. - Q: anther-cap, front view. - R: pollinia. A (habit) drawn from *Richards* 2497 (holotype) and B–R from *Giles* 964A by Susanna Stuart-Smith. Scale: single bar = 1 mm; double bar = 1 cm.

36. DENDROCHILUM ANOMALUM Carr

Dendrochilum anomalum *Carr* in Gard. Bull. Straits Settlem. 8: 87 (1935). Type: Borneo, Sarawak, Ulu Koyan, Mt. Dulit, *native collector* in *Richards* 2497 (holotype SING, isotype K).

Epiphyte. Rhizome abbreviated, 2–3 mm in diameter. ***Roots*** 1–1.5 mm in diameter, smooth. ***Cataphylls*** 4, 1–7 cm long, obtuse to acute, finely veined, rather fleshy, dark brown to purplish, finely speckled darker brown, becoming fibrous. ***Pseudobulbs*** 2.3–4 × 0.7–2 cm, crowded on rhizome, ovoid, smooth to finely wrinkled, yellowish-olive. ***Leaf-blade*** 9.3–15.5 × 3.3–4.5 cm, oblong-elliptic, abruptly narrowed below, obtuse, shortly apiculate, main veins 9, coriaceous, margin slightly recurved; petiole 1.3–2.5 cm long, to 2.9 mm wide, sulcate, tough. ***Inflorescence*** from within the unexpanded leaf, many-flowered, lax to subdense; flowers borne 3–4 mm apart, laterally arranged, but twisting to form a spiral; peduncle 16–25 cm long, 1–2 mm in diameter, terete, rather stout, yellowish-green; rachis 17.5–27 cm long, 1–1.5 mm in diameter, quadrangular, pale brownish-green, sparsely finely setose, with 3–4 imbricate, 5–6 mm long adpressed basal sterile bracts; floral bracts 5–5.5 × 4.8–6 mm, broadly ovate, obtuse or subacute, concave, multi-veined, finely setose, yellow-green or pink, speckled brown. ***Flowers*** described as dull orange-yellow with an orange lip with central parts red or as sepals greenish cream or pale green to yellowish green, tipped brownish pink, the lip dark orange-buff, pale brownish orange or pale reddish brown and column pale reddish brown or orange, with an orange to salmon-pink anther-cap. ***Pedicel*** with ***ovary*** 2.5–3.7(–4) mm long, clavate, curved, minutely papillose, ovary slightly finely setose. ***Sepals*** sparsely finely setose on abaxial surface. ***Dorsal sepal*** 5–6.3 × 1.5–1.9 mm, narrowly oblong, acute, veins 3. ***Lateral sepals*** 6–7 × 1.6–2.1 mm, obliquely oblong-elliptic, shortly acuminate, acute, inconspicuously sigmoidly curved, dorsally carinate near apex, veins 3. ***Petals*** 4.8–6 × 1–1.2 mm, linear-ligulate or narrowly oblong-elliptic, acute, sometimes subfalcate, very sparsely finely setose on abaxial surface, veins 3. ***Lip*** 3-lobed from the middle, 5–6 mm long when flattened, 2–3 mm wide across side lobes, 3 mm wide across mid-lobe, immobile, erect and parallel with column, very minutely papillose, especially beneath; side lobes clasping column, triangular, subacute, margins minutely papillose towards apex, posterior margin erect, roundly dilate below apex, recurved towards apex with a short transverse fold; mid-lobe abruptly and strongly reflexed, ovate, subacute, fleshy, *c.* 3 mm long, minutely transversely rugose, margins minutely erose, provided with three obscure broad rounded keels extending from base to below apex; disc between side lobes provided with a fleshy V-shaped keel from above, the base to *c.* $^{1}/_{4}$ below base of mid-lobe, with a thinner keel, dilate towards apex, on either side. ***Column*** 4–4.5 × 1–1.1 mm, papillose, slightly curved; foot 0.6–1 mm long, papillose; apical hood transversely oblong above the triangular base, truncate, minutely erose; stigma transversely oblong, anterior margin elevate; stelidia borne at or just below stigma, narrowly triangular-falcate, upcurved, acute, papillose, *c.* 0.7 mm long; anther-cap 0.5–0.6 × 1 mm, ovate, cucullate, minutely papillose; pollinia four. Plate 9B.

HABITAT AND ECOLOGY: Heath forest; hill forest with *Gymnostoma sumatrana* on ultramafic substrate; "kerangas"-like forest with 15–20 metre high trees with thin boles, close together, light, with little undergrowth, many light green moss cushions on ground, soil sandy. Alt. 600 to 1000 m. Flowering observed in January, August and November.

DISTRIBUTION IN BORNEO: KALIMANTAN TIMUR: Apokayan. SABAH: Mt. Kinabalu; Tenom District. SARAWAK: Mt. Dulit.

GENERAL DISTRIBUTION: Endemic to Borneo.

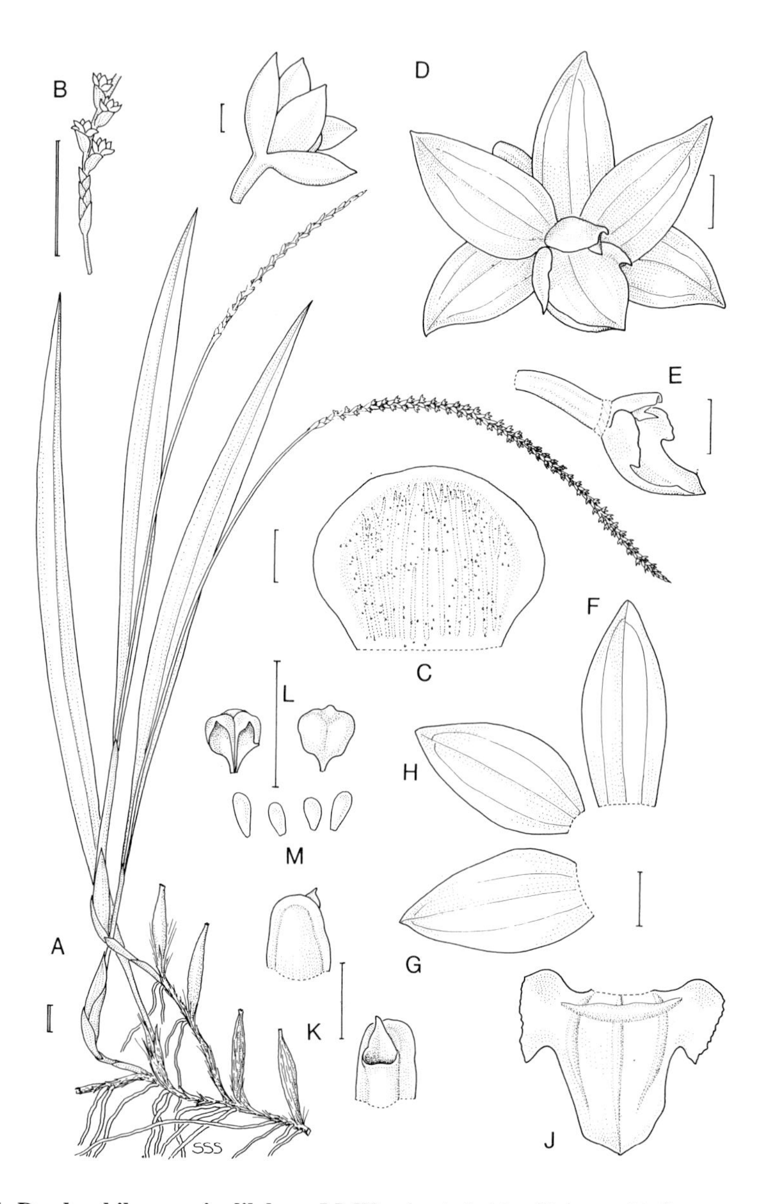

Figure 37. Dendrochilum auriculilobum J.J. Wood. - A: habit. - B: base of inflorescence. - C: floral bract, flattened. - D: flower, front view. - E: pedicel with ovary, lip and column, side view. - F: dorsal sepal. - G: lateral sepal. - H: petal. - J: lip, flattened. - K: column, front and back views. - L: anther-cap, front and back views. - M: pollinia. All drawn from *Vermeulen & Duistermaat* 1057 (holotype) by Susanna Stuart-Smith. Scale: single bar = 1 mm; double bar = 1 cm.

NOTES: A quite distinct species which appears to have no close allies. It seems to prefer forest types with a rather open canopy allowing higher light levels to penetrate.

DERIVATION OF NAME: The generic name is derived from the Greek *dendron*, tree, and *cheilos*, lip, or *chilos*, green food, alluding to either the epiphytic habit and the often conspicuous lip, or to the epiphytic habit and the consequent deriving of food on trees. The specific epithet is derived from the Latin *anomalus*, diverging from the usual or abnormal, referring to the isolated position of this species within the genus.

37. DENDROCHILUM AURICULILOBUM J.J. Wood

Dendrochilum auriculilobum *J.J. Wood* in Wood & Cribb, Checklist Orchids Borneo: 165, fig. 22 H & J (1984). Type: Borneo, Sabah, Sipitang District, Sungai Rurun headwaters, *Vermeulen & Duistermaat* 1057 (holotype L, isotype K).

Terrestrial. Pseudobulbs 2.8–3 × 0.6–0.8 cm, cylindrical, 2–3 cm apart on rhizome. ***Cataphylls*** 3.5–8 cm long, pale fawn speckled pale brown, enclosing young pseudobulbs and basal part of peduncle, becoming fibrous. ***Leaf blade*** 20–25 × 1.2–1.3 cm, linear-lanceolate, acute, thin-textured; petiole 5–6 cm long. ***Inflorescence*** many-flowered; peduncle 20–25 cm long, filiform, wiry; rachis 18 cm long, curving, with about 6 imbricate sterile bracts up to 1 mm long at base; floral bracts 3–3.2 × 4–4.5 mm, broadly ovate, obtuse, margin hyaline, prominently veined, finely setose, involute, entirely enclosing pedicel with ovary. ***Flowers*** arranged in two ranks, each flower borne 2–2.5 mm apart; somewhat fragrant; very pale greenish; lip green. ***Pedicel*** with ***ovary*** 2.3–2.4 mm long, straight. ***Sepals*** and ***petals*** spreading. ***Dorsal sepal*** 4 × 1.7–1.8 mm, ovate-elliptic, acute. ***Lateral sepals*** 4 × 2–2.1 mm, ovate, acute. ***Petals*** 3.9–4 × 1.7–1.8 mm, elliptic, acute. ***Lip*** 2.5 mm long, 3 mm wide across side lobes, concave, cup-like; side lobes 1.1 × 0.2–0.3 mm, auriculate, acute; mid-lobe 1.5–1.6 × 1.5 mm, oblong-ovate, obtuse, with a small obtuse apical mucro, 3-veined, with a fleshy, transverse basal ridge. ***Column*** 0.8 × 1 mm, oblong; stelidia, foot and apical hood absent; anther-cap *c.* 0.6–0.7 × 0.6–0.7 mm, cucullate, apiculate.

HABITAT AND ECOLOGY: Low and open mossy ridge forest with a dense undergrowth of bamboo and rattan palms. Alt. 1700 m. Flowering observed in December.

DISTRIBUTION IN BORNEO: SABAH: Sipitang District.

GENERAL DISTRIBUTION: Endemic to Borneo.

NOTES: *D. auriculilobum* is related to *D. hologyne* Carr, described from Sarawak and also recorded from Sipitang District in South-west Sabah. *D. auriculilobum* is distinguished by its much shorter, thicker pseudobulbs, laxer inflorescences, larger flowers and lip with auriculate side lobes and a fleshy transverse basal ridge.

DERIVATION OF NAME: The specific epithet is derived from the Latin *auriculatus*, with ear-like appendages, and *lobus*, lobe, referring to the side lobes of the lip.

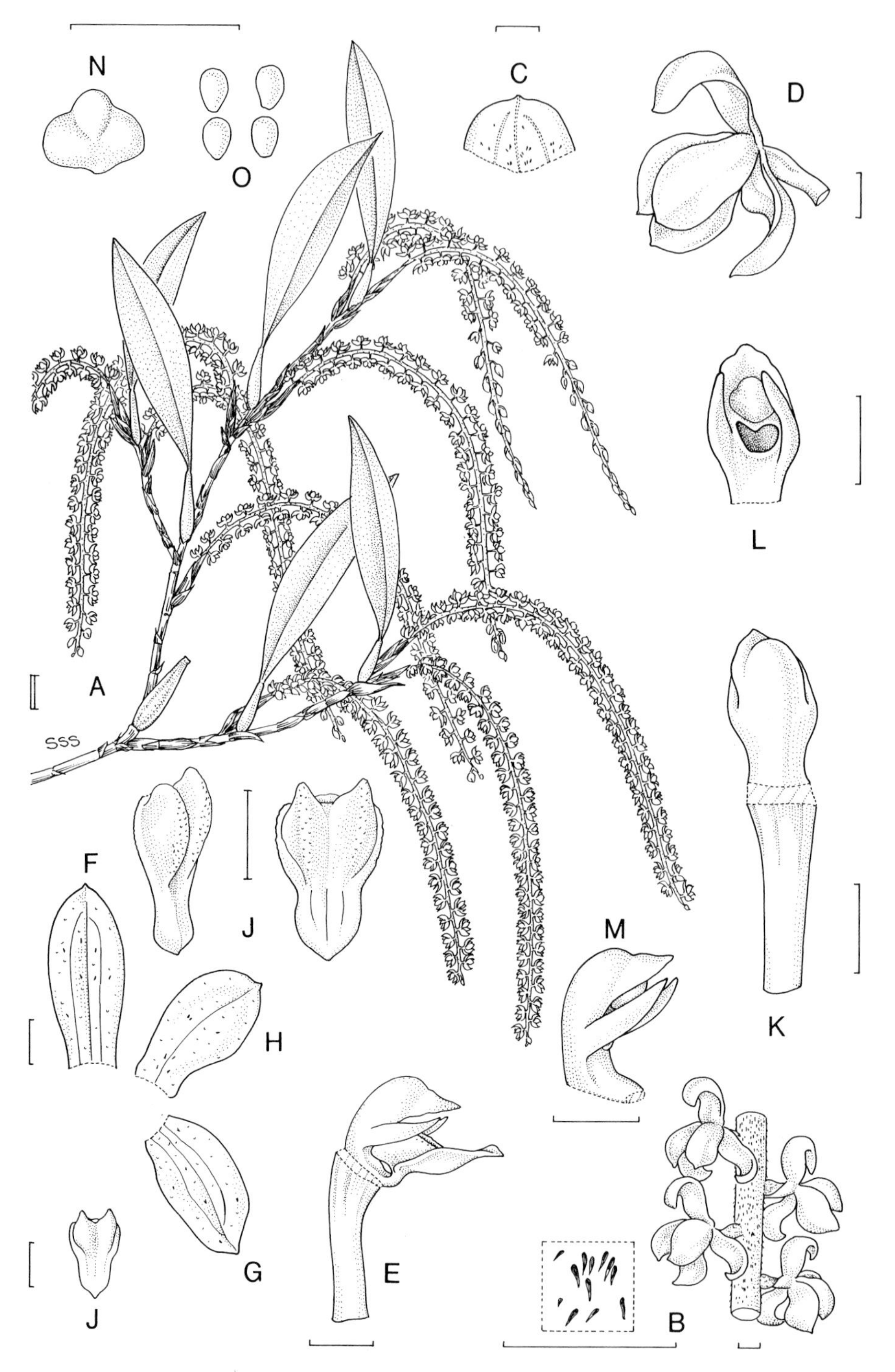

Figure 38. Dendrochilum crassum Ridl. - A: habit. - B: portion of inflorescence with detail showing trichomes. - C: floral bract, flattened. - D: flower, side view. - E: pedicel with ovary, lip and column, side view. - F: dorsal sepal. - G: lateral sepal. - H: petal. - J: lip, front and side views. - K: pedicel with ovary and column, back view. - L: column, front view. - M: column, side view. - N: anther-cap. - O: pollinia. A (habit) drawn from *Vermeulen & Duistermaat* 693 and B–O from *Wood* 733 by Susanna Stuart-Smith. Scale: single bar = 1 mm; double bar = 1 cm.

38. DENDROCHILUM CRASSUM Ridl.

Dendrochilum crassum *Ridl.* in J. Linn. Soc., Bot. 32: 288 (1896). Type: Peninsular Malaysia, Perak, Hermitage Hill, *Ridley* s.n. (not located).

Sprawling ***epiphyte***, sometimes ***lithophytic***. ***Rhizome*** 30–50 cm long, 0.4–0.6 cm wide, branching, creeping or straggling through undergrowth, internodes 0.5–2.5 cm long, orange. ***Roots*** 1–1.2 mm in diameter, wiry, smooth. ***Pseudobulbs*** 1.8–4.5 × 0.5–0.8 cm, cylindrical, ridged, olive to orange. ***Leaf blade*** 5–10 × 1.2–4 cm, elliptic or oblong-elliptic, obtuse, minutely mucronate, coriaceous; petiole 2–8 mm long, sulcate. ***Inflorescence*** heteranthous, densely many-flowered; flowers borne 2–3 mm apart; peduncle and rachis covered in a brown furfuraceous indumentum; peduncle 1–2 cm long, terete, almost entirely enclosed by several acute imbricate sheaths up to 2 cm long; rachis 8–16 cm long, quadrangular, often becoming decurved; floral bracts 2–2.1 × 2.1–2.2 mm, broadly ovate, obtuse, apiculate, covered with scattered brown trichomes. ***Flowers*** sweetly scented; sepals and petals green with whitish margins and apex; lip pale green or whitish, lower half often bright green; column pale green; anther-cap cream. ***Pedicel*** with ***ovary*** 3.5–3.6 mm long, narrowly clavate, ovary sparsely furfuraceous. ***Sepals*** and ***petals*** fleshy, minutely papillose, 3-veined, incurved. ***Dorsal sepal*** 4.5 × 2 mm, oblong to oblong-spathulate, obtuse to subacute, with scattered brown furfuraceous trichomes on abaxial surface and towards base on adaxial surface. ***Lateral sepals*** 4 × 2–2.1 mm, obliquely elliptic, obtuse to subacute, with similarly distributed trichomes. ***Petals*** 3.9–4 × 2 mm, spathulate, obovate, obtuse, 3-veined, erect and incurved, sparsely covered with furfuraceous trichomes on abaxial surface. ***Lip*** 2.4–2.5 mm long, 1.2–1.3 mm wide proximally, 0.8–0.9 mm wide distally, narrowly pandurate, apex ovate, obtuse, sessile, minutely papillose; disc with two fleshy keels extending from base to centre of lip. ***Column*** 1.8–1.9 mm long; foot 0.2–0.3 mm long; apical hood ovate, margin entire or slightly irregular; stelidia borne centrally each side of the broadly rounded stigmatic cavity, linear-lanceolate, somewhat falcate, acute, slightly shorter than apical hood; anther-cap ovate, cucullate. Plate 9C & D.

HABITAT AND ECOLOGY: Hill forest; steep rocky roadside banks with *Arundina graminifolia*, *Gahnia*, *Lycopodium*, *Melastoma*, *Nepenthes fusca*, ferns, etc., on sandstone and shale outcrops, often in exposed sunny sites. Alt.1200 to 1500 m. Flowering observed from March to May, October and December.

DISTRIBUTION IN BORNEO: SABAH: Crocker Range (Mt. Alab, Kimanis road, Sinsuron road); Mt. Kinabalu; Mt. Trus Madi; Nabawan.

GENERAL DISTRIBUTION: Peninsular Malaysia and Borneo.

NOTE: *D. crassum* belongs to subgenus *Dendrochilum* which has its centre of distribution in Sumatra.

DERIVATION OF NAME: The specific epithet is derived from the Latin *crassus*, thick, referring to the thick, leathery leaves and fleshy floral parts.

Figure 39. Dendrochilum cruciforme J.J. Wood var. **cruciforme**. - A: habit. - B: flower, front view. - C: pedicel with ovary and column, back view. - D: column, oblique view. - E: habit. - F: floral bract, flattened. - G: flower, side view. - H: dorsal sepal. - J: lateral sepal. - K: petal. - L: lip, flattened. - M: lip, natural position, side view. - N: pedicel with ovary and column, side view. - O: pedicel with ovary and column, back view. - P: column, front view. - Q: anther-cap, back and front views. - R: pollinia. A–D drawn from *J. & M.S. Clemens* 40134 and E–R from *J. & M.S. Clemens* s.n. (holotype) by Susanna Stuart-Smith. Scale: single bar = 1 mm; double bar = 1 cm.

39. DENDROCHILUM CRUCIFORME J.J. Wood var. CRUCIFORME

Dendrochilum cruciforme *J.J. Wood* in Wood & Cribb, Checklist Orchids Borneo: 169, fig. 24A & B (1994). Type: Borneo, Sabah, Mt. Kinabalu, Penibukan, *J. & M.S. Clemens* s.n. (holotype K, isotypes AMES, BO, E, HBG).

var. **cruciforme**

Clump-forming ***epiphyte. Pseudobulbs*** 0.5–2.5 × 0.2–0.4 cm, cylindrical or narrowly fusiform, enclosed in sheaths when young. ***Leaves*** narrowly linear, subacute to acute, narrowed into a slender petiole; blade 5–11.5 × 0.1–0.3(0.4) cm; petiole 0.5–2 cm long. ***Inflorescence*** gently curving, densely many-flowered; flowers borne 1–2 mm apart; peduncle 2.5–6.5 cm long; rachis 3.5–5(–6.5) cm long, quadrangular; floral bracts 1–1.5 × 1 mm, ovate, acute. ***Flowers*** having a musty scent according to some; variously described as pinkish cream with purple spots, creamy green, creamy yellow, cream or pure white; lip usually with a dark purple-brown blotch at base of side lobes and on apex of keels, rarely pale greenish white with a greenish lip. ***Pedicel*** with ***ovary*** 0.5–1(–2) mm long, clavate. ***Dorsal sepal*** 2.2–2.6(–3) × 0.6–0.9 mm, oblong-elliptic, acute. ***Lateral sepals*** 2.1–2.5(–2.8) × 0.8–1 mm, ovate-elliptic, acute. ***Petals*** 2–2.1 × 0.4–0.6 mm, narrowly elliptic, acute. ***Lip*** 2 mm long, 1.8–2 mm wide across side lobes, cruciform, 3-veined, with two rounded basal keels joined at the base by a transverse ridge; side lobes triangular or triangular-acute, acute or subacute, often somewhat falcate; mid-lobe narrowly triangular-acuminate, cuspidate. ***Column*** 1–1.2 × 0.2–0.3 mm; hood acute, acuminate or bifid; stelidia basal, 0.9–1.1 mm long, linear-ligulate, obtuse; foot absent; anther-cap cucullate. Plate 10A.

HABITAT AND ECOLOGY: *Leptospermum/Dacrydium* forest on sandstone and shale; lower to upper montane forest. Alt. 900 to 2000 m. Flowering observed in January, March, June, August, October, November and December.

DISTRIBUTION IN BORNEO: SABAH: Mt. Alab; Mt. Kinabalu; Maliau Basin.

GENERAL DISTRIBUTION: Endemic to Borneo.

NOTES: *D. cruciforme* is the type species of section *Cruciformia* which is endemic to Borneo. The habit and floral structure resembles *D. dolichobrachium* (Schltr.) Merr., but the lip of *D. cruciforme* has shorter keels, broader triangular or triangular-ovate side lobes and a much smaller, narrowly triangular-acuminate, cuspidate mid-lobe. It may also be distinguished from *D. devogelii* J.J. Wood by the shorter pedicel with ovary, cruciform lip and shorter stelidia. It differs from *D. gibbsiae* Rolfe by its much dwarfer stature, shorter, narrower grass-like leaves, shorter inflorescence with slightly smaller flowers, a three, rather than five-lobed lip and longer column.

Clemens 40134 (AMES, BM, E) is a more robust specimen with pseudobulbs up to 2.5 × 0.4 cm, leaf blades up to 11.5 × 0.4 cm, and pure white flowers lacking the characteristic dark blotch on the lip. *De Vogel* 8661 (K, L) is a variant with a longer filiform pedicel, greenish white flowers, again without a dark blotch on the lip, less pronounced side lobes and a deeply bifid apical column hood.

The degree of lobing of the lip in *D. cruciforme* appears quite variable, some specimens having more pronounced side lobes than others. This is also true of the widespread *D. gibbsiae*. The shape of the apical column hood also ranges from acute to acuminate and entire to deeply bifid (see fig. 39a, H–M), but is nearly always longer than the stelidia. The full range of variation is difficult to ascertain given the limited material available.

DERIVATION OF NAME: The specific epithet is derived from the Latin *cruciformis*, cross-shaped, in reference to the lip.

39a. DENDROCHILUM CRUCIFORME J.J. Wood var. LONGICUSPE J.J. Wood

Dendrochilum cruciforme *J.J. Wood* var. **longicuspe** *J.J. Wood* in Wood & Cribb, Checklist Orchids Borneo: 170, fig. 24 C & D (1994). Type: Borneo, Sabah, Mt. Kinabalu, Kadamaian River, *Carr* 3675, SFN 28004 (holotype K, isotype SING).

Floral bracts *c.* 3.5 × 2 mm. ***Flowers*** unscented; white or pale yellow, base of lip greenish-yellow. ***Dorsal sepal*** 3 × 0.9 mm. ***Lateral sepals*** 2.9–3 × 0.9–1 mm. ***Petals*** 2.5–2.6 × *c.* 0.7 mm. ***Lip*** 2.5 mm long, 1.5 mm wide across side lobes; side lobes narrow, somewhat falcate; mid-lobe 1 mm long, narrowly triangular, acuminate. ***Column*** 1 × 0.2–0.3 mm; hood tridentate, shorter than stelidia; stelidia 1–2 mm long, ligulate, obtuse; foot absent.

HABITAT AND ECOLOGY: Twig epiphyte in upper montane forest. Alt. 2000–2500 m. Flowering observed in August.

DISTRIBUTION IN BORNEO: SABAH: Mt. Kinabalu.

GENERAL DISTRIBUTION: Endemic to Borneo.

NOTES: Differing from the typical variety in having a lip with a longer, acuminate mid-lobe and narrower, somewhat falcate side lobes. The stelidia are longer than the tridentate apical hood.

DERIVATION OF NAME: The varietal epithet is derived from the Latin *longus*, long, and *cuspis*, a sharp, rigid point, and refers to the mid-lobe of the lip.

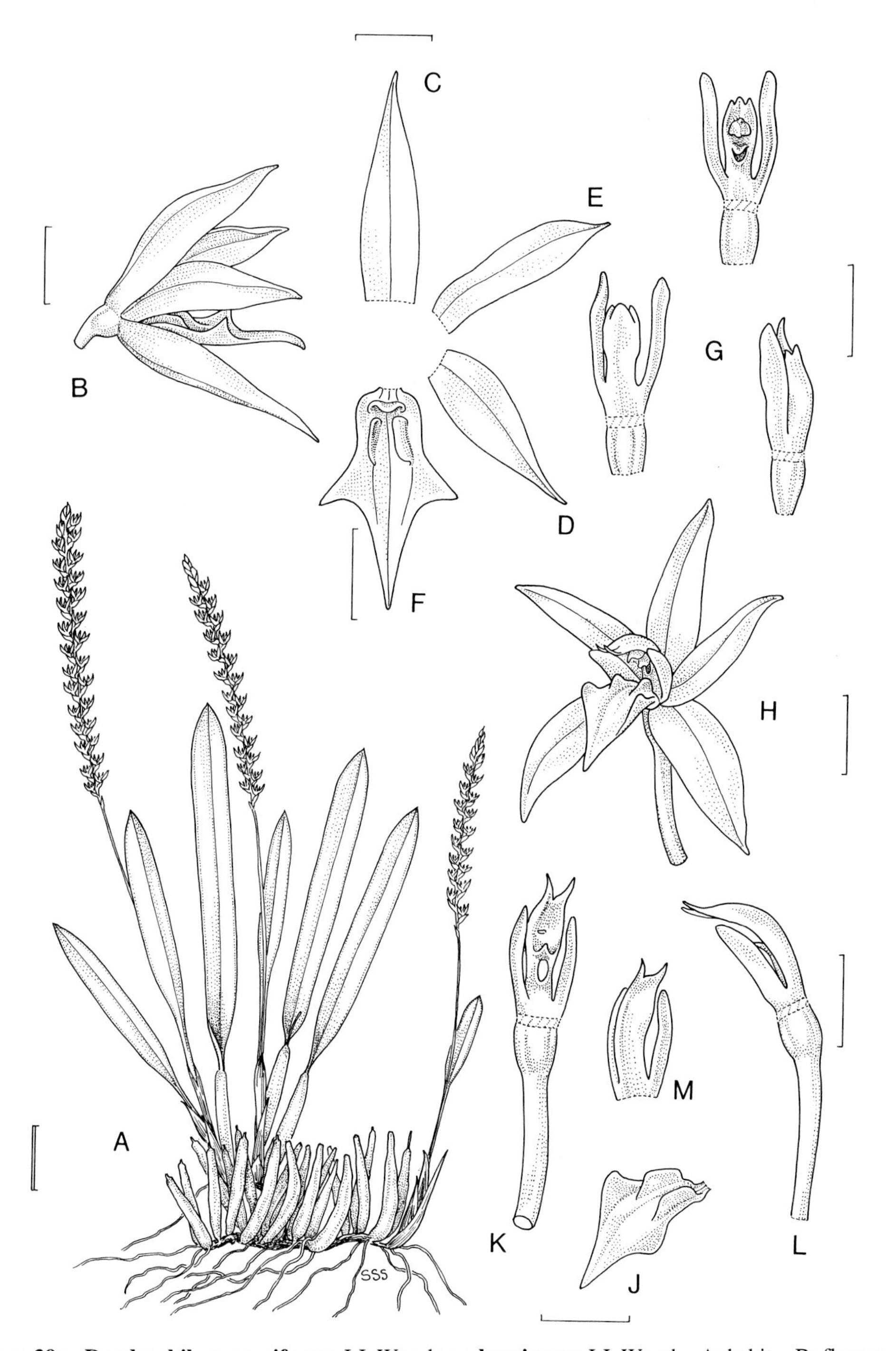

Figure 39a. Dendrochilum cruciforme J.J. Wood var. **longicuspe** J.J. Wood. - A: habit. - B: flower, side view. - C: dorsal sepal. - D: lateral sepal. - E: petal. - F: lip, flattened. - G: pedicel with ovary and column, front, back and side views. **D. cruciforme** J.J. Wood var. **cruciforme** (variant). - H: flower, front view. - J: lip, oblique view. - K: pedicel with ovary and column, front view. - L: pedicel with ovary and column, side view. - M: column, back view. A–G drawn from *Carr* 3675, SFN 28004 (holotype) and H–M from *de Vogel* 8661 by Susanna Stuart-Smith. Scale: single bar = 1 mm; double bar = 1 cm.

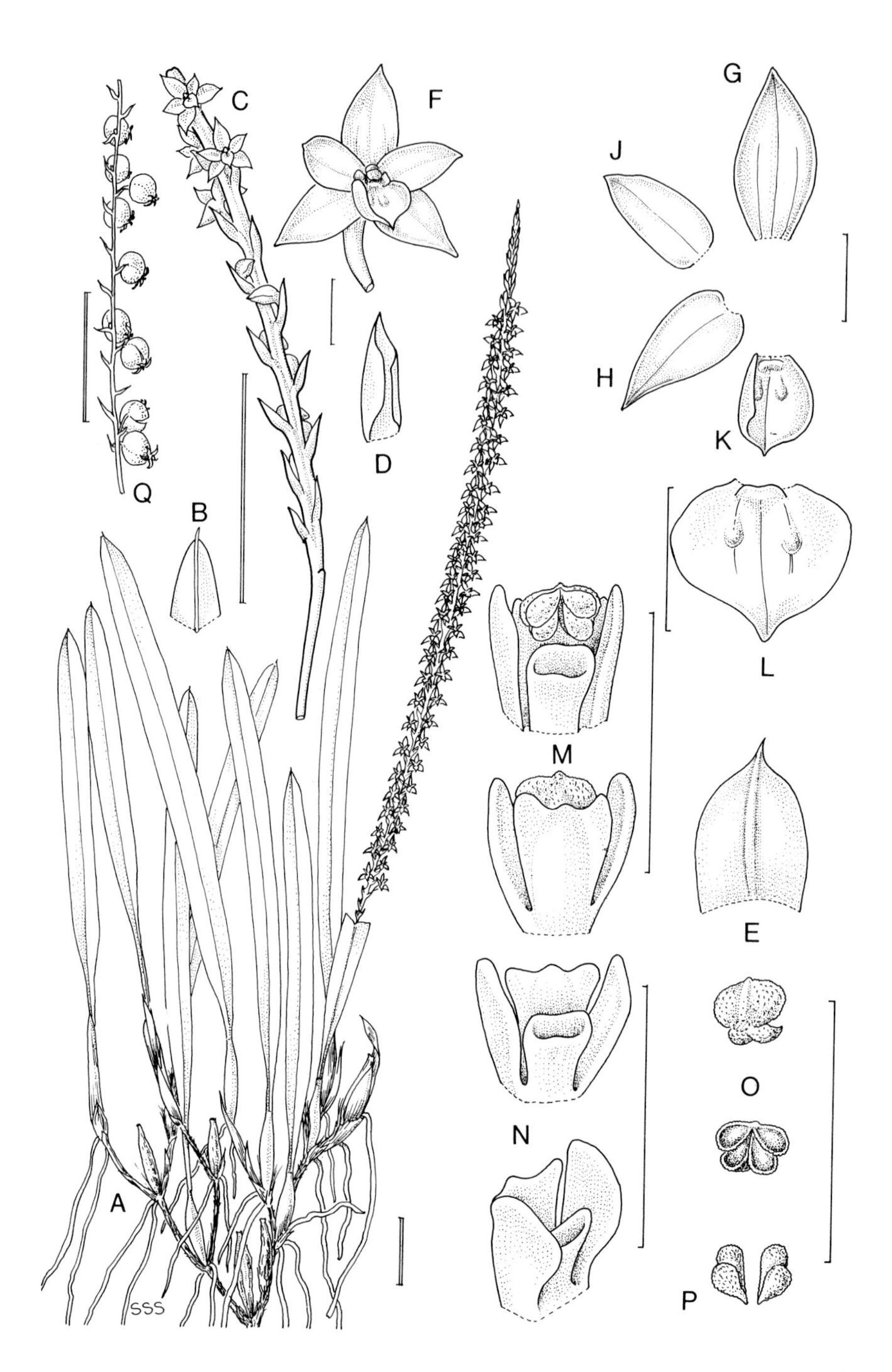

Figure 40. Dendrochilum cupulatum J.J. Wood. - A: habit. - B: leaf apex. - C: base of inflorescence. - D: floral bract, natural position. - E: floral bract, flattened. - F: flower, front view. - G: dorsal sepal. - H: lateral sepal. - J: petal. - K: lip, natural position. - L: lip, flattened. - M: column with anther-cap, front and back views. - N: column, front and oblique views. - O: anther-cap, front and back views. - P: pollinia. - Q: infructescence. A (habit) drawn from *Hansen* 498, B–P drawn from *de Vogel* 1013 and Q from *Kitayama* 893 by Susanna Stuart-Smith. Scale: single bar = 1 mm; double bar = 1 cm.

40. DENDROCHILUM CUPULATUM J.J. Wood

Dendrochilum cupulatum *J.J. Wood* in Wood & Cribb, Checklist Orchids Borneo: 171, fig. 23 J–L (1994). Type: Borneo, Sarawak, Mt. Mulu National Park, above Camp 4, *Lamb* MAL 12 (holotype K).

Creeping, clump-forming ***epiphyte. Rhizome*** tough, branching profusely, producing numerous wiry roots. ***Pseudobulbs*** 0.9–2.3 × 0.4–0.8 cm, cylindrical or elliptic, borne 0.3–2.4 cm apart on rhizome. ***Cataphylls*** pale brown, unspotted, concealing young pseudobulbs at first, but soon becoming fibrous. ***Leaves*** erect, blade 4–10 × 0.4–0.5(–0.8) cm, linear-ligulate, conduplicate at base, apex slightly carinate, apiculate; petiole 0.1–0.5 cm long. ***Inflorescence*** erect, densely many-flowered; peduncle 1.5–3 cm long, greenish-yellow; rachis 6–14 cm long, quadrangular, greenish yellow; floral bracts 2–2.5 × 1.1–1.2 mm, ovate, apiculate, carinate. ***Flowers*** 3 mm across, borne in 2 ranks, each flower 1 mm apart; with a slightly spicy scent; greenish yellow, very pale greenish with a brighter green lip or white with a greenish white lip. ***Pedicel*** with ***ovary*** 1.8 mm long, slightly curved. ***Sepals*** and ***petals*** spreading. ***Dorsal sepal*** 2 × 1 mm, ovate-elliptic, acute. ***Lateral sepals*** 2 × 1 mm, ovate-elliptic, acute. ***Petals*** *c.* 1.6–1.7 × 0.8–0.9 mm, ovate-elliptic, acute. ***Lip*** 1.1 × 1.5–1.6 mm, broadly ovate, concave, cupulate, apex apiculate, with a basal ridge and two small rounded central calli. ***Column*** 0.2 × 0.2–0.3 mm, cuneate, truncate; stelidia 0.5 mm long, basal, oblong, somewhat truncate; foot absent; anther-cap minute, cucullate. Plate 10B.

HABITAT AND ECOLOGY: Upper montane forest; upper montane shrubbery on exposed ridges; recorded as a branch epiphyte in ridge-top forest on sandstone. Alt. 1400 to 2100 m. Flowering observed in March, April and December.

DISTRIBUTION IN BORNEO: SABAH: Crocker Range, Kimanis road. SARAWAK: Mt. Mulu National Park; Hose Mountains.

GENERAL DISTRIBUTION: Endemic to Borneo.

NOTES: *D. cupulatum* is a member of section *Eurybrachium* and is related to *D. corrugatum* (Ridl.) J.J. Sm. (syn. *D. fimbriatum* Ames) from Sabah but distinguished by its well-spaced pseudobulbs, linear-ligulate leaves and slightly smaller flowers with a shorter dorsal sepal, non-falcate lateral sepals, entire petals, an entire, strongly concave, cupulate lip with two tiny rounded central calli, and a shorter column. It may also be distinguished from *D. alatum* Ames, also from Sabah, by the well-spaced pseudobulbs, denser inflorescence with slightly smaller flowers with entire petals, a shorter, strongly concave, cupulate lip with an apiculate apex and tiny calli, and a shorter, truncate column. The spikes of tiny crowded flowers look remarkably like those of *D. sublobatum* Carr, a species belonging to section *Platyclinis,* so far only recorded from Sarawak. This, however, has crowded pseudobulbs and longer leaves. On closer inspection, the flowers of *D. sublobatum* have a decurved, slightly lobed, ovate-acuminate lip with a large, horseshoe-shaped callus and a column with a conspicuous hood and long linear stelidia.

DERIVATION OF NAME: The specific epithet is derived from the Latin *cupula*, a cup, in reference to the concave, cup-like lip.

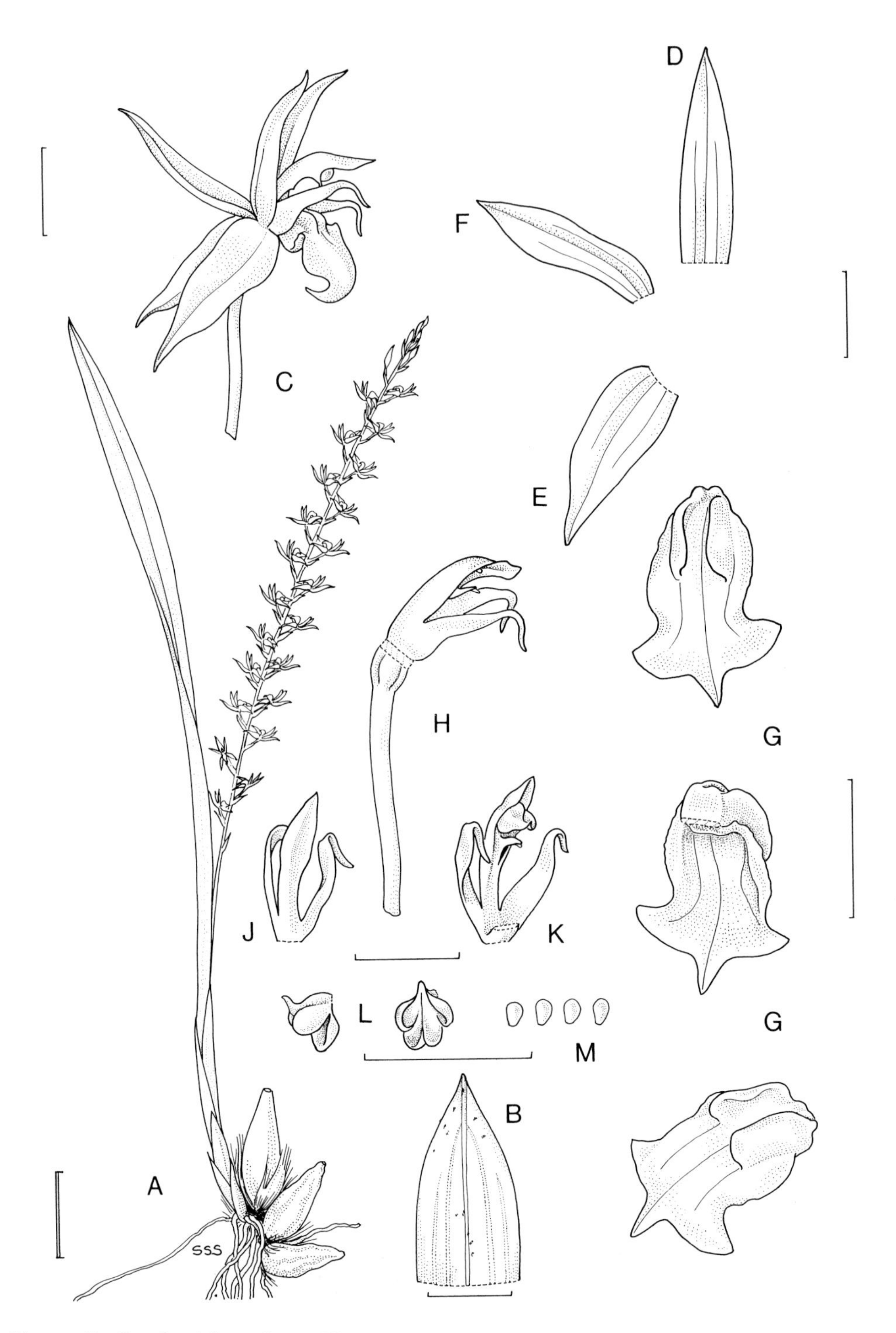

Figure 41. Dendrochilum devogelii J.J. Wood. - A: habit. - B: floral bract, flattened. - C: flower, side view. - D: dorsal sepal. - E: lateral sepal. - F: petal. - G: lip, front, back and oblique views. - H: pedicel with ovary and column, side view. - J: column, back view. - K: column, front view. - L: anther-cap, side and back views. - M: pollinia. All drawn from *de Vogel* 8376 (holotype) by Susanna Stuart-Smith. Scale: single bar = 1 mm; double bar = 1 cm.

41. DENDROCHILUM DEVOGELII J.J. Wood

Dendrochilum devogelii *J.J. Wood* in Wood & Cribb, Checklist Orchids Borneo: 173, fig. 24 E & F (1994). Type: Borneo, Sabah, Sipitang District, ridge between Maga River and Pa Sia River, *de Vogel* 8376 (holotype L, isotype K).

Tufted ***epiphyte. Pseudobulbs*** 1–1.5 × 0.4–0.5 cm, cylindrical. ***Leaves*** 9.5 × 0.3 cm, linear-ligulate, acute. ***Inflorescence*** laxly many-flowered; peduncle 2.2 cm long; rachis 6.5 cm long, with a 3.5 mm long sterile bract at base; floral bracts 1–2 mm long, ovate, acute. ***Flowers*** pale green. ***Pedicel*** with ***ovary*** 3 mm long, very narrowly clavate. ***Dorsal sepal*** 2.5 × 0.8 mm, oblong, acute. ***Lateral sepals*** 2–2.1 × 0.8–0.9 mm, oblong, slightly falcate, acute. ***Petals*** 2 × 0.5–0.6 mm, linear, acute, margins a little uneven. ***Lip*** 1–1.1 × 0.9–1 mm, pandurate, with a fleshy, U-shaped keeled basal callus; side lobes rounded to subacute; mid-lobe acute, apiculate, mucronate. ***Column*** 2 mm long; stelidia basal, 2–2.1 mm long, ligulate, obtuse, tips decurved; apical hood entire, shorter than stelidia; anther-cap cucullate.

HABITAT AND ECOLOGY: Rather dense primary forest dominated by *Agathis* and *Lithocarpus* up to 30 metres high, on poor sandy soil, bedrock sandstone, much leaf litter, little undergrowth. Alt. 1450 m. Flowering observed in October.

DISTRIBUTION IN BORNEO: SABAH: Sipitang District.

GENERAL DISTRIBUTION: Endemic to Borneo.

NOTES: *D. devogelii* is only known from the type collection which consists of only two pseudobulbs, one leaf and one inflorescence. It is related to *D. cruciforme* but is distinguished by the pandurate lip with a proportionately longer basal unlobed portion, stelidia longer than the apical hood and decurved at the apex, and the longer pedicel.

DERIVATION OF NAME: The specific epithet honours Dr E.F. de Vogel, orchid taxonomist at the Rijksherbarium, Leiden, The Netherlands, who collected the type.

42. DENDROCHILUM EXASPERATUM Ames

Dendrochilum exasperatum *Ames,* Orchidaceae 6: 50 (1920). Type: Borneo, Sabah, Mt. Kinabalu, Marai Parai Spur, *J. Clemens* 396 (holotype AMES).

Epiphyte, rarely ***terrestrial. Rhizome*** 1–8 cm long, *c.* 0.4 cm in diameter. ***Roots*** produced in a large, dense mass, smooth, *c.* 1 mm in diameter. ***Cataphylls*** usually 4, 1–5 cm long, finely veined, reddish-brown, unspotted, soon becoming fibrous. ***Pseudobulbs*** 1–6 × 0.4–0.6 cm, closely caespitose, narrowly cylindrical or fusiform, somewhat flattened, finely wrinkled when dry. ***Leaf blade*** 5–16 × (1.1–)1.5–3.5 cm, oblong-elliptic to elliptic, obtuse and minutely mucronate, sometimes somewhat constricted *c.* 2–2.5 cm

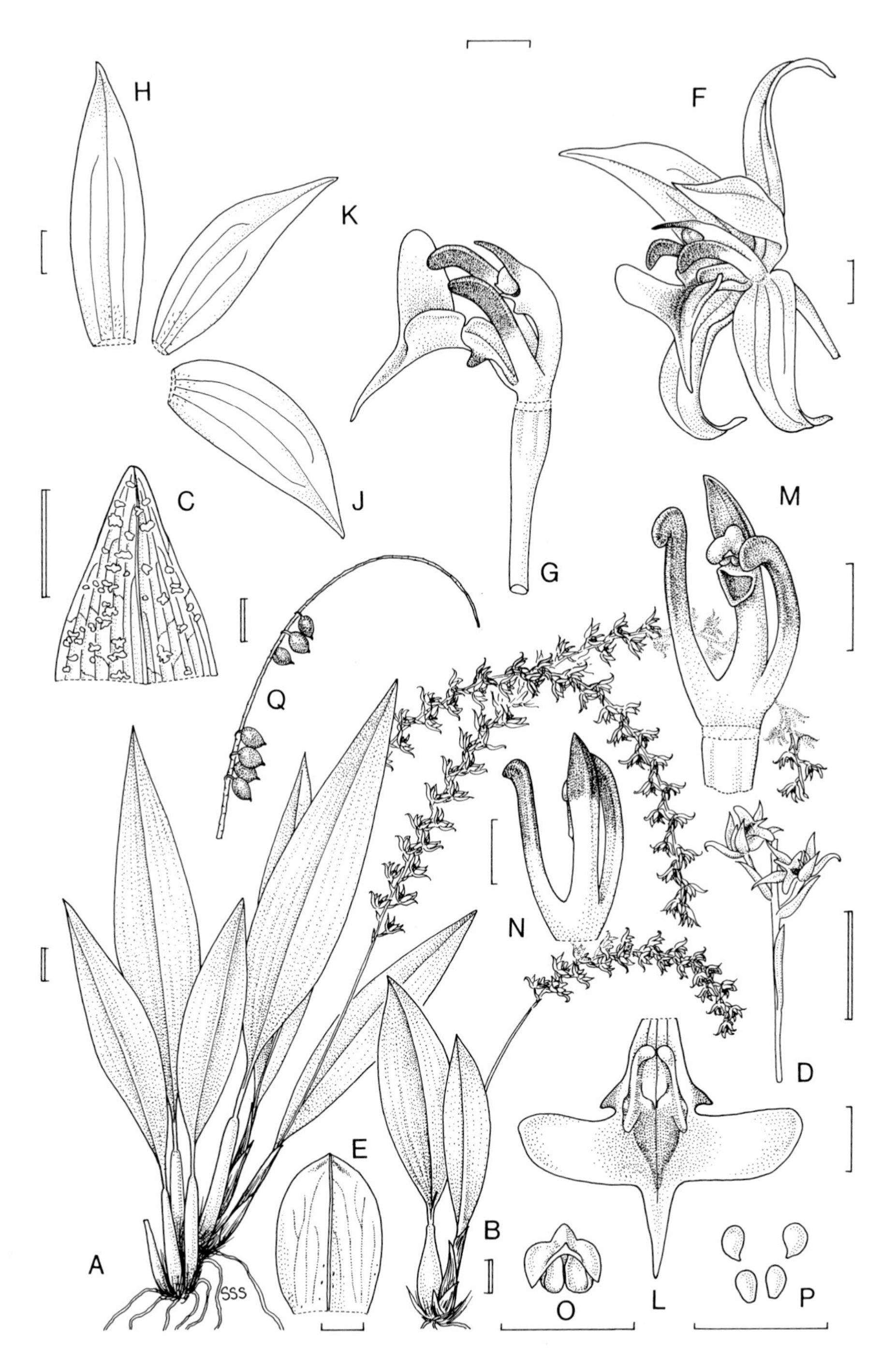

Figure 42. Dendrochilum exasperatum Ames. - A & B: habit. - C: leaf apex showing calcium oxalate crystals. - D: floral bract. - E: flower, oblique view. - F: pedicel with ovary, lip and column, side view. - G: dorsal sepal. - H: lateral sepal. - J: petal. - K: lip, flattened. - L: column, oblique view. - M: column, back view. - N: anther-cap and pollinia. - O: infructescence. A (habit) drawn from *Vermeulen & Duistermaat* 668 & 1038, B (habit) from *J. & M.S. Clemens* 50246, C–N from *Vermeulen & Duistermaat* 668 and O from *Carr* 3668, SFN 28029 by Susanna Stuart-Smith. Scale: single bar = 1 mm; double bar = 1 cm.

below apex, main veins 7(–9), with numerous fine transverse veins, thin and parchment-like to somewhat leathery-textured, frequently with numerous small irregular calcium oxalate bodies similar to *D. gibbsiae*, particularly noticeable in dried material; petiole 0.5–4 cm long, sulcate. ***Inflorescence*** gently curving, many-flowered, subdense; flowers borne 2–4 mm apart, buds appearing curved and acuminate; peduncle 5–15 cm long, terete; rachis 8–25 cm long, quadrangular, with a solitary basal sterile bract 4–9 mm long; floral bracts 2–6 × 2 mm, oblong-ovate to ovate, obtuse, apiculate, longer than pedicel with ovary, veins prominent and raised on abaxial surface. ***Flowers*** scented or unscented; variously described as greenish with some ochre, the lip brown in centre; pale reddish ochre, the lip dark red in centre and with reddish brown side lobes, and a pinkish mid-lobe with a white tip; pale greenish yellow, with a triangular yellow-brown spot at base of each mid-lobe side lobule; cream-yellow with a dark speck on lip; sepals pale ochre-red, the petals yellowish and the lip ochre, with two brown central spots; pale brown with tips of upper sepals and lip appendages reddish brown. ***Pedicel*** with ***ovary*** 1.8–3 mm long, clavate. ***Dorsal sepal*** 6 × 1.5 mm, narrowly oblong-elliptic, acute, veins 3. ***Lateral sepals*** 4.6–5 × 1.7–1.8 mm, slightly obliquely oblong-elliptic, acute, veins 3. ***Petals*** 4–5 × 1.1–1.5 mm, narrowly elliptic, acute, somewhat falcate, veins 2–3. ***Lip*** 5-lobed, 2.5–3 mm long, 1.8–2 mm wide across mid-lobe; veins 3; side lobes reduced, triangular, rounded; mid-lobe cruciform, expanded into spreading to erect oblong, rounded to truncate, sometimes slightly falcate side lobules and a triangular, acuminate, cuspidate, horizontal to decurved terminal lobule; side lobules 1–1.1 × 0.9–1 mm, veins 4; terminal lobule 1–1.1 mm long; disc with two lamellate keels joined at base to form a horseshoe-shaped U, arising from base and terminating at base of mid-lobe, with a low septum level with side lobes between. ***Column*** 2.8–3 mm long, slender, very obscurely dorsally carinate; foot absent; apical hood elongate, 1 mm long, oblong-triangular, obtuse to acute, entire; rostellum small, triangular, acute; stelidia basal, ligulate, apex obtuse, rather spathulate or hamate, 2 mm long, slightly shorter than apical hood; anther-cap 0.4–0.5 × 0.6 mm, ovate, cucullate, glabrous. ***Fruit*** 5–7 × 3.5–4 mm, ovoid-elliptic, lines of dehiscence 3.

HABITAT AND ECOLOGY: Lower montane ridge forest on sandstone soils; mossy forest, recorded growing in 60% sunlight high up in canopy; low and rather open, wet, somewhat podsolic forest, with a very dense undergrowth of *Pandanus* and rattan palms; cleared, steeply sloping roadsides on sandstone and shale outcrops partially covered with grass and small bushes, fully exposed to the sun. Alt. 900 to 2400 m. Flowering observed in January, April, August, September, November and December.

DISTRIBUTION IN BORNEO: SABAH: Crocker Range, Mt. Alab, Kimanis road; Mt. Kinabalu; Sipitang District, Mt. Lumaku, Sungai Rurun headwaters; Tenom District, Mt. Anginon. SARAWAK: Mt. Dulit.

GENERAL DISTRIBUTION: Endemic to Borneo.

NOTES: *D. exasperatum* is a close ally of the commoner *D. gibbsiae*, with which it shares similar lip and column characters. The leaves, however, are generally shorter, broader, more obtuse, and have a distinctive fine transverse venation, while the floral bracts and flowers are larger and the lip side lobes are less developed.

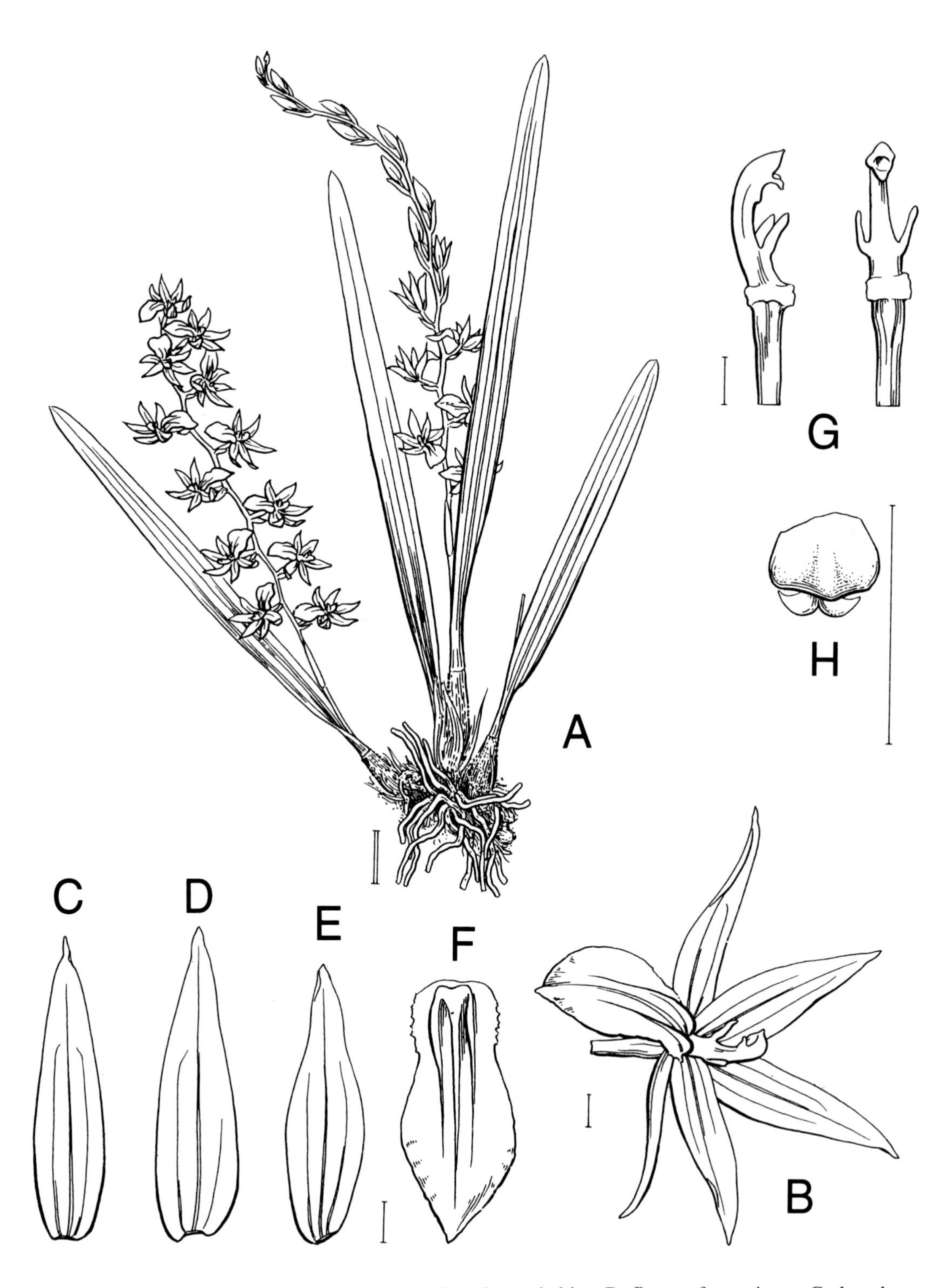

Figure 43. Dendrochilum galbanum J.J. Wood. - A: habit. - B: flower, front view. - C: dorsal sepal. - D: lateral sepal. - E: petal. - F: lip. - G: pedicel with ovary and column, side and front views. - H: anther-cap, back view. All drawn from *Burtt & Martin* B. 5328 (holotype) by Eleanor Catherine. Scale: single bar = 1 mm; double bar = 1 cm.

DERIVATION OF NAME: The specific epithet is derived from the Latin *exasperatus*, covered with short hard points, roughened, presumably referring to the surface of the leaves which frequently contain numerous small calcium oxalate bodies producing a rough texture, particularly in dried material.

43. DENROCHILUM GALBANUM J.J. Wood

Dendrochilum galbanum *J.J. Wood* in Orchid Rev. 102 (1197): 147, fig. 80 (1994). Type: Borneo, Sarawak, route from Bakelalan to Mt. Murud, SW of Camp III, *Burtt & Martin* B. 5328 (holotype E, isotypes K, SAR).

Clump-forming ***epiphyte. Pseudobulbs*** 0.5–1 × 0.4–0.5 cm, ovoid, enclosed by sheaths when young. ***Leaves*** 4–7 × 0.3–0.6 cm, linear-ligulate, obtuse and minutely mucronate, attenuate into a short sulcate petiole 2–5 mm long. ***Inflorescence*** 12- to 20-flowered, longer than leaves, erect; flowers borne 3–5 mm apart; peduncle 0.3–1.8 cm long, with a couple of apical sterile bracts 3–4 mm long; rachis 4–9.5 cm long, quadrangular, orange-coloured; floral bracts 3.5–4 mm long, ovate-elliptic, acute to acuminate, concave, light brown. ***Flowers*** greenish yellow or bright yellow; lip yellow, brown at the middle, the callus red; column green; anther-cap yellow. ***Pedicel*** with ***ovary*** 3 mm long, slender. ***Dorsal sepal*** 7.5–8 × 2 mm, narrowly elliptic, acute. ***Lateral sepals*** 7.9–8 × 2 mm, narrowly ovate-elliptic, acute. ***Petals*** 7 × 2 mm, narrowly elliptic, subacute. ***Lip*** 6.2–6.3 × 2.9–3 mm, elliptic, acute, the side lobes absent, erose towards base, the disc with two fleshy keels and a prominent median vein, all joined at the base and terminating a little above erose portion. ***Column*** 3 mm long, slender, greatly curved, with a short foot; hood entire; stelidia 1 mm long, borne a short distance above base; anther-cap minute, cucullate.

HABITAT AND ECOLOGY: Upper montane mossy forest. Alt. 1800 to 2250 m. Flowering observed in September.

DISTRIBUTION IN BORNEO: SARAWAK: Mt. Murud.

GENERAL DISTRIBUTION: Sumatra and Borneo.

NOTE: An attractive species closely related to *D. graminoides* Carr, described from Mt. Kinabalu in Sabah, but distinguished by the much shorter, broader leaves, erect inflorescences and larger flowers with a longer, broader lip, and longer column with slightly shorter stelidia.

DERIVATION OF NAME: The specific epithet is derived from the Latin *galbanus*, the colour of gum galbanum, which is greenish-yellow, and refers to the flower colour.

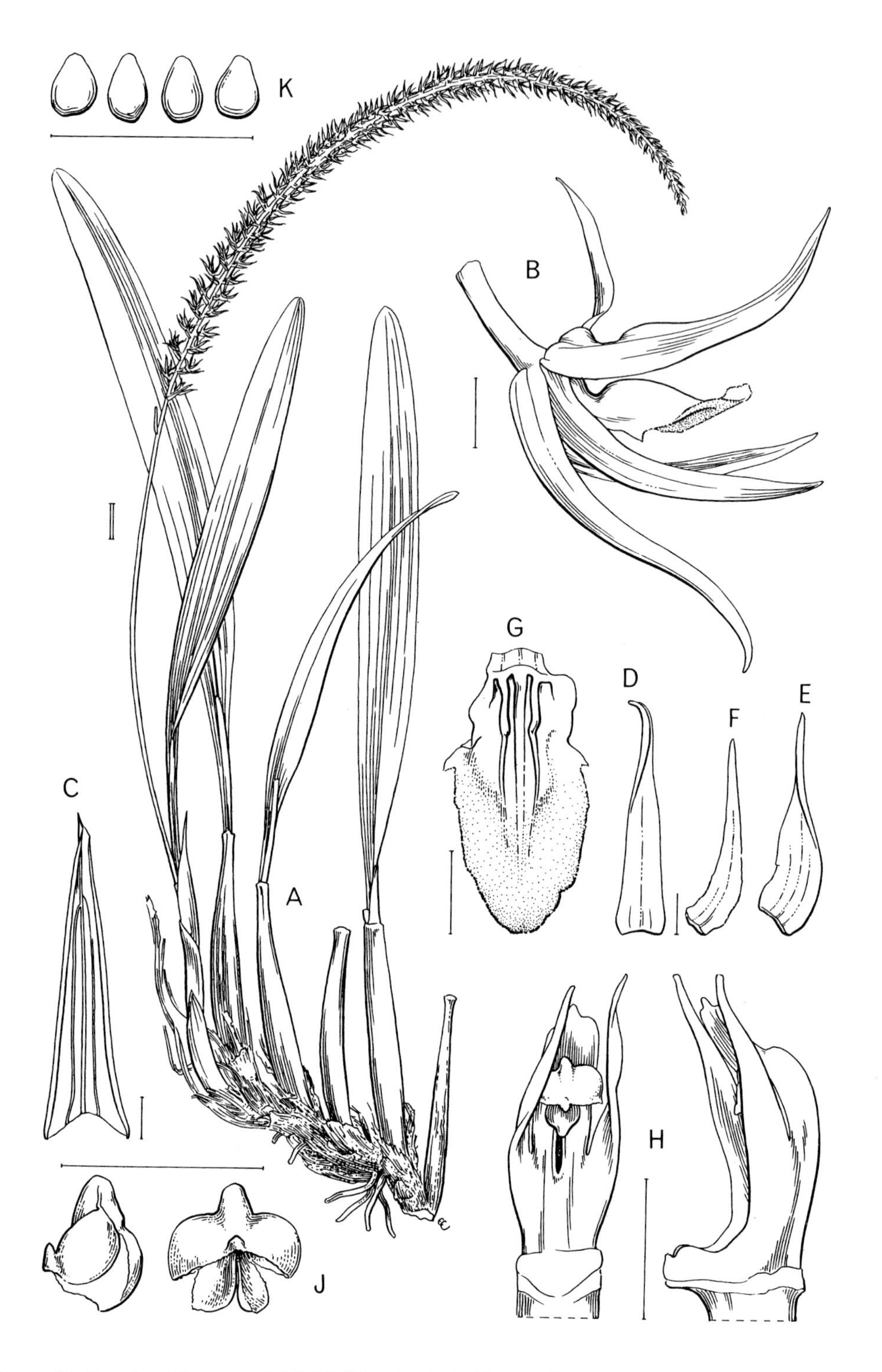

Figure 44. Dendrochilum geesinkii J.J. Wood. - A: habit. - B: flower, side view. - C: floral bract. - D: dorsal sepal. - E: lateral sepal. - F: petal. - G: lip, flattened. - H: column, front and side views. - J: anther-cap, side and back views. - K: pollinia. All drawn from *Geesink* 9180 by Eleanor Catherine. Scale: single bar = 1 mm; double bar = 1 cm.

44. DENDROCHILUM GEESINKII J.J. Wood

Dendrochilum geesinkii *J.J. Wood* in Wood & Cribb, Checklist Orchids Borneo: 175, fig. 17 (1994). Type: Borneo, Kalimantan Timur, base of Mt. Tapa Sia, between Long Bawan and Panado, *Geesink* 9180 (holotype L, isotypes BO, K, KYO).

Epiphyte with a creeping rhizome enclosed in fibrous sheaths. ***Pseudobulbs*** 6–8 × 0.6–0.8 cm, each about 0.5–0.6 cm apart, cylindrical, enclosed by three 1.5–4 cm long, acute to acuminate ***cataphylls*** when immature. ***Leaves*** 12.6–23.5 × 1.4–2.2 cm, oblong-elliptic, obtuse and mucronate to subacute, narrowed and conduplicate at base, coriaceous. ***Inflorescence*** densely many-flowered; peduncle 13–17 cm long, naked; rachis up to 23 cm long; floral bracts 5–7 mm long, 1–1.1 mm wide at base, subulate, acuminate. ***Flowers*** pale green. ***Pedicel*** with ***ovary*** 2 mm long. ***Dorsal sepal*** 6 mm long, 1 mm wide (across base), linear-lanceolate, acuminate. ***Lateral sepals*** 5.5 mm long, 1 mm wide (across base), lanceolate, slightly falcate, acute. ***Petals*** 5 mm long, *c.* 0.8 mm wide (across base), linear-lanceolate, slightly falcate, acute, minutely erose. ***Lip*** 3.8–4 × 2 mm, oblong-elliptic, obtuse, with tiny auriculate side lobes, the margin minutely erose-papillose, the disc with three fleshy keels near base, grading into prominent veins which terminate towards apex. ***Column*** 2 mm long, with a foot; stelidia borne from near base, 2.1–2.2 mm long, acute, slightly exceeding hood; hood *c.* 0.2 mm wide, slightly irregularly toothed; anther-cap cucullate.

HABITAT AND ECOLOGY: Hill forest on sandstone; mixed hill-dipterocarp forest on ridges. Alt. 1290 to 1400 m. Flowering observed in July and August.

DISTRIBUTION IN BORNEO: KALIMANTAN TIMUR: Mt. Tapa Sia. SARAWAK: Mt. Batu Lawi area.

GENERAL DISTRIBUTION: Endemic to Borneo.

NOTE: This species is related to *D. crassifolium* Ames but may be distinguished by the much longer pseudobulbs, slightly longer floral bracts, and slightly smaller flowers with narrower petals, obscure, auriculate lip side lobes and stelidia exceeding the apical column hood.

DERIVATION OF NAME: Named in honour of the late Rob Geesink of the Rijksherbarium, Leiden, The Netherlands, who collected the type during a joint Indonesian-Japanese expedition sponsored by LIPI, Jakarta and the Ministry of Education, Tokyo.

Figure 45. Dendrochilum gibbsiae Rolfe. - A-C: habits. - D: floral bract, flattened. - E: flower, oblique view. - F: pedicel with ovary, lip and column, front view. - G: dorsal sepal. - H: lateral sepal. - J: petal. - K: lip. - L: column, oblique view. - M: column, side view. - N: column, back view. - O: anther-cap, oblique view. - P: pollinia. - Q–U: lips. - V: infructescence. A (habit) drawn from *Carr* 3620, SFN 27884, B (habit) from *Carr* 3134, SFN 26745, C–P from *Wood* 578, Q from *Gibbs* 4085 (holotype of *D. kinabaluense* Rolfe), R from *J. Clemens* 361 (holotype of *D. quinquelobum* Ames), S from *Bailes & Cribb* 722, T from *Gibbs* 4087 (holotype of *D. gibbsiae* Rolfe), U from *Wood* 621 and V from *Carr* SFN 27970 by Susanna Stuart-Smith. Scale: single bar = 1 mm; double bar = 1 cm.

45. DENDROCHILUM GIBBSIAE Rolfe

Dendrochilum gibbsiae *Rolfe* in J. Linn. Soc., Bot. 42: 147 (1914). Type: Borneo, Sabah, Mt. Kinabalu, Marai Parai Spur, *Gibbs* 4087 (holotype BM, isotype K).

Dendrochilum kinabaluense Rolfe in J. Linn. Soc., Bot. 42: 148 (1914). Type: Borneo, Sabah, Mt. Kinabalu, Marai Parai Spur, 2100 m, Feb. 1910, *Gibbs* 4085 (holotype BM, isotype K).

D. quinquelobum Ames, Orchidaceae 6: 63–64, pl. 82, II, 2 (1920). Type: Borneo, Sabah, Mt. Kinabalu, Kiau, 30 November 1915, *J. Clemens* 361 (holotype AMES, isotypes BM, K, SING).

Epiphyte or ***terrestrial. Rhizome*** 1–10 (rarely up to 16) cm long, 0.2–0.3 cm in diameter. ***Roots*** produced in a large, dense mass, smooth, 0.5–1.8 mm in diameter. ***Cataphylls*** 3–4, 0.5–2.5 cm long, finely veined, greyish-brown or reddish-brown, sometimes finely speckled, persistent and taking a long time to become fibrous. ***Pseudobulbs*** 1.2–4.8 × 0.3–1 cm, closely caespitose, rarely distant and up to 2.5 cm apart, cylindrical, narrowly oblong, ovoid-oblong or narrowly fusiform, smooth or slightly wrinkled, always green. ***Leaf-blade*** (3.5–)10–28 × (0.3–)0.6–1.8(–2.1) cm, linear-lanceolate or oblong-lanceolate, subacute or acute, thin-textured, main veins 5–7, with numerous parallel secondary veins, transverse veins absent, containing scattered to densely placed crystalline calcium oxalate bodies, often covering the entire surface, variable in shape, from simple to intricate and dendritic in shape, particularly noticeable in dried material; petiole (1–)1.5–3 cm long, sulcate. ***Inflorescence*** an ascending to gently curving, many-flowered, narrow subdense raceme; flowers borne 2–3 mm apart; peduncle terete, (3.5–)8–13 cm long; rachis quadrangular, (5.5–)10–16(–25) cm long, basal sterile bract absent; floral bracts 1.2–3 × 0.1–1.2 mm, ovate-triangular or oblong-elliptic, acute, spreading. ***Flowers*** strongly scented of cinnamon or oranges, or unscented; variously described as cream, flesh-coloured, salmon, cream-green, pale yellowish-green, pale greenish, greenish yellow, creamy yellow, yellow, straw-coloured or pale khaki with a pink pedicel, pale green lip, orange-brown near base and pale ochre at apex, or orange, with 2 pink, purple, maroon, brown or orange-brown spots at base of mid-lobe; calli yellowish-green, brown at tip; column pale green or brownish green, reddish at base; stelidia greenish white; anther cream. ***Pedicel*** with ***ovary*** 1–3 mm long, narrowly clavate. ***Dorsal sepal*** 2.2–4 × 0.8–1.1 mm, oblong-elliptic, acute, concave, minutely papillose-hairy at base inside, veins 3. ***Lateral sepals*** 2–4 × 1.2–1.5 mm, oblong-ovate, acute, minutely papillose-hairy inside, veins 3. ***Petals*** 1.6–3.6 × 0.8–1 mm, narrowly elliptic or oblong-elliptic, acute, minutely papillose-hairy inside, veins 1–2. ***Lip*** 5-lobed, 1.6–3 mm long, 2.5–4 mm wide across mid-lobe, veins 3, median prominent; slightly decurved at side lobes; side lobes 0.2–1 × 0.2–0.6 mm, oblong, obtuse, truncate or slightly erose, sometimes reduced to triangular, falcate teeth, spreading or erect; mid-lobe cruciform, expanded into spreading or retrorse, oblong and rounded to oblong-falcate, obtuse side lobules and a cuspidate, acuminate, deflexed or ascending terminal lobule; side lobules 1–2 × 0.5–1.1 mm, veins 3; terminal lobule 0.8–1 mm long; disc with two fleshy lamellate keels joined at base to form a horseshoe-shaped U, arising from just above base and terminating at base of mid-lobe, and usually with an obscure, slightly raised transverse septum level with side lobes between. ***Column*** 1.4–2 mm long, slender; foot absent; apical hood short, narrowly ovate, obtuse or subacute; rostellum small, ovate-triangular, obtuse or subacute; stelidia (1–)1.5–2.1 mm long, basal, oblong-linear, apex obtuse, hamate, equalling or longer than apical hood; anther-cap ovate, cucullate, minute, glabrous. ***Fruit*** *c.* 5 × 4.5–5 mm, ovoid. Plate 10C & D.

HABITAT AND ECOLOGY: Lower montane oak-laurel forest; mossy forest on ultramafic substrate with *Gymnostoma* and undergrowth of climbing bamboo; low mossy and xerophyllous scrub forest on extreme ultramafic substrate; steep, open roadside banks on sandstone and shale outcrops, associated with *Melastoma*, *Nepenthes fusca*, *Gahnia* spp., *Lycopodium* spp., ferns, etc.; limestone boulders in primary forest; recorded as epiphytic on tree trunks and lower branches, rarely growing above two metres up. Alt. *c.* 870 to 2400 m. Flowering observed throughout the year.

DISTRIBUTION IN BORNEO: BRUNEI: Mt. Pagon. KALIMANTAN TIMUR: Apokayan. SABAH: Crocker Range; Mt. Kinabalu; Mt. Tawai. SARAWAK: Mt. Ampungan; Mt. Mulu National Park; Mt. Murud area; Hose Mountains.

GENERAL DISTRIBUTION: Endemic to Borneo.

NOTES: *D. gibbsiae* is the most widespread and one of the most frequently collected members of section *Cruciformia*. Plants are variable both vegetatively and in the dimensions and shape of the floral parts, unlike the majority of species in the genus. It is usually found growing abundantly in dense or large populations.

Variants with narrow leaves, racemes and floral parts have been called *D. kinabaluense* Rolfe and those with strongly recurved falcate lip mid-lobe side lobules assigned to *D. quinquelobum* Ames. Examination of the large number of collections now available shows every gradation from short and straight to strongly falcate lip mid-lobe side lobules. Recognition of these taxa seems unjustified.

A characteristic feature of *D. gibbsiae* and some related species, eg. *D. dolichobrachium* (Schltr.) Schltr. and *D. hastilobum* J.J. Wood, are the numerous crystalline calcium oxalate bodies present in the leaf blades. These may be few in number or, more commonly, densely distributed over the entire surface and clearly visible in dried and spirit material. Their shape may be simple, or complicated, resembling an asymmetrical snowflake.

DERIVATION OF NAME: The specific epithet honours Miss Lilian Gibbs, the collector of the type.

46. DENDROCHILUM GRANDIFLORUM (Ridl.) J.J. Sm.

Dendrochilum grandiflorum (*Ridl.*) *J.J. Sm.* in Recueil Trav. Bot. Néerl. 1: 66 (1904). Type: Borneo, Sabah, Mt. Kinabalu, *Haviland* 1142 (holotype K).

Platyclinis grandiflora Ridl. in Stapf in Trans. Linn. Soc. London, Bot. 2, 4: 233 (1894).

Acoridium grandiflorum (Ridl.) Rolfe in Orchid Rev. 12: 220 (1904).

Epiphyte, occasionally ***terrestrial. Rhizome*** up to 10 cm long, 0.4–0.6 cm in diameter, branches 2–4 cm long, tough, clothed in fibrous sheaths. ***Roots*** much branched, smooth, 0.5–1 mm in diameter. ***Cataphylls*** 3–4, 1.5–5 cm long, finely veined, greyish-brown to brown, finely speckled, persistent, becoming fibrous. ***Pseudobulbs*** (1.8–) 2.5–6 × 0.4–0.6 cm, caespitose, up to 1.5 cm apart, narrowly cylindrical, terete, with narrow longitudinal furrows, usually appearing deeply sulcate when dried, surface densely minutely rugose and quite distinctive in dried material; red, orange, sometimes yellow. ***Leaf-blade*** 6–18 × (0.8–)1.3–2.5(–3.3) cm, narrowly elliptic to elliptic, acute, apiculate, dark green and glossy above, slightly paler beneath, coriaceous, margin slightly constricted 1.5–2.5 cm below apex, main veins 5, with numerous small transverse veins forming a reticulate pattern, most distinctive in dried material, abaxial and adaxial surface distributed with minute brown trichomes, petiole 1–3 cm long, sulcate below blade, with minute brown trichomes. ***Infloresence*** several- to many-flowered, gently curving, subdense to lax; flowers borne 4–7 mm apart; peduncle 7–16 cm long, terete, glabrous; rachis 8–15 cm long, quadrangular, sulcate, minutely finely setose, with a 4–6 mm long basal sterile bract; floral bracts 3–6.8 × 4 mm, ovate, shortly apiculate, margin minutely erose, veins prominent, raised, surface finely setose. ***Flowers*** usually unscented, with a salmon-pink pedicel with ovary; sepals and petals flesh-coloured or pinkish brown; lip deep flesh-coloured with darker keels; column dark olive with pink base and stelidia, the rostellum yellow; anther-cap cream suffused rose towards apex; flowers also variously described as yellow, light brown, old rose-pink and bright flesh-coloured. ***Pedicel*** with ***ovary*** 4 mm long, clavate, curved, finely setose. ***Dorsal sepal*** 7–7.5 × 2.5–2.6 mm, oblong-elliptic to ovate-elliptic, acute, somewhat concave, minutely papillose-hairy at base inside, veins 3. ***Lateral sepals*** 6.9–7 × 2.6–2.8 mm, obliquely ovate-elliptic, dorsally carinate at apex, acute, adaxial surface sparsely minutely papillose-hairy, veins 3. ***Petals*** 5.5–6.5 × 2 mm, oblong-elliptic, acute, very minutely papillose at base inside, veins 3. ***Lip*** tridentate, 5 mm long, 5–5.1 mm wide across side lobes, 1.5–1.6 mm wide at base, immobile; veins 3, median simple, two each side branching into two to three secondary veins on side lobes; side lobes 2–2.1 mm long, triangular, acute, spreading; mid-lobe 1.5–1.6 mm long, cuspidate, acute; disc with an M-shaped callus composed of two elevated, fleshy, semi-circular, wing-like keels 0.3–0.4 mm high, terminating *c.* halfway along lip and becoming low and fleshy at base where they are joined, between which is a short, fleshy, blunt basal boss. ***Column*** 2.8–3 mm long, *c.* 0.8 mm wide, slightly narrowed at middle, gently curved; foot absent; apical hood ovate, rounded, entire; rostellum triangular, obtuse; stelidia 2 × 0.2 mm, basal, narrowly linear, obtuse, equalling rostellum to slightly shorter than apical hood; anther-cap *c.* 0.5 × 0.5 mm, ovate, cucullate, glabrous. Plate 11A & B.

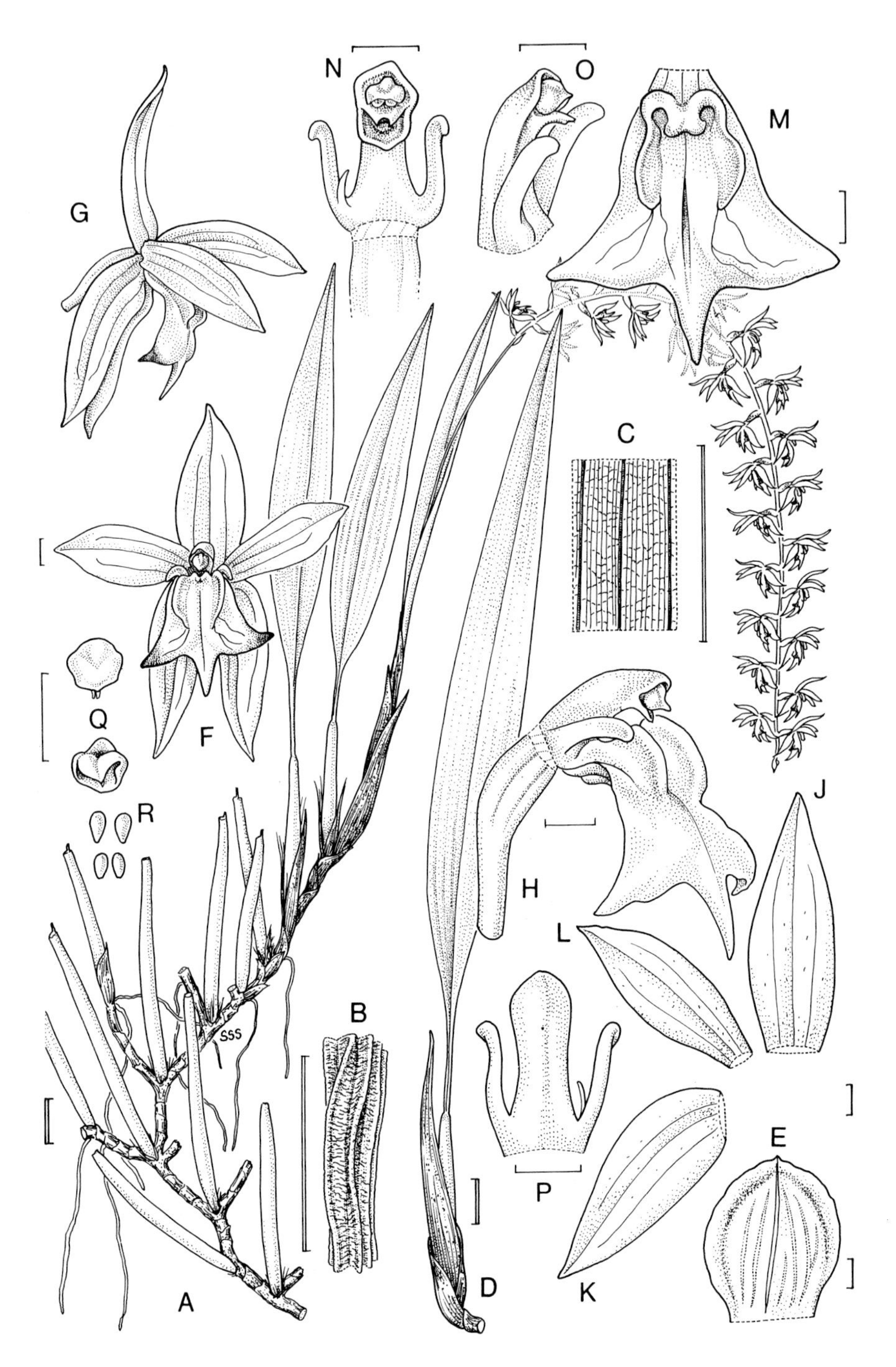

Figure 46. Dendrochilum grandiflorum (Ridl.) J.J. Sm. - A: habit. - B: portion of old pseudobulb, close-up. - C: close-up detail of leaf venation. - D: habit. E: floral bract, flattened. - F: flower, front view. - G: flower, side view. - H: pedicel with ovary, lip and column, side view. - J: dorsal sepal. - K: lateral sepal. - L: petal. - M: lip, flattened. - N: upper portion of ovary and column with anther-cap, front view. - O: column, side view. - P: column, back view. - Q: anther-cap, back and front views. - R: pollinia. A–C drawn from *Sinclair* 9180, D from *Carr* 3476, SFN 27430 and E-R from *Wood* 608 by Susanna Stuart-Smith. Scale: single bar = 1 mm; double bar = 1 cm.

HABITAT AND ECOLOGY: Upper montane forest, often in open sites and preferring forest developed on ultramafic substrate; recorded as terrestrial in *Leptospermum recurvum* forest; epiphytic in thick moss cushions on the trunks and branches of dead trees and on rotting stumps; rocky wooded stream banks. Alt. 900 to 3800 m. Flowering observed in January, February, March, May, June, July, August, November and December.

DISTRIBUTION IN BORNEO: SABAH: Mt. Kinabalu.

GENERAL DISTRIBUTION: Endemic to Borneo.

NOTES: *D. grandiflorum* is an easily recognised species in section *Cruciformia,* so far only recorded on Mt. Kinabalu from where we have numerous collections from a wide altitudinal range along the well-trodden summit trail. Populations are particularly common above 2400 m where it is more commonly a terrestrial, often forming huge clumps. It is curious that it has never been found in suitable habitats on, for example, Mt. Murud in Sarawak or Mt. Trus Madi in Sabah.

Sterile herbarium specimens can easily be identified by the finely rugose older pseudobulbs and reticulate vein pattern on the leaf blades.

DERIVATION OF NAME: The specific epithet refers to the relatively large flowers compared to its near relatives.

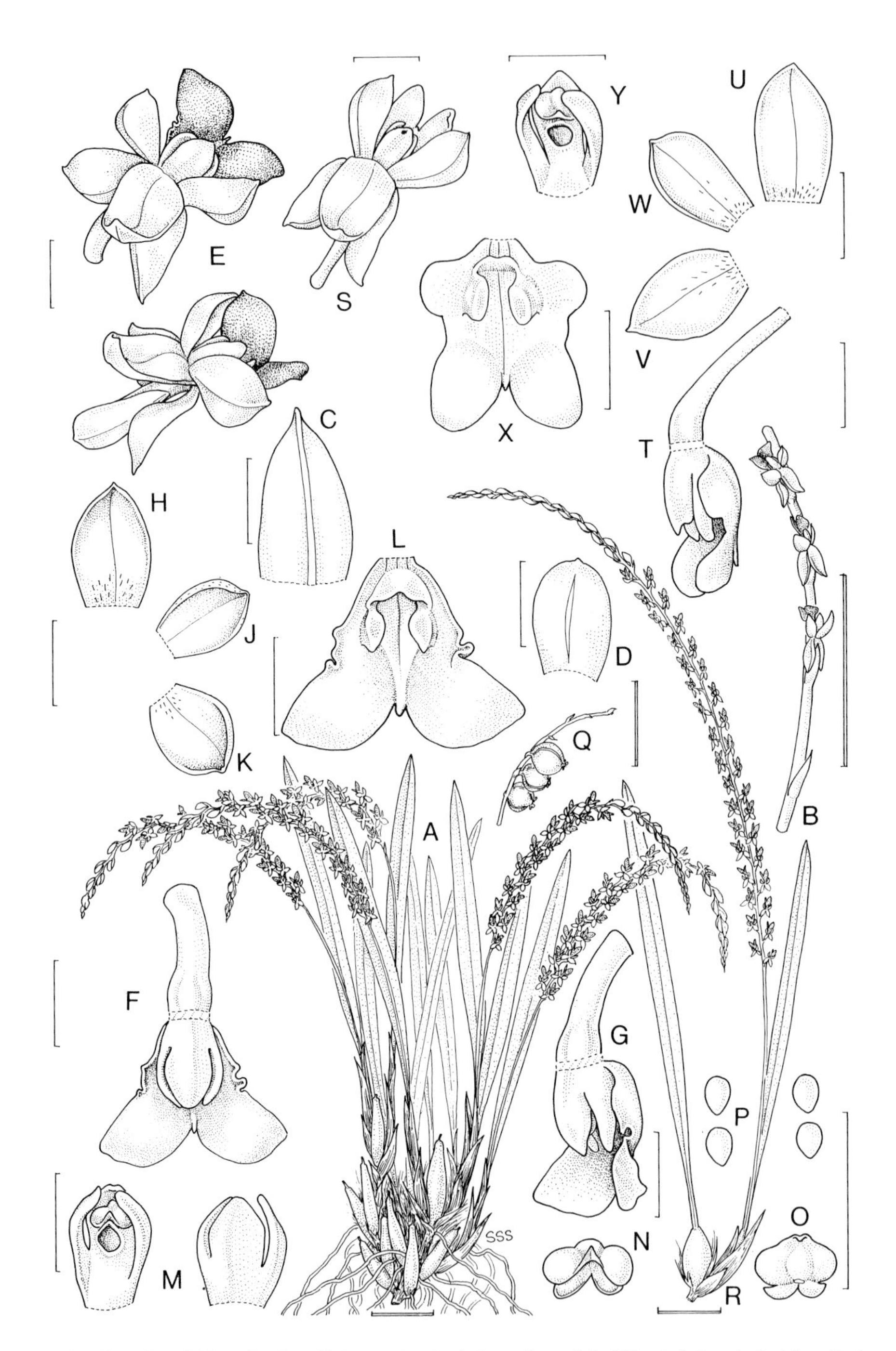

Figure 47. Dendrochilum haslamii Ames (typical form from Mt. Kinabalu). - A: habit. - B: base of inflorescence. - C & D: floral bracts, flattened. - E: flowers. - F: pedicel with ovary, lip and column, front view. - G: pedicel with ovary, lip and column, side view. - H: dorsal sepal. - J: lateral sepal. - K: petal. - L: lip, flattened. - M: column, front and back views. - N: anther-cap, front view. - O: anther-cap, back view. - P: pollinia. - Q: infructescence. (Variant from Mt. Mulu). - R: habit. - S: flower. - T: pedicel with ovary, lip and column, side view. - U: dorsal sepal. - V: lateral sepal. - W: petal. - X: lip, flattened. - Y: column, front view. A (habit) drawn from *Collenette* 21535, B–P from *Gunsalam* 3, Q from *J. & M.S. Clemens* 31663 and R–Y from *G. Lewis* 369 by Susanna Stuart-Smith. Scale: single bar = 1 mm; double bar = 1 cm.

47. DENDROCHILUM HASLAMII Ames

Dendrochilum haslamii Ames, Orchidaceae 6: 53, pl. 85 (1920). Type: Borneo, Sabah, Mt. Kinabalu, *Haslam* s.n. (holotype AMES).

Epiphyte. Rhizome abbreviated, up to 8 cm long, usually much shorter, 2 mm in diameter, clothed in fibrous sheaths. ***Roots*** 0.5–1 mm in diameter, flexuous, produced in a large mass, smooth. ***Cataphylls*** 3–4, 0.5–1.5 cm long, reddish-brown, unspotted, distinctly papillose, persistent, slowly becoming fibrous. ***Pseudobulbs*** 0.5–2 × 0.4–0.6 cm, caespitose, crowded on to rhizome, subfusiform or ovoid, smooth to slightly rugose, strongly rugose in dried material, epidermal cells often clearly defined at high magnification in dried material, surface appearing almost papillose; always green. ***Leaf-blade*** 2.5–6 × 0.2–0.6 cm, coriaceous, linear-ligulate to linear-lanceolate, obtuse, minutely apiculate, main veins 3, median prominent, containing scattered to densely distributed crystalline calcium oxalate bodies similar to *D. gibbsiae*; petiole 2–7 mm long, sulcate. ***Inflorescence*** 15- to 30-flowered, curving, subdense; flowers borne 1.8–2.5 mm apart; peduncle 1.2–5(–9) cm long, terete, slender, dull red; rachis 3–7 cm long, quadrangular, somewhat sulcate, becoming fractiflex, dull red, with or without a 3 mm long sterile bract; floral bracts 2 × 1.2–1.3 mm, ovate, apiculate, median vein prominent. ***Flowers*** unscented; sepals and petals creamy yellow, yellowish green or yellow; lip bright chestnut-brown or brownish orange, apical cusp creamy yellow; column olive-brown, stelidia pale brown, tipped yellowish brown. ***Pedicel*** with ***ovary*** 2–3 mm long, clavate, gently curved, ovary 0.5 mm long, rotund. ***Sepals*** strongly concave and reflexed. ***Dorsal*** and ***lateral sepals*** 1.5–1.75 × 1.1 mm, oblong-ovate, obtuse to subacute, 1-veined, slightly carinate. ***Petals*** 1.5 × 1 mm, ovate to oblong-ovate, obtuse to subacute, strongly concave, slightly spreading, directed forward. ***Lip*** 4-lobed, 2 mm long, 1.5 mm wide below middle, usually 3 mm wide across terminal lobules; veins 3, indistinct; side lobes obscure, rounded, undulate; mid-lobe expanded into two ovate to oblong, obtuse, falcate, divaricate lobules between which is a short triangular, acute cusp or mucro; disc with a U-shaped fleshy callus extending from near base and terminating at middle of disc. ***Column*** 1 mm long, straight; foot absent; apical hood entire, obtuse; stelidia, 1–1.1 mm long, basal, linear, obtuse; anther-cap cucullate, glabrous. ***Fruit*** ovoid. Plate 11C.

HABITAT AND ECOLOGY: Upper montane ridge-top forest, most frequently on ultramafic substrate; usually found as a twig epiphyte on mossy branches, sometimes growing low down on trunks and branches, often in exposed sites; recorded as epiphytic on *Leptospermum*. Alt. 2400 to 3100 m. Flowering observed from July to December.

DISTRIBUTION IN BORNEO: SABAH: Mt. Kinabalu. SARAWAK: Mt. Mulu National Park.

GENERAL DISTRIBUTION: Endemic to Borneo.

NOTES: *D. haslamii* is easily identifiable by its diminutive size and chestnut-brown bilobed lip contrasting with clear yellow sepals and petals. It often forms large clusters of pseudobulbs, often numbering over fifty on one plant.

DERIVATION OF NAME: The specific epithet honours George Haslam who collected the type.

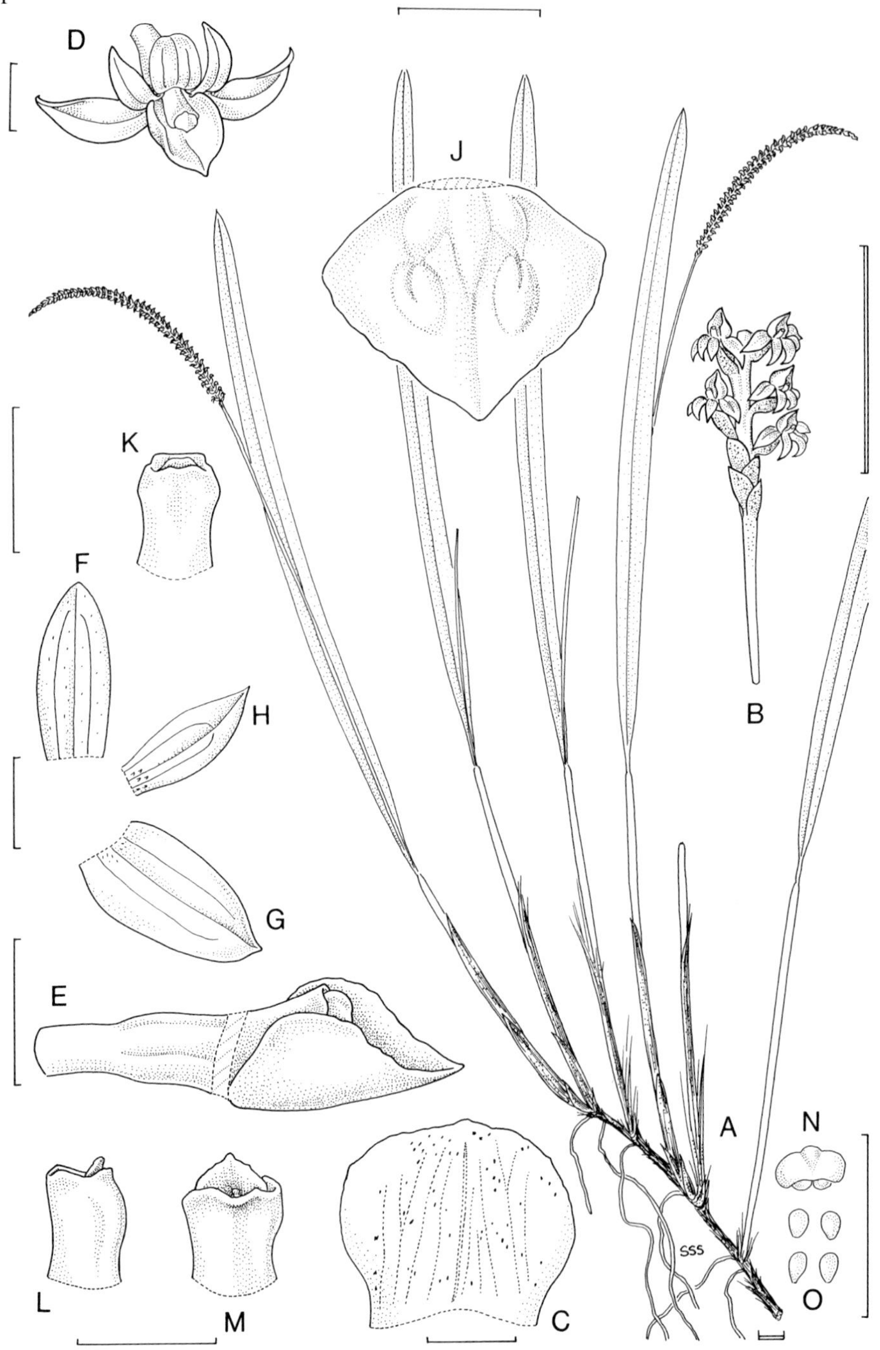

Figure 48. Dendrochilum hologyne Carr. - A: habit. - B: base of inflorescence. - C: floral bract. - D: flower, front view. - E: pedicel with ovary, lip and column, side view. - F: dorsal sepal. - G: lateral sepal. - H: petal. - J: lip, flattened. - K: column, front view. - L: column, side and back views. - M: anther-cap and pollinia. A (habit) drawn from *de Vogel* 8339 and B–M from *Wood* 646 by Susanna Stuart-Smith. Scale: single bar = 1 mm; double bar = 1 cm.

48. DENDROCHILUM HOLOGYNE Carr

Dendrochilum hologyne *Carr* in Gard. Bull. Straits Settlem. 8: 89 (1935). Type: Borneo, Sarawak, Dulit Ridge, 1400 m, 17 September 1932, *P.M. Synge* S.513 (holotype SING, isotype K).

A clump forming ***epiphyte*** or ***terrestrial. Rhizome*** long-creeping, 0.2–0.3 cm in diameter, internodes 1.5–7 cm long, rooting at nodes, enclosed in brown, finely speckled darker brown imbricate cataphylls. ***Pseudobulbs*** (5.5–)10–14.5 × 0.2–0.3 cm, narrowly cylindrical, wrinkled, completely enclosed by tubular, imbricate, membranous, acute cataphylls at first, partially covered when mature. ***Cataphylls*** slightly wrinkled, brown with fine darker brown speckling. ***Leaves*** 13–23 × 0.6–1.2 cm, linear-ligulate, acute, thinly coriaceous, veins prominent below, abruptly narrowed into a grooved petiole 1.2–1.8 cm long. ***Inflorescence*** apical from the almost mature pseudobulb, densely many-flowered, equal to or longer than leaves; peduncle 15–20 cm long, terete, filiform, with 2–3 apical imbricate, adpressed sterile bracts below rachis, otherwise naked; rachis 8–12 cm long, quadrangular, concave alternately on each side above the flower; floral bracts 2–2.7 × 2.5–3.2 mm, oblong-ovate or broadly triangular-ovate, obtuse, prominently 5-veined. ***Flowers*** with pale brown or fawn sepals and petals; lip greenish brown, apricot at base. ***Ovary*** 1.5 mm long, sessile. ***Sepals*** and ***petals*** spreading to reflexed. ***Dorsal sepal*** 2 × 9 mm, oblong-elliptic, narrowly obtuse. ***Lateral sepals*** 2.2–2.5 × 1–1.1 mm, oblong-ovate, obtuse to subacute, slightly keeled near apex. ***Petals*** 2–2.2 × 0.5–0.8 mm, oblong or narrowly oblong-elliptic, acute. ***Lip*** 1.8–2 × 2 mm when flattened, entire, triangular-ovate or obliquely subquadrate, obtuse or acute, concave, margins erect, obscurely 3-veined, with two obscure central swellings each side of median vein. ***Column*** 0.8–1 mm long; stelidia absent; rostellum triangular; anther-cap cucullate. Plate 11D.

HABITAT AND ECOLOGY: Lower montane forest; ridge forest with *Agathis*, small rattans, etc.; podsol forest; often growing on mossy tree boles or as a terrestrial in leaf litter. Alt. 100 to 1600 m. Flowering observed in September, October and December.

DISTRIBUTION IN BORNEO: SABAH: Sipitang District. SARAWAK: Mt. Dulit.

GENERAL DISTRIBUTION: Endemic to Borneo.

NOTES: Carr in his original description misleadingly describes the lip as having three rounded basal keels. His description was based on dried material and no such keels could be found on flowers preserved in alcohol at Kew. Only two obscure central swellings are present on the lip.

DERIVATION OF NAME: The specific epithet is derived from the Greek *holo,* meaning entire, complete, whole or undivided, and *gyno*, female or pertaining to female organs, in reference to the column which lacks stelidia.

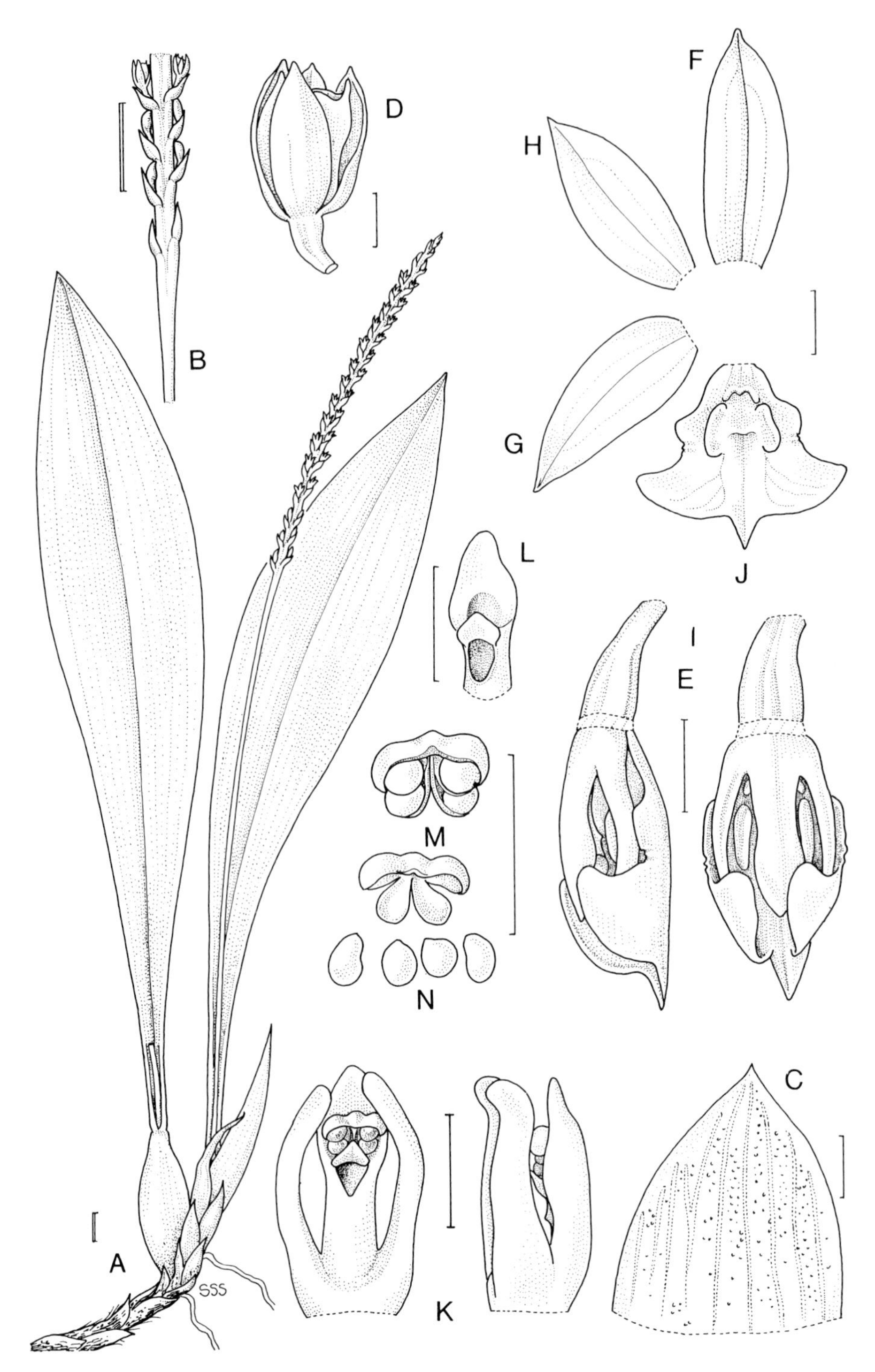

Figure 49. Dendrochilum hosei J.J. Wood. - A: habit. - B: base of inflorescence. - C: floral bract, flattened. - D: flower, side view. - E: pedicel with ovary, lip and column, side and back views. - F: dorsal sepal. - G: lateral sepal. - H: petal. - J: lip, flattened. - K: column, front and side views. - L: column apex. - M: anther-cap with pollinia, front view. - N: pollinia. All drawn from *de Vogel* 1244 (holotype) by Susanna Stuart-Smith. Scale: single bar = 1 mm; double bar = 1 cm.

49. DENDROCHILUM HOSEI J.J. Wood

Dendrochilum hosei *J.J. Wood* in Wood & Cribb, Checklist Orchids Borneo: 179, fig. 23 D & E (1994). Type: Borneo, Sarawak, Kapit District, northern Hose Mountains, base of ridge leading to Bukit Batu, *de Vogel* 1244 (holotype L, isotype K).

Epiphyte. Pseudobulbs up to 5 × 2 cm, elliptic, slightly wrinkled, enclosed in brown cataphylls up to 8 cm long when young. ***Leaves*** tough, coriaceous; blade 22–25 × 5.3–5.7 cm, elliptic, acute, cuneate at base; petiole 5–6 cm long, deeply sulcate. ***Inflorescence*** many-flowered, rather rigid; flowers borne 5–6 mm apart; peduncle 22 cm long, 2–2.8 mm wide, tough; rachis 14 cm long, 2–3 mm wide, fleshy; floral bracts 3–5 × 2 mm, triangular-ovate, acute. ***Flower*** colour not recorded. ***Pedicel*** with ***ovary*** *c.* 1.8 mm long, clavate, curved. ***Dorsal sepal*** 4 × 1.9 mm, oblong-elliptic to ovate-elliptic, apiculate. ***Lateral sepals*** 4 × 1.9 mm, ovate-elliptic, apiculate. ***Petals*** 3.5–3.6 × 1.6–1.7 mm, elliptic, acute. ***Lip*** 2.9–3 mm long, 4 mm wide across side lobes, pandurate, margins slightly erose towards base; side lobes obscure, oblong, margins rather uneven to erose; mid-lobe with erect oblong-ovate, obtuse, slightly falcate lateral lobules and a tooth-like apiculate, mucronate central lobule; disc with a basal fleshy callus and two separate fleshy crest-like keels either side which almost touch the basal ridge at their base. ***Column*** 2.1 mm long, slightly carinate dorsally, foot absent; stelidia basal, 2 mm long, ligulate, obtuse, decurved at apex; apical hood ovate, entire; anther-cap cucullate.

HABITAT AND ECOLOGY: Lower montane forest about 25 metres high on sandstone. Alt. 1200 m. Flowering observed in December.

DISTRIBUTION IN BORNEO: SARAWAK: Hose Mountains.

GENERAL DISTRIBUTION: Endemic to Borneo.

NOTES: The lip shape and column structure of this rather unattractive species belonging to section *Cruciformia* bear a striking resemblance to *D. gibbsiae* Rolfe and *D. grandiflorum* (Ridl.) J.J. Sm. The broad leaves of *D. hosei* are quite different however, and similar to those of *D. longifolium* Rchb.f., while the rather rigid inflorescence is not unlike *D. longipes* J.J. Sm. The two keels on the disc are aligned very close to, but are not united with the transverse basal ridge. In many species, including *D. gibbsiae* and *D. grandiflorum*, these are united to form a U- or M-shaped structure.

DERIVATION OF NAME: The specific epithet refers to the Hose Mountains in Sarawak which were named after the Reverend George F. Hose (1838–1922), Bishop of Singapore, Labuan and Sarawak from 1881 until 1908.

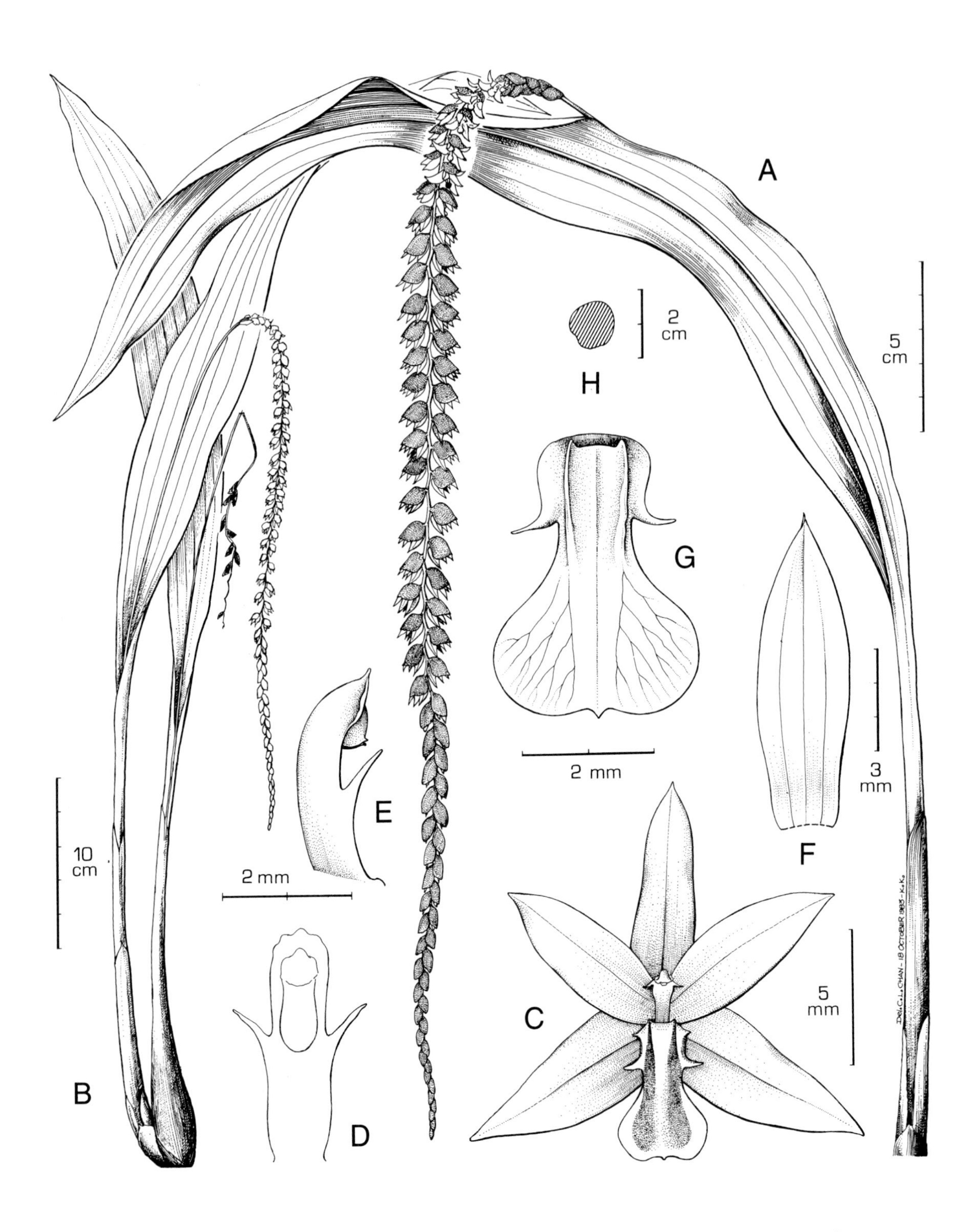

Figure 50. Dendrochilum imbricatum Ames. - A: inflorescence and leaf. - B: habit. - C: flower, front view. - D: column, front view. - E: column, side view. - F: dorsal sepal. - G: lip, front view. - H: pseudobulb, transverse section. All drawn from *Phillipps & Lamb* s.n. by C.L. Chan.

50. DENDROCHILUM IMBRICATUM Ames

Dendrochilum imbricatum *Ames*, Orchidaceae 6: 54, pl. 82, fig. 1 (1920). Type: Borneo, Sabah, Mt. Kinabalu, Kiau, *J. Clemens* 179 (holotype AMES, isotypes BO, K, SING).

Robust ***epiphyte. Rhizome*** abbreviated, tough, covered in sheaths. ***Roots*** 1–2 mm in diameter, wiry, smooth. ***Pseudobulbs*** 4.5–8.5 cm long, 1.5 cm wide at base, 0.5–0.6 cm wide at apex, aggregated on rhizome, contiguous, cylindrical to narrowly conical, smooth, when immature concealed by *c.* 4 imbricate tubular brownish cataphylls, the uppermost large and elongated, 12–18 cm long, which also encloses the petiole and part of the peduncle. ***Leaves*** large, thin-textured, chartaceous when dry; blade 33–40 × 3.5–5 cm, oblong-elliptic, acute, cuneate at base, often gradually and asymmetrically constricted 8–13 cm below apex; petiole 12–18 × 0.4 cm, tough, sulcate. ***Inflorescence*** synanthous, densely many-flowered; flowers borne up to 5 mm apart; peduncle and rachis minutely sparsely brown furfuraceous; peduncle 18–38 cm long, terete, erect to porrect; rachis 38–40 cm long, distinctly quadrangular, angles slightly alate, pendulous, with several imbricate sterile bracts at base; floral bracts 8–13 × 8 mm, ovate-elliptic, rounded, deeply convex, imbricate as the raceme expands, a little shorter than or equal to the flowers when the raceme is mature, brownish. ***Flowers*** sweetly scented; sepals and petals yellowish-green or pale green; side lobes of lip pale yellow with dark brown margins and apex, the mid-lobe creamy with a large brown central patch, keels brown; column white, flushed yellow beneath. ***Pedicel*** with ***ovary*** 2 mm long, minutely furfuraceous. ***Sepals*** 3-veined, glabrous. ***Dorsal sepal*** 7–9 × 3 mm, oblong-elliptic to triangular-lanceolate, acute. ***Lateral sepals*** 7–9 × 3 mm, as dorsal sepal, slightly oblique. ***Petals*** 7–8.1 × 2.9 mm, oblong to narrowly elliptic, acute, 3-veined. ***Lip*** 4–4.2 mm long, 3-lobed, reflexed, *c.* 2.5 mm wide across side lobes; free portion of side lobes 0.9–1 mm long, triangular, apex acute and subsetaceous; mid-lobe 2.5–3 × 2–3.1 mm, cuneate at base, expanded into a rounded flabellate or suborbicular blade, sometimes obscurely apiculate; disc with two low, fleshy keels extending from base and terminating on base of mid-lobe; mid-lobe sulcate, especially on base of mid-lobe. ***Column*** 3–3.2 mm long; foot 0.1–0.2 mm long; apical hood oblong, rounded, obscurely retuse or irregularly erose; stelidia 1 mm long, subulate, median, arising just below stigmatic cavity, reaching just beyond the ovate, acute rostellum; anther-cap ovate, cucullate. Plate 12A & B.

HABITAT AND ECOLOGY: Hill and lower montane forest; kerangas vegetation. Alt. 600 to 1550 m. Flowering observed in August, October and November.

DISTRIBUTION IN BORNEO: KALIMANTAN: Locality unknown. SABAH: Mt. Kinabalu. SARAWAK: Ulu Limbang, route to Batu Lawi.

GENERAL DISTRIBUTION: Endemic to Borneo.

NOTE: *D. imbricatum* belongs to section *Platyclinis* and is easily distinguished by the large bracts which nearly conceal the flowers.

DERIVATION OF NAME: The specific epithet is derived from the Latin *imbricatus*, overlapping like roof tiles, and refers to the floral bracts.

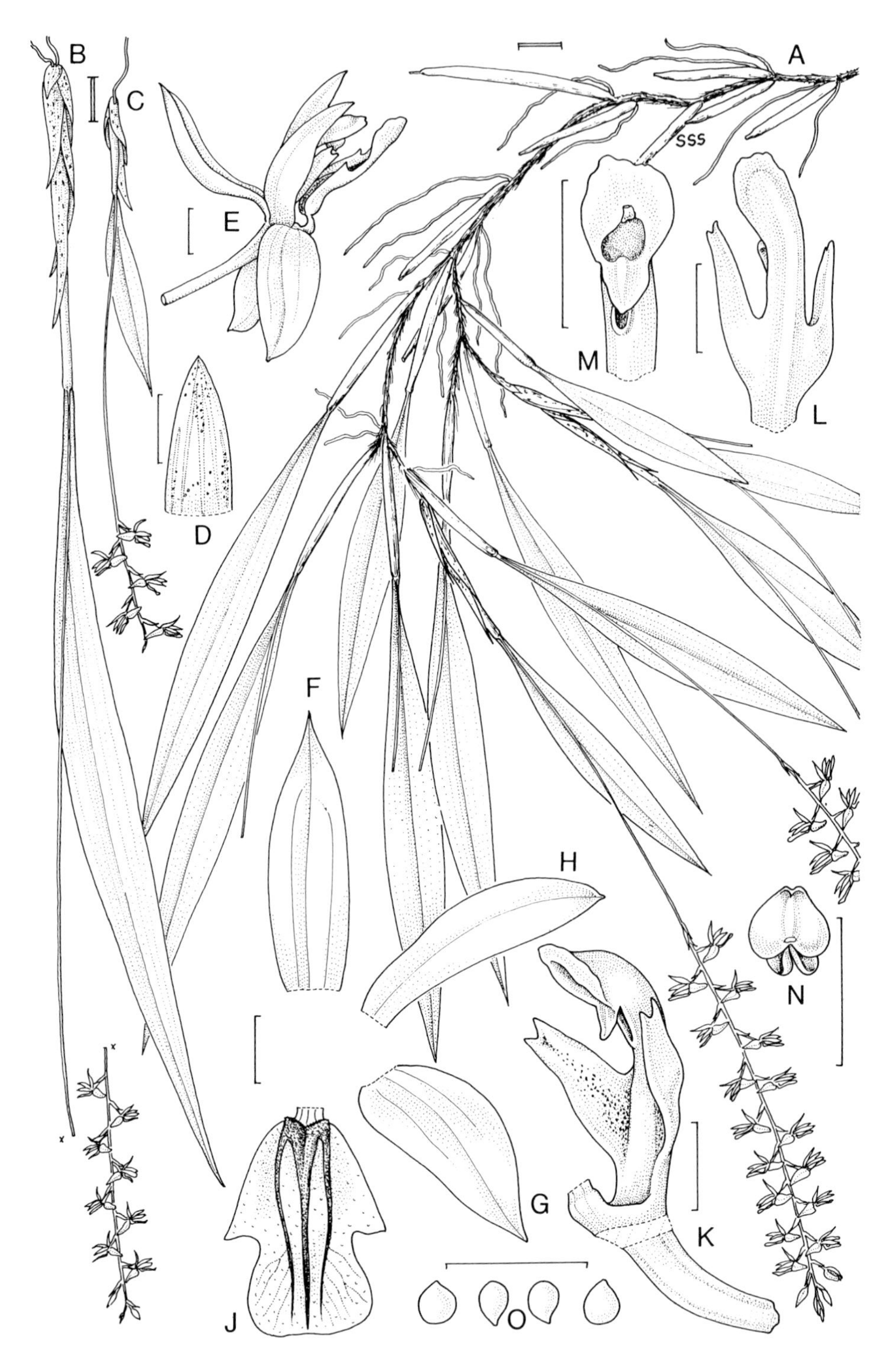

Figure 51. Dendrochilum imitator J.J. Wood. - A, B & C: habits. - D: floral bract, flattened. - E: flower, side view. - F: dorsal sepal. - G: lateral sepal. - H: petal. - J: lip, flattened. - K: pedicel with ovary and column, oblique view. - L: column, back view. - M: column apex. - N: anther-cap, back view. - O: pollinia. A and D–O drawn from *de Vogel* 8451 (holotype), B from *Beaman* 10347 and C from *de Vogel* 8426 by Susanna Stuart-Smith. Scale: single bar = 1 mm; double bar = 1 cm.

51. DENDROCHILUM IMITATOR J.J. Wood

Dendrochilum imitator *J.J. Wood* in Wood & Cribb, Checklist Orchids Borneo: 179, fig. 22G (1994). Type: Borneo, Sabah, Sipitang District, trail from Long Pa Sia to Long Samado, *c.* 4 km SSW of Long Pa Sia, *de Vogel* 8451 (holotype L, isotype K).

Pendulous ***epiphyte. Rhizome*** branching, up to 18 cm long, rooting at nodes. ***Pseudobulbs*** 1.2–3.5(–4.5) × 0.2–0.3 cm, narrowly cylindrical or narrowly fusiform, enclosed in several brown-spotted acute cataphylls 1.5–3.5 cm long. ***Leaves*** thin-textured; blade 4–10 × 0.8–1.4 cm, elliptic, acute, with numerous small transverse veins; petiole 0.3–0.6 cm long, sulcate. ***Inflorescence*** 6- to many-flowered, lax; flowers borne 4–5 mm apart; peduncle 4.5–9 cm long, filiform, minutely furfuraceous at base; rachis 2.5–10.5 cm long, quadrangular, minutely furfuraceous, with 1 or 2 ovate, acute, basal sterile bracts; floral bracts 2–2.5 mm long, ovate, acute. ***Flowers*** pale green or dull pale greenish-brown, lip with a large central brownish blotch, median keel pale green; column cream. ***Pedicel*** with ***ovary*** 4 mm long, clavate, slightly curved. ***Dorsal sepal*** 4.5–5 × 1–1.2 mm, lanceolate, acute, concave. ***Lateral sepals*** 4–4.5 × 1.5–1.9 mm, obliquely ovate-elliptic, acute, reflexed. ***Petals*** 4–5 × 0.9–1 mm, ligulate, subacute, slightly falcate. ***Lip*** 3.2–4 mm long, 1.6–2.1 mm wide across mid-lobe, stipitate to column-foot by a narrowly oblong claw; side lobes triangular, acute, irregularly erose-denticulate; mid-lobe obovate, obtuse, sometimes with a small apical mucro; disc with three fleshy papillose keels, the outer contiguous at base and extending to at or just below base of mid-lobe, median keel broad at base, narrowing at the middle and usually becoming pronounced again at the apex, extending almost to apex of mid-lobe. ***Column*** 3.5 mm long, arcuate; foot 0.5–0.8 mm long; apical hood ovate, cucullate; stelidia 1.5 mm long, arising just above base, ligulate, bifid at apex; anther-cap cucullate, with a dorsal tooth.

HABITAT AND ECOLOGY: Recorded as a trunk epiphyte at the type locality which is dry, rather low and open ridge forest on soil probably derived from sandstone and shales. This forest is interspersed with small open patches of grass and *Gleichenia* ferns. It also grows as a branch epiphyte overhanging water in dense primary riverine forest up to 30 metres high on soil derived from sandstone. *Vermeulen & Duistermaat* 962 (K, L) was collected in open podsol forest. Alt. 1300 to 1400 m. Flowering observed in October and December.

DISTRIBUTION IN BORNEO: SABAH: Sipitang District.

GENERAL DISTRIBUTION: Endemic to Borneo.

NOTES: *D. imitator* is distinguished from the closely related *D. lacteum* Carr because of the spotted rhizome sheaths, floral bracts shorter than the pedicel with ovary, smaller flowers with a longer pedicel with ovary, shorter and a little narrower, acute rather than acuminate sepals and petals, reflexed lateral sepals, a shorter lip with a much more pronounced median keel, and bifid stelidia on the column.

De Vogel 2079 (K, L, spirit material only), collected at 1300 m on Mt. Pagon in Brunei, is similar to *D. imitator* but has minute hairs at the base of broader sepals and petals, non-reflexed lateral sepals as in *D. lacteum*, and a broader lip. The pronounced median keel on the lip and the bifid column stelidia are, however, typical of *D. imitator*. Further collections are necessary before its status can be clarified.

DERIVATION OF NAME: The specific epithet is derived from the Greek *imitator*, to imitate, in reference to the close resemblance of this species to *D. lacteum*.

52. DENDROCHILUM JOHANNIS-WINKLERI J.J. Sm.

Dendrochilum johannis-winkleri *J.J. Sm.* in Mitt. Inst. Allg. Bot. Hamburg 7: 36, t. 5, fig. 26 (1927). Type: Borneo, Kalimantan Barat, Bukit Tilung, *Winkler* 1495 (holotype HBG).

Epiphyte. Rhizome short, creeping, branched, covered when young with tubular, dark brown-spotted cataphylls *c.* 3 cm long which disintegrate and become fibrous with age. ***Pseudobulbs*** 0.8–1 × 0.4–0.5 cm, oblong-ovoid. ***Leaves*** erect; blade 4.75–11 × 0.55–0.8 cm, linear-subspathulate, broadest distally, apex equally shallowly retuse, rounded; petiole 0.6–1.4 cm long, canaliculate. ***Inflorescence*** erect, laxly many-flowered; flowers borne 2 mm apart; peduncle 8.5–10 cm long, filiform, naked; rachis *c.* 6 cm long, narrow, angular; floral bracts 2 × 2 mm long, oblong-ovate, broadly obtuse. ***Flowers*** greenish white. ***Pedicel*** with ***ovary*** 1–1.1 mm long, clavate. ***Dorsal sepal*** 4 × 1.5 mm, oblong, subacute or acute, sometimes suberose at apex, concave, 3-veined. ***Lateral sepals*** 3.7–3.8 × 1.5 mm, obliquely subovate-oblong, somewhat falcate, acute, barely apiculate, concave, 3-veined. ***Petals*** 3.2–3.3 × 1.2–1.3 mm, obliquely oblong, somewhat falcate, acute, slightly concave, minutely erose, 3-veined. ***Lip*** obscurely 3-lobed, clawed at base, recurved, convex, 2.2 mm long, 1 mm wide across side lobes; side lobes obscure, rounded, erose; mid-lobe *c.* 1.2 × 0.7–0.8 mm, oblong-trullate, acute; disc with three fleshy minutely papillose ridges joined at base, outer two broader, median narrow, all three terminating on lower half of mid-lobe. ***Column*** 1.7–1.8 mm long, gently curved; stelidia borne at the middle, 2 mm long, obliquely linear-oblong, obtuse, slightly concave and minutely papillose distally, porrect; apical hood rounded, sometimes crenate, minutely papillose; foot 0.4 mm long, incurved, truncate; rostellum conspicuous, convex, rounded and abruptly contracted and apiculate; anther-cap 0.5 mm wide, cucullate, transversely triangular, acuminate or acute, apex recurved.

HABITAT AND ECOLOGY: Not recorded. Alt. 800–1500 m. Flowering observed in February.

DISTRIBUTION IN BORNEO: BRUNEI: Bukit Retak. KALIMANTAN BARAT: Bukit Tilung.

GENERAL DISTRIBUTION: Endemic to Borneo.

NOTE: *D. johannis-winkleri*, which belongs to section *Platyclinis*, is only known from the type and one other recent collection from Brunei.

DERIVATION OF NAME: Named in honour of Professor Hans Winkler who collected the type.

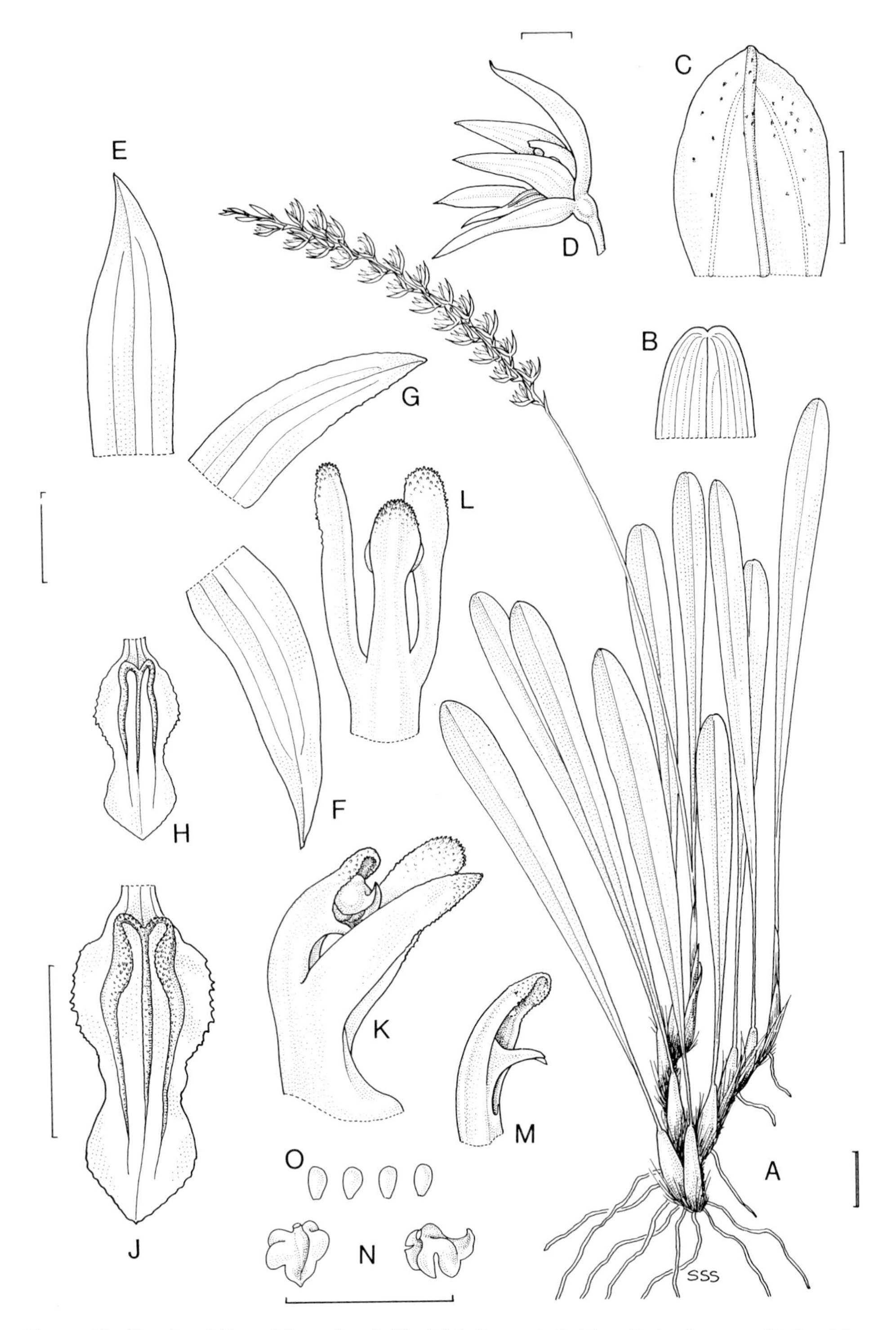

Figure 52. Dendrochilum johannis-winkleri J.J. Sm. - A: habit. - B: leaf apex. - C: floral bract, flattened. - D: flower, side view. - E: dorsal sepal. - F: lateral sepal. - G: petal. - H & J: lip, flattened. - K: column, side view. - L: column, back view. - M: column apex, side view. - N: anther-cap, back and side views. - O: pollinia. All drawn from *Winkler* 1495 (holotype) by Susanna Stuart-Smith. Scale: single bar = 1 mm; double bar = 1 cm.

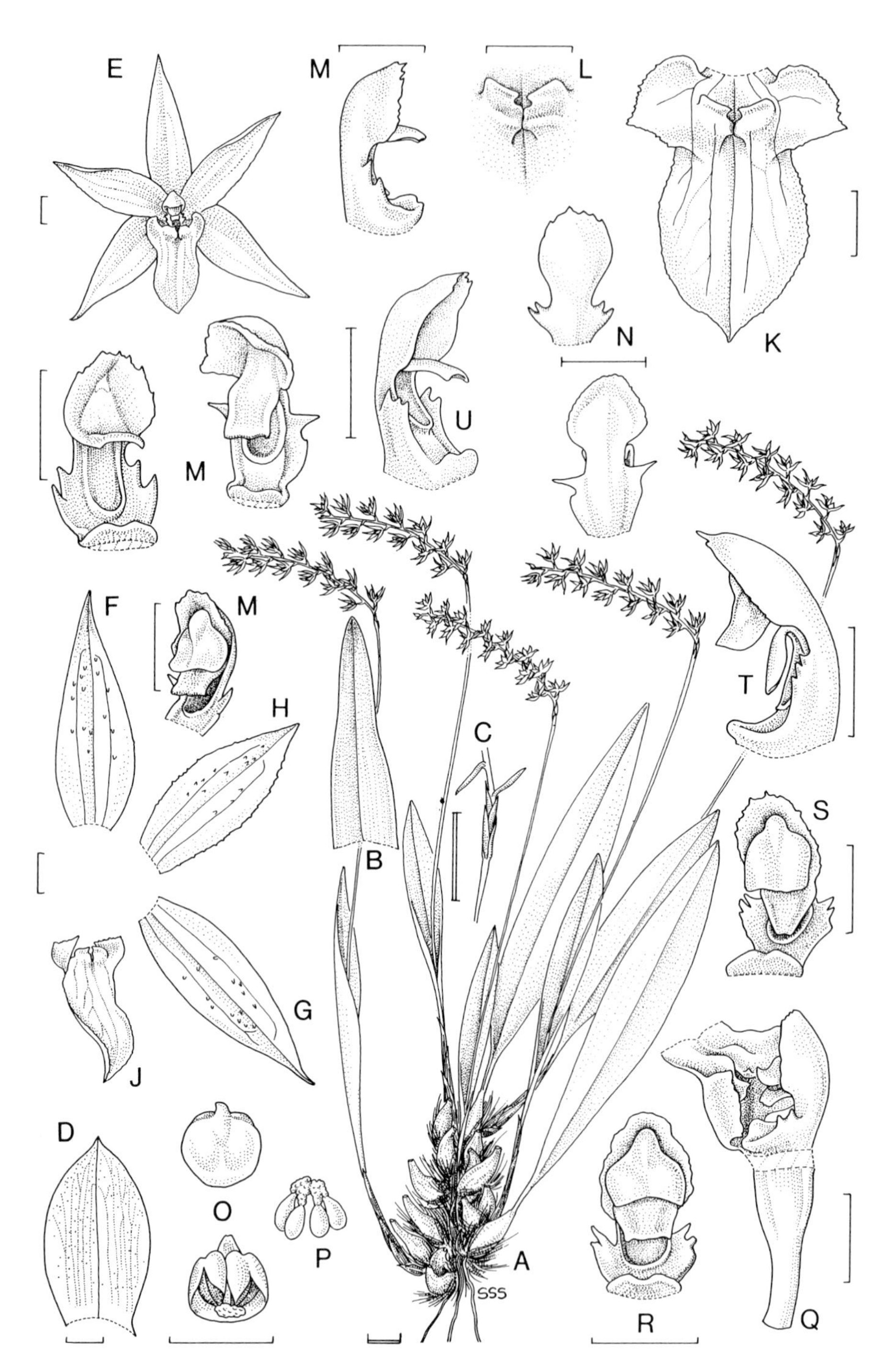

Figure 53. Dendrochilum kingii (Hook. f.) J.J. Sm. - A: habit. - B: leaf apex. - C: base of inflorescence. - D: floral bract, flattened. - E: flower, front view. - F: dorsal sepal. - G: lateral sepal. - H: petal. - J: lip, natural position. - K: lip, flattened. - L: basal calli on disc of lip. - M: columns, front, side and oblique views. - N: columns, back view. - O: anther-cap, back and front views. - P: pollinia. - Q: pedicel with ovary, part of lip and column, side view. - R: column with anther-cap, front view. - S: column with anther-cap, front view. - T: column with anther-cap, side view. - U: column without anther-cap, oblique view. A–C drawn from *Ridley* s.n., D–P from *Cribb* 89/65, Q & R from *Lamb* s.n., S & T from *Lamb* AL 359/85 and U from *de Vogel* 919 by Susanna Stuart-Smith. Scale: single bar = 1 mm; double bar = 1 cm.

53. DENDROCHILUM KINGII (Hook. f.) J.J. Sm.

Dendrochilum kingii (*Hook. f.*) *J.J. Sm.* in Recueil Trav. Bot. Néerl. 1: 76 (1904). Type: Peninsular Malaysia, Perak, *Scortechini* s.n. (lectotype K).

Platyclinis kingii Hook. f., Hooker's Icon. Pl. XXI: plate 2015 (1880); Fl. Brit. Ind. 5: 708 (1890).

P. sarawakensis Ridl. in J. Linn. Soc., Bot. 31: 267 (1896). Type: Borneo, Sarawak, *Biggs* s.n., cult. Hort. Bot. Penang (not located).

Dendrochilum sarawakense (Ridl.) J.J. Sm. in Recueil Trav. Bot. Néerl. 1: 66 (1904).

Acoridium kingii (Hook. f.) Rolfe in Orchid Rev. 12: 220 (1904).

A. sarawakense (Ridl.) Rolfe in Orchid Rev. 12: 220 (1904).

Dendrochilum palawanense Ames, Orchidaceae 2: 103, fig. s.n. (1908). Type: Philippines, Palawan, Mt. Pulgar, 1250 m, *Foxworthy* in Bur. Sci. 553 (holotype AMES, isotype K).

D. bicallosum J.J. Sm. in Bull. Dép. Agric. Indes Néerl. 22: 17 (1909), *nom. illeg.* (non Ames).

D. bigibbosum J.J. Sm. in Bull. Dép. Agric. Indes Néerl. 45: 13 (1911). Types: Borneo, Kalimantan, Sungai Keriboeng, *Hallier* 1312 (syntype BO, isosyntypes K, L). Sarawak, Mt. Bengkaum, *Brooks* s.n. (syntype BO, isosyntypes K, L, SING).

Clump forming ***epiphyte*** or ***terrestrial. Rhizome*** 1–10 cm long, covered in fibrous sheaths. ***Roots*** 0.5–1 mm in diameter, wiry, smooth. ***Pseudobulbs*** 1.8–4 × 0.5–1.2 cm, oblong to ovoid, often ampulliform, smooth to wrinkled, aggregated to 1.5 cm apart on rhizome, enclosed when immature by cataphylls. ***Cataphylls*** 3–4, imbricate, tubular, acute, minutely furfuraceous-punctate, 2–7.5 cm long. ***Leaves*** thin to rather thick and leathery-textured, shiny; blade (8–)12–40 × 1–2.8(–3.5) cm, linear-lanceolate to narrowly elliptic, ligulate, sometimes distally constricted *c.* 3 cm below apex, obtuse to acute; petiole (0.3–)1.5–8 cm long, sulcate. ***Inflorescence*** synanthous, laxly *c.* 6- to 30-flowered; flowers borne 3–7 mm apart; peduncle 12–33 cm long, terete, erect to curving, glabrous or sparsely minutely furfuraceous-punctate; rachis (6–)9–11(–13) cm long, somewhat fractiflex, quadrangular, sulcate, decurved, minutely furfuraceous-punctate to almost glabrous, with 1–3 sterile basal acute bracts 5–7 mm long; floral bracts 5–8 × 4–5 mm, linear-lanceolate, ovate-elliptic or ovate, acute to shortly acuminate, convolute. ***Flowers*** 1–1.1 cm across, sweetly scented or unscented; sepals greenish orange, pale brown, pinkish brown, ochre-brown, translucent lemon-yellow, flushed green or pale apple-green and pale pink at base; petals straw-coloured or any combination of the above; lip orange-brown, yellow, salmon-pink, dark pinkish brown or greenish ochre, the calli

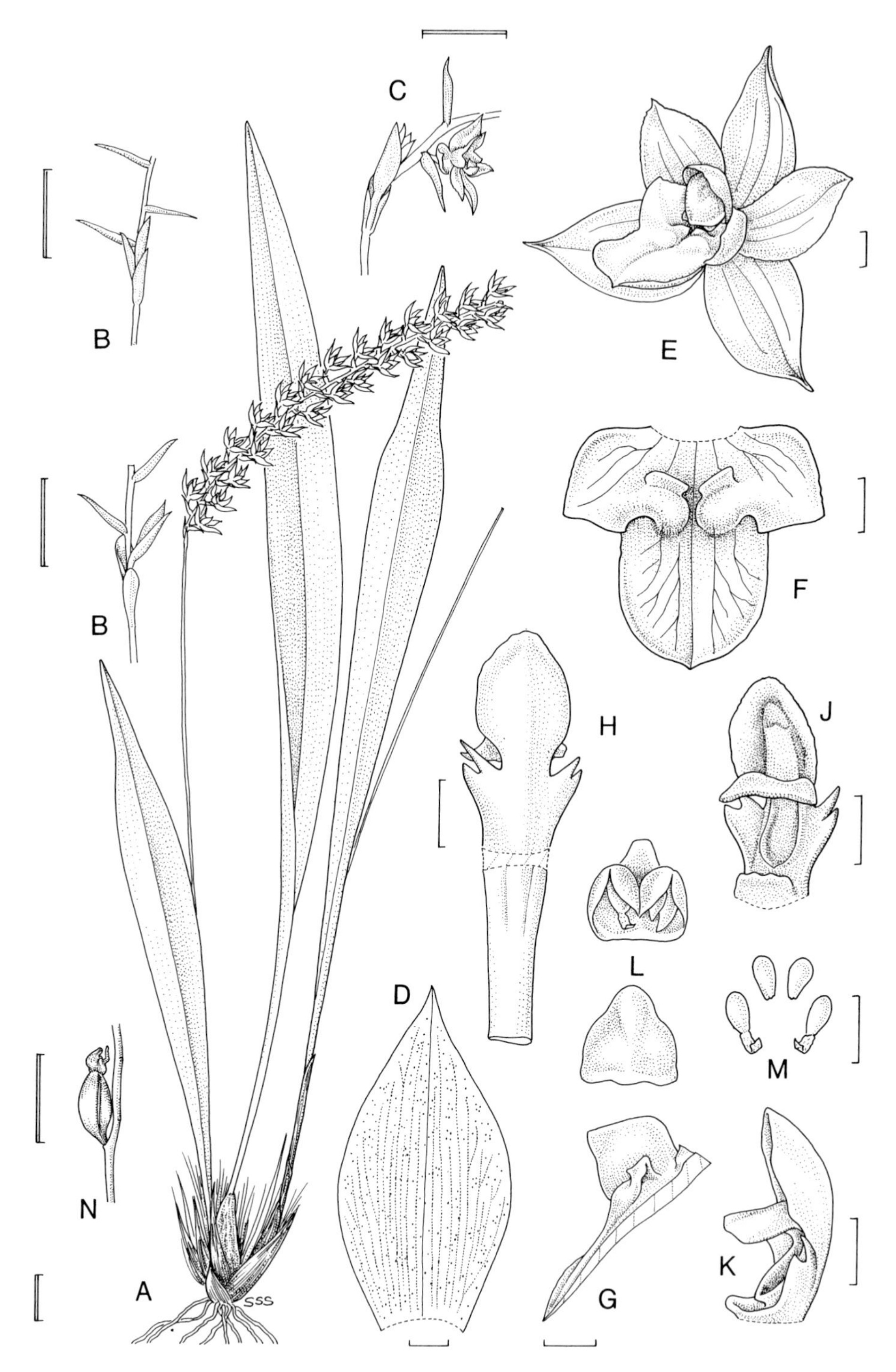

Figure 53a. Dendrochilum kingii (Hook.f.) J.J. Sm. - A: habit. - B & C: base of inflorescences. - D: floral bract, flattened. - E: flower, oblique view. - F: lip, flattened. - G: lip, longitudinal section. - H: pedicel with ovary and column, back view. - J: column, front view. - K: column, side view. - L: anther-cap, front and back views. - M: pollinia. - N: fruit. A & B drawn from *Hallier* 1312 (syntype of *D. bigibbosum* J.J.Sm.), C-M from *de Vogel* 1584 and N from *Ahmad Talip* SAN 70988 by Susanna Stuart-Smith. Scale: single bar = 1 mm; double bar = 1 cm.

olive-green, sometimes with a green median line on the mid-lobe; column yellow, pink or orange-brown, stelidia often cream or white. ***Pedicel*** with ***ovary*** 2–4.5 mm long, clavate. ***Sepals*** and ***petals*** spreading, 3-veined. ***Dorsal sepal*** 5.5–8 × 2–3 mm, oblong-elliptic, ovate-elliptic to narrowly elliptic, acute to acuminate. ***Lateral sepals*** 6–8 × 2–3.1 mm, oblong-elliptic or ovate-elliptic, slightly oblique, acuminate to narrowly acuminate. ***Petals*** 5–6.6 × 2–3 mm, obliquely ovate-elliptic, oblong-elliptic or narrowly elliptic, shortly narrowed at base, subacute or acute, less often acuminate, margin often minutely erose. ***Lip*** 4–5.5 mm long, 2.6–5 mm wide across side lobes when flattened; side lobes 0.8–1.8 × 1–2 mm wide at base, erect, auriculate, subquadrate or obliquely rounded, obtuse, crenulate to erose; mid-lobe 2.6–4 × 2–3 mm, somewhat deflexed, oblong-ovate, broadly ovate to sub-elliptic, obtuse, obtuse and shortly cuspidate or acute to acuminate, sometimes erose at base; disc with 2 fleshy, rounded, depressed flap-like calli between side lobes, their forward ends converging and often touching, with a sulcate area between and often extending a short way on to the swollen base of the mid-lobe. ***Column*** 2–3.5 mm long, decurved; foot short, but distinct; apical hood large, ovate, suborbicular, entire or minutely unevenly denticulate, *c.* 0.9–1.5 mm wide; stelidia broad, quadrate, wing-like, extending from base to below apical hood, regularly or irregularly and acutely bidentate or tridentate, the teeth shorter than or just reaching base of apical hood, sometimes less wing-like and reduced to two subulate narrowly triangular teeth; stigmatic cavity margin prominent and fleshy; rostellum large, 0.6–0.9 × 0.6–1.4 mm, narrowly or broadly rectangular or quadrate, sometimes triangular, incurved; anther-cap 1.4–1.8 × 1.1–1.5 mm, galeate, umbonate. Plate 13A.

HABITAT AND ECOLOGY: Hill-dipterocarp forest; podsol forest on very wet sandy soil, with dipterocarps and *Dacrydium*, etc.; hill forest on limestone; recorded as growing on *Eugenia rejangense* on a stream bank; epiphytic on the trunks and boles of trees, more rarely in the crown; also observed as a terrestrial in podsol forest. Alt. 300 to 900 m. Flowering observed from June to December.

DISTRIBUTION IN BORNEO: KALIMANTAN: Sungai Keribung. KALIMANTAN TIMUR: Sangkulirang. SABAH: Lahad Datu District, Ulu Segama; Nabawan area; Pun Batu; Sipitang District, Long Pa Sia area. SARAWAK: Bukit Woen; Bukit Semako; Iban River, Belaga; Kenaban River, Upper Plieran; Mt. Matang; Mt. Temabok, Upper Baram Valley; Ulu Lawas; Hose Mountains.

GENERAL DISTRIBUTION: Peninsular Malaysia, Borneo, Philippines (Palawan).

NOTES: A variable lowland species belonging to section *Platyclinis* and superficially resembling *D. rufum* (Rolfe) J.J. Sm. (section *Mammosa*). Although the column structure is similar to *D. rufum*, the lip is quite different, lacking the saccate hypochile and prominent calli of that species. Vernacular names include 'Bunga Tupan' (Kayan dialect) and 'Dar-chang' (Murut dialect).

DERIVATION OF NAME: The specific epithet honours Sir George King (1840–1909) who, with Robert Pantling, produced the authoritative two-volume work entitled The Orchids of the Sikkim Himalaya.

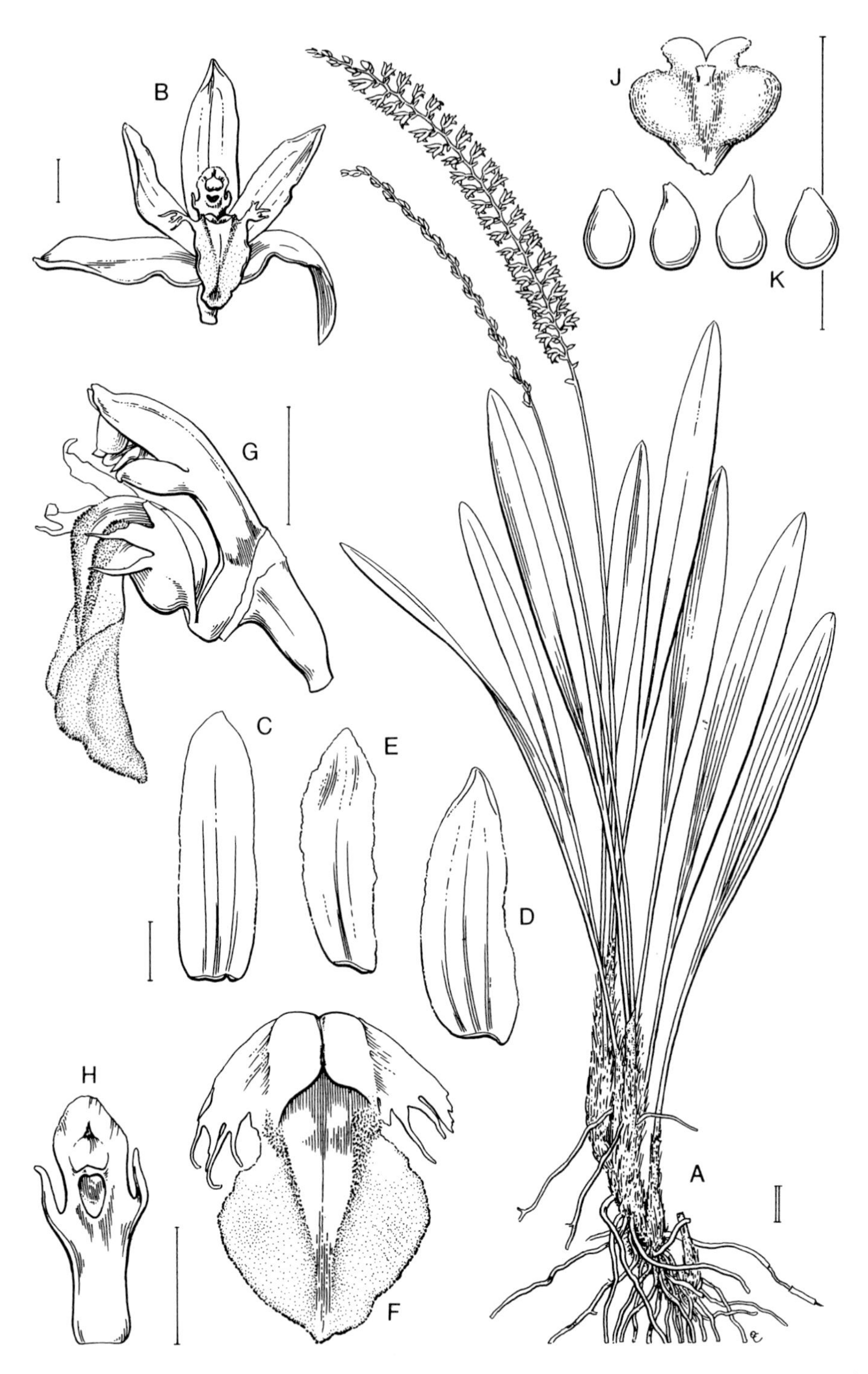

Figure 54. Dendrochilum lacinilobum J.J. Wood & A. Lamb. - A: habit. - B: flower, front view. - C: dorsal sepal. - D: lateral sepal. - E: petal. - F: lip, flattened. - G: pedicel with ovary, lip and column, side view. - H: column, front view. - J: anther-cap, back view. - K: pollinia. All drawn from *Lamb & Surat* in *Lamb* AL 1390/91 (holotype) by Eleanor Catherine. Scale: single bar = 1 mm; double bar = 1 cm.

54. DENDROCHILUM LACINILOBUM J.J. Wood & A. Lamb

Dendrochilum lacinilobum *J.J. Wood & A. Lamb* in Wood & Cribb, Checklist Orchids Borneo: 181, fig. 19 (1994). Type: Borneo, Sabah, Crocker Range, Ulu Apin Apin, *Lamb & Surat* in *Lamb* AL 1390/91 (holotype K).

Tufted ***epiphyte*** forming large clumps. ***Pseudobulbs*** 1.5–4 × 0.3–0.5 cm, narrowly fusiform, covered in pale brown, darker brown mottled fibrous cataphylls. ***Leaf blade*** 8.5–11(–17) × 1–1.8 cm, narrowly elliptic or ligulate-elliptic, gradually tapering into petiole, apex obtuse and mucronate, thin-textured; petiole (2–)4–5 cm long. ***Inflorescence*** dense, with over 50 flowers in two ranks; peduncle (9–)15–16 cm long, terete, naked, pale green; rachis 10–13 cm long, quadrangular, yellowish; floral bracts 2–3 mm long, 2 mm wide when flattened, ovate-oblong, obtuse, concave, involute, prominently veined, brown. ***Flowers*** 0.9 cm across; sweetly scented; sepals and petals creamy white; lip creamy white or very pale green with yellowish cinnamon or ochre keels; column pink. ***Pedicel*** with ***ovary*** 1 mm long. ***Dorsal sepal*** 4.5 × 1.5 mm, oblong, acute. ***Lateral sepals*** 4.5 × 1.5–1.6 mm, oblong-elliptic, acute. ***Petals*** 4 × 1.1–1.2 mm, oblong-elliptic, acute, minutely erose-papillose. ***Lip*** 3 mm long when flattened, 3-lobed; side lobes spreading, 0.8–1 mm long, apex irregularly laciniate; mid-lobe strongly recurved in natural position, 2 mm long, 1 mm wide at base, expanding into a 1.5 mm wide ovate-elliptic, rounded or subacute blade, minutely toothed at base, surface minutely papillose, with two low, raised keel-like and minutely papillose ridges extending from base and terminating near apex, with a furrow between; disc with two flange-like basal keels, each about 0.4 mm wide, which curve toward the middle and meet, tapering off and terminating at base of mid-lobe. ***Column*** 2 mm long, with a foot; stelidia subacute, borne near apex, either side of stigmatic cavity, extending as high as rostellum; apical hood ovate, entire; anther-cap 0.2 × 0.2 mm, ovate, cucullate. Plate 13B.

HABITAT AND ECOLOGY: Epiphytic in thick moss on the trunks and branches of trees in hill and lower montane forest on sandstone and shale ridges. Alt. 900 to 1600 m. Flowering observed in April, October, November and December.

DISTRIBUTION IN BORNEO: SABAH: Crocker Range, Kimanis road, Ulu Apin Apin; Tenom District, Kallang Waterfall, Mt. Anginon.

GENERAL DISTRIBUTION: Endemic to Borneo.

NOTES: *D. lacinilobum* is similar in habit to *D. gramineum* (Ridl.) Holttum from Peninsular Malaysia and *D. kamborangense* Ames from Borneo. It differs from both in having white sepals and petals, distinct irregularly laciniate lip side lobes, two flange-like basal keels each about 0.4 mm wide and apical stelidia. It is further distinguished from *D. gramineum* by its wider petals and slightly broader lip, and from *D. kamborangense* by its smaller flowers with narrower petals and lip.

DERIVATION OF NAME: The specific epithet is derived from the Latin *laciniatus*, deeply divided into narrow divisions with tapering pointed incisions, and *lobus*, lobe, referring to the distinctive side lobes of the lip.

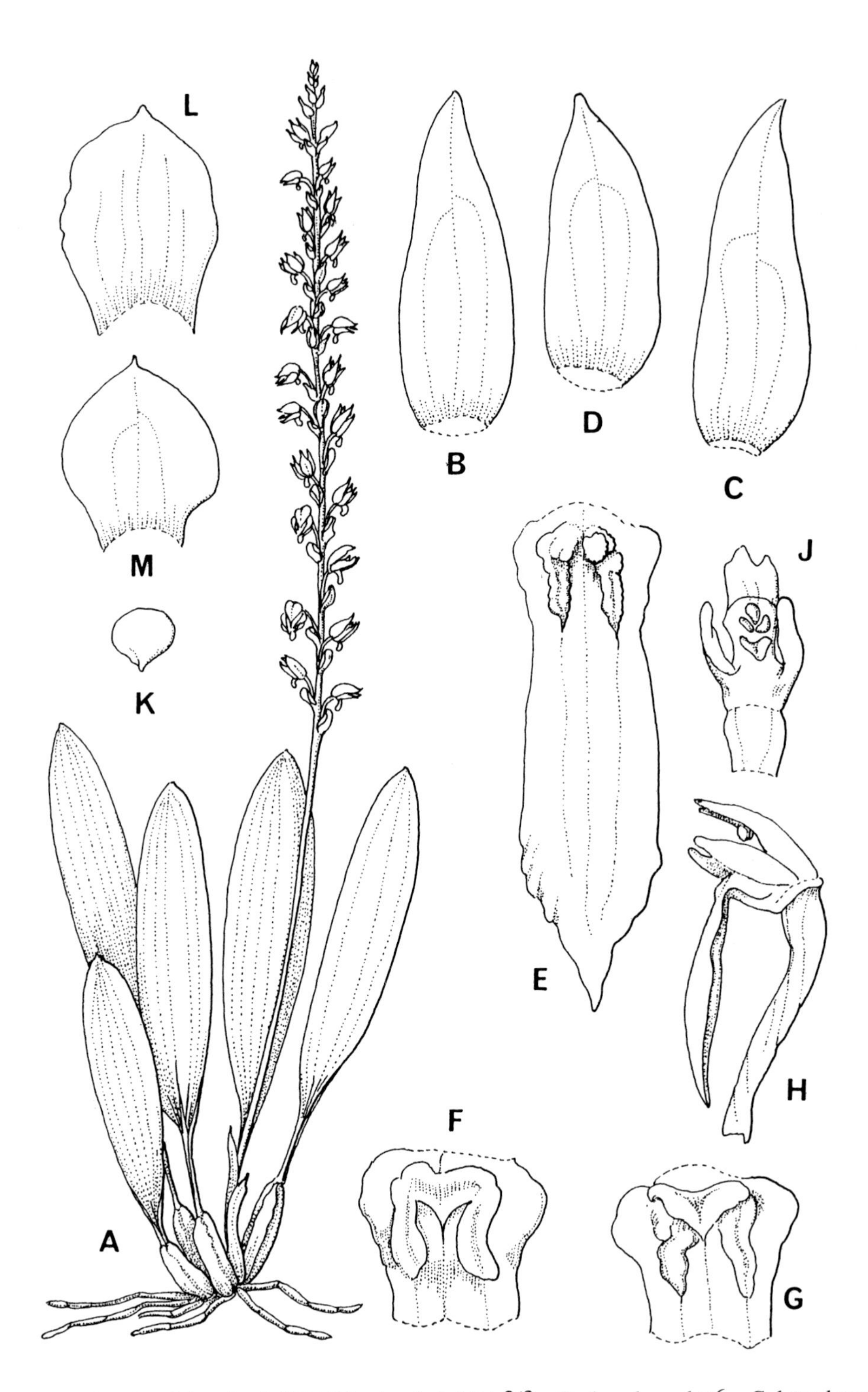

Figure 55. Dendrochilum lewisii J.J. Wood. - A: habit ×2/3. - B: dorsal sepal ×6. - C: lateral sepal ×6. - D: petal ×6. - E: lip, flattened ×12. - F & G: close-ups of basal calli on disc of lip ×14. - H: pedicel with ovary, lip and column, side view ×6. - J: column, front view ×8. - K: anther-cap, back view ×14. - L: lower floral bract, flattened ×6. - M: upper floral bract, flattened ×6. All drawn from *G. Lewis* 366 (holotype) by Mair Swann.

55. DENDROCHILUM LEWISII J.J. Wood

Dendrochilum lewisii *J.J. Wood* in Kew Bull. 39(1): 78, fig. 4 (1984). Type: Borneo, Sarawak, Mt. Mulu, *G. Lewis* 366 (holotype K).

Erect, tufted ***epiphyte. Rhizome*** short. ***Pseudobulbs*** crowded together, 1–2 × 0.5 cm, fusiform, enclosed by 2 closely adpressed, brown cataphylls to 1.8 cm long when young. ***Leaves*** erect, pale green; blade 3.2–5 × 0.9–1.3 cm, oblong-elliptic, obtuse and mucronate, coriaceous; petiole 0.5–1 cm long, sulcate. ***Inflorescence*** erect to gently curving, densely many-flowered; flowers borne 3–3.5 mm apart; peduncle 4–4.5 cm long, glabrous; rachis 16–18 cm long, quadrangular, canaliculate on two sides, glabrous, with 2 small ovate basal sterile bracts; floral bracts 2 × 2.5 mm, ovate, obtuse and mucronate, margins revolute. ***Flowers*** cream. ***Pedicel*** with ***ovary*** 4 mm long, slender. ***Sepals*** 6 × 1.5–2 mm, ovate-elliptic, acute, curved forward. ***Petals*** 5 × 2.1 mm, ovate to ovate-elliptic. ***Lip*** 3.8–4 × 1.2–1.5 mm, entire, lanceolate, acute to acuminate, with small erose basal auricles, sharply deflexed at base, margin erose towards apex; disc with a roughly M-shaped basal callus. ***Column*** 2 mm long, gently curved, apex usually obscurely tridentate; foot absent; stelidia 1.5 mm long, basal, linear, subacute; rostellum small, ovate; anther-cap ovoid.

HABITAT AND ECOLOGY: Upper montane ridge-top forest. Alt. 2200 to 2300 m. Flowering observed in October.

DISTRIBUTION IN BORNEO: SARAWAK: Mt. Mulu National Park.

GENERAL DISTRIBUTION: Endemic to Borneo.

NOTES: *D. lewisii* belongs to section *Eurybrachium* and is only known from the type. It is reported to be locally common on Mt. Mulu.

DERIVATION OF NAME: Named in honour of Gwilym Lewis, an expert on legumes at Kew, who collected the type.

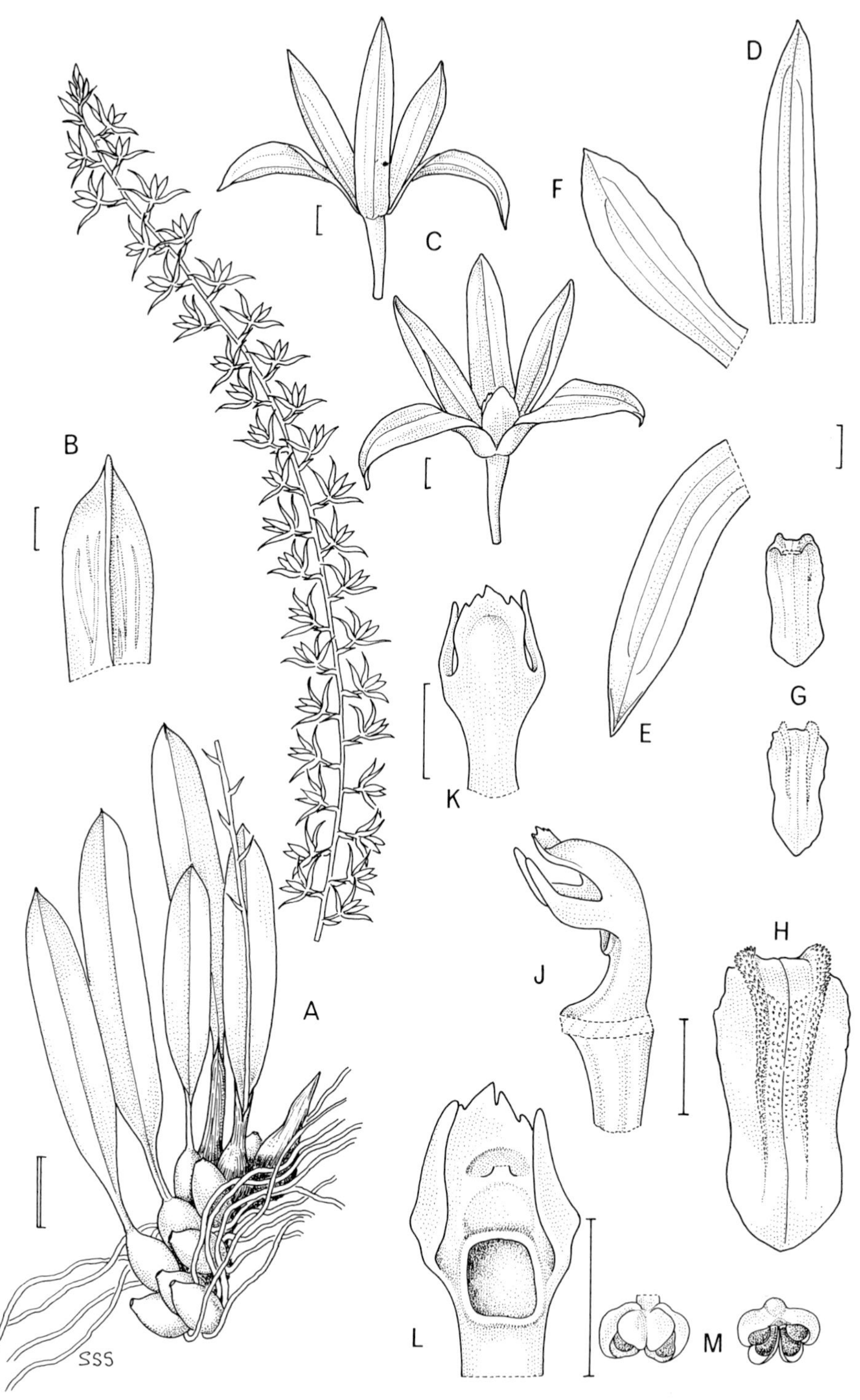

Figure 56. Dendrochilum lumakuense J.J. Wood. - A: habit. - B: floral bract, flattened. - C: flower, front and back views. - D: dorsal sepal. - E: lateral sepal. - F: petal. - G: lip, front and back views. - H: lip, front view. - J: pedicel with ovary and column, side view. - K: column, back view. - L: column, front view. - M: anther-cap with pollinia, front and back views. All drawn from *Comber* 108 (holotype) by Susanna Stuart-Smith. Scale: single bar = 1 mm; double bar = 1 cm.

56. DENDROCHILUM LUMAKUENSE J.J. Wood

Dendrochilum lumakuense *J.J. Wood* in Wood & Cribb, Checklist Orchids Borneo: 184, fig. 24 G & H (1994). Type: Borneo, Sabah, Sipitang District, Mt. Lumaku, *J.B. Comber* 108 (holotype K).

Clump-forming ***epiphyte. Pseudobulbs*** 0.8–1.5 × 0.6–1 cm, ovoid, yellowish, wrinkled in dried material, enclosed in greyish-fawn black-spotted cataphylls when young, caespitose. ***Leaves*** erect; blade 3–5 × 0.7–0.8 cm, ligulate, subacute, minutely apiculate, conduplicate at base; petiole 0.5–1 cm long, sulcate. ***Inflorescence*** erect to gently curving, densely many-flowered; flowers borne 2–2.5 mm apart; peduncle 3.5–4 cm long, naked; rachis 8–18 cm long, both brown; floral bracts 3 mm long, subulate, acute. ***Flowers*** open simultaneously, unscented, green. ***Pedicel*** with ***ovary*** 3 mm long, narrow. ***Dorsal sepal*** 7–8 × 1–1.2 mm, linear-lanceolate, acute. ***Lateral sepals*** 6.5–7 × 1–1.2 mm, linear-lanceolate, acute, slightly falcate. ***Petals*** 6–6.8 × 1.8–2 mm, spathulate, acute or subacute. ***Lip*** 3–3.1 × 1.1–1.2 mm, oblong-pandurate, obtuse, with two papillose basal keels, curved in the basal portion. ***Column*** 2 mm long, gently curved; hood irregularly toothed; stelidia borne between stigma and hood, subulate, acute, equalling apex of hood; anther-cap cucullate.

HABITAT AND ECOLOGY: Growing in 80% sun at the top of a 15 metre high tree in montane forest. Further habitat details not provided. Alt. 1800 m. Flowering observed in December.

DISTRIBUTION IN BORNEO: SABAH: Mt. Lumaku.

GENERAL DISTRIBUTION: Endemic to Borneo.

NOTES: *D. lumakuense* resembles certain forms of *D. dewindtianum* var. *dewindtianum* in habit, but can be distinguished by the green flowers with an oblong-pandurate lip lacking side lobes. The flower structure is remarkably similar to many of the species in subgenus *Dendrochilum*, but the synanthous inflorescence is typical of subgenus *Platyclinis* section *Platyclinis*.

DERIVATION OF NAME: The specific epithet records the type locality, Mt. Lumaku (1966 m) in South-west Sabah.

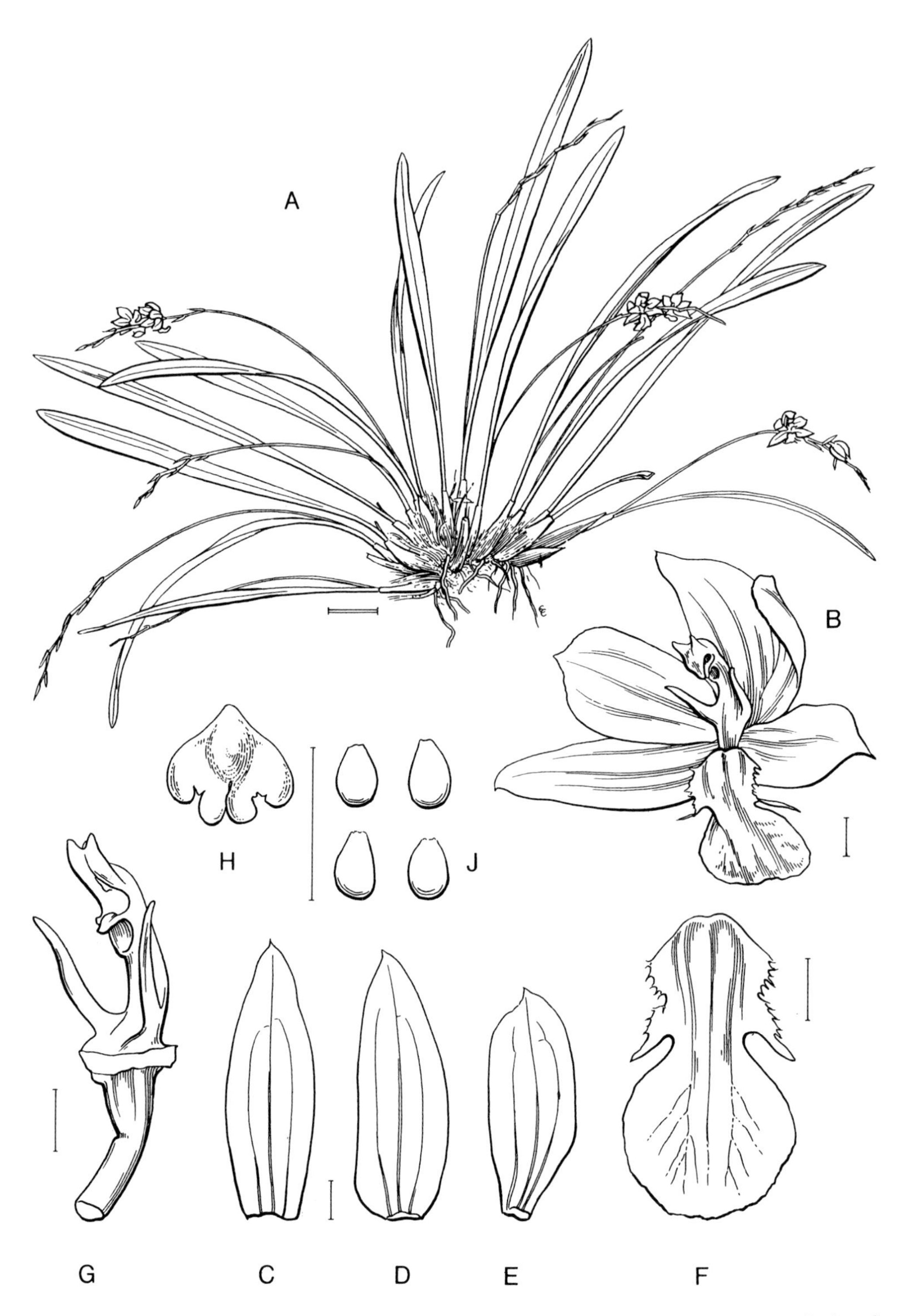

Figure 57. Dendrochilum magaense J.J. Wood. - A: habit. - B: flower, oblique view. - C: dorsal sepal. - D: lateral sepal. - E: petal. - F: lip, flattened. - G: pedicel with ovary and column, oblique view. - H: anther-cap, back view. - J: pollinia. All drawn from *Wood* 657 (holotype) by Eleanor Catherine. Scale: single bar = 1 mm; double bar = 1 cm.

57. DENDROCHILUM MAGAENSE J.J. Wood

Dendrochilum magaense *J.J. Wood* in Orchid Rev. 102 (1197): 147, fig. 81 (1994). Type: Borneo, Sabah, Sipitang District, Ulu Long Pa Sia, above Maga River 8 km NW of Long Pa Sia, *Wood* 657 (holotype K).

Clump-forming ***epiphyte. Pseudobulbs*** 0.8–1.2 × 0.2–0.5 cm, ovoid-elliptic, olive-green. ***Leaf blade*** 3–7.5 × 0.3–0.4 cm, ligulate, obtuse; petiole 0.6–1.8 cm long, slender. ***Inflorescence*** pendulous, 4- to 8-flowered; flowers borne 3–4 mm apart; peduncle 3–5 mm long, filiform, usually with 2 or 3 apical sterile bracts; rachis 2–3.8 cm long, quadrangular, thicker than peduncle; floral bracts 4–5 mm long, ovate to ovate-elliptic, apiculate, involute. ***Flowers*** wide-opening; salmon-pink to buff or cream with an orange to pale ochre centre to the sepals and petals, and a pale orange to ochre centre to the lip. ***Pedicel*** with ***ovary*** 2.5–2.8 mm long, clavate. ***Dorsal sepal*** 7.2–7.3 × 2.2–2.3 mm, oblong-elliptic, acute. ***Lateral sepals*** 7.5 × 2.4–2.5 mm, oblong-elliptic, acute. ***Petals*** 6.5 × 2.5–2.6 mm, elliptic, obtuse and mucronate, margin very minutely erose. ***Lip*** 4.9–5 mm long, 2 mm wide across base of side lobes; mid-lobe 2.5–2.6 × 2.8–2.9 mm, broadly elliptic from a cuneate base, obtuse; side lobes triangular, acuminate, irregularly serrate to lacerate, free portion *c.* 1 mm long; disc thickened, with two low fleshy ridges joined at base and terminating at base of mid-lobe. ***Column*** 3.5 mm long, with a foot *c.* 0.5–0.6 mm long, gently curved; hood prominent, slightly recurved, bifid, *c.* 0.7–0.8 mm wide; stelidia 2 mm long, acicular, borne at base; anther-cap 0.9 × 0.9 mm, ovate, cucullate.

HABITAT AND ECOLOGY: Lower montane oak/chestnut ridge-top forest with *Agathis alba*; open, low, dry stunted forest 5–10 metres high, with a dense field layer of terrestrial orchids, etc. on a narrow sandstone ridge. Alt. 1400 to 1500 m. Flowering observed in October.

DISTRIBUTION IN BORNEO: SABAH: Sipitang District.

GENERAL DISTRIBUTION: Endemic to Borneo.

NOTE: *D. magaense* is closely related to *D. tenompokense* Carr, described from Mt. Kinabalu, but is distinguished by its inflorescence which has a shorter rachis, the slightly larger flowers which are cream or salmon-pink flushed with orange or ochre, the broader, obtuse lip mid-lobe and longer column with a prominent, slightly recurved bifid hood and acicular basal stelidia.

DERIVATION OF NAME: The specific epithet refers to the Maga River in South-west Sabah, in the forest above which the type was collected.

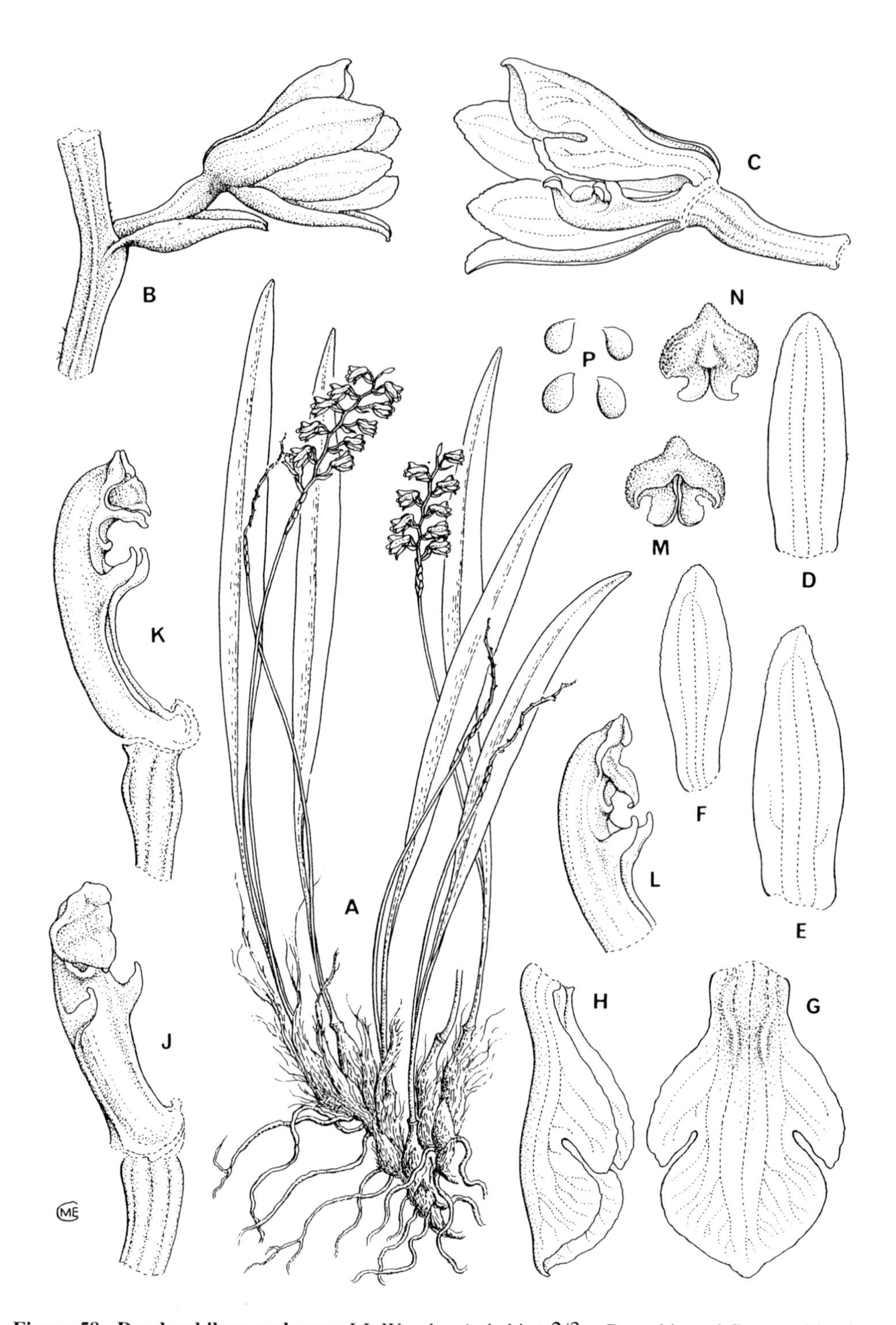

Figure 58. Dendrochilum muluense J.J. Wood. - A: habit ×2/3. - B: rachis and flower, side view ×6. - C: flower with half of dorsal sepal, and lateral sepal and petal removed, side view ×8. - D: dorsal sepal ×8. - E: lateral sepal ×8. - F: petal ×8. - G: lip, flattened ×10. - H: lip, side view ×10. - J: column with anther-cap, oblique view ×14. - K: column with anther-cap removed, side view ×14. - L: column apex with anther-cap removed, side view ×14. - M: anther-cap, back view ×20. - N: anther-cap, front view ×20. - P: pollinia ×20. All drawn from *Nielsen* 143 (holotype) by Maureen Church.

58. DENDROCHILUM MULUENSE J.J. Wood

Dendrochilum muluense *J.J. Wood* in Kew Bull. 39(1) : 80, fig. 5 (1984). Type: Borneo, Sarawak, Mt. Mulu National Park, Mt. Mulu, ridge at Camp 4, *Nielsen* 143 (holotype AAU, isotype K).

Erect ***epiphyte. Rhizome*** 2–18 cm long, *c.* 2 mm in diameter, branching, tough, clothed in fibrous sheaths. ***Roots*** 0.2–0.3 mm in diameter, numerous, filiform, wiry, smooth. ***Cataphylls*** 3–4, 0.5–4.5 cm long, brown to reddish–brown with darker brown spotting, becoming fibrous. ***Pseudobulbs*** 0.8–2(–2.5) × 0.3–0.8 cm, crowded on rhizome, narrowly conical or fusiform, erect, finely rugose when dried. ***Leaf blade*** (0.4–) 5–15.5 × 0.4–1.3 cm, thin-textured, linear–lanceolate to narrowly elliptic, obtuse and mucronate, main veins 5, with numerous small transverse veins, scattered calcium oxalate bodies sometimes present; petiole 1–4.5 cm long, narrow, sulcate. ***Inflorescence*** 8- to 17-flowered, lax to subdense; flowers borne 1.8–2 mm apart; peduncle 6–14.5 cm long, terete, filiform, erect, porrect or gently curving, glabrous or minutely furfuraceous above; rachis 1–5–3.5 cm long, quadrangular, held at a sharp angle to the peduncle, straight to slightly fractiflex, minutely furfuraceous with 2–3, 1–3 mm long basal sterile bracts; floral bracts 2.5–3 mm long, ovate-obtuse or obtuse and mucronate. ***Flowers*** non–resupinate; sepals and petals white, cream or whitish green; lip white or yellowish orange, with a brownish spot at the base, often yellowish at tip; column pale yellow, sometimes with a brown spot, foot red. ***Pedicel*** with ***ovary*** 1–2.5 mm long, narrowly clavate, deflexed and turning orange at anthesis. ***Dorsal sepal*** 4–4.5 × 1.5 mm, oblong to oblong-elliptic, obtuse, veins 3. ***Lateral sepals*** 4–5 × 1.8 mm, slightly obliquely oblong to oblong-elliptic, obtuse, veins 3. ***Petals*** 3.5–4.5 × 1.2–1.5 mm, oblong-elliptic, obtuse, margin erose when magnified, veins 3. ***Lip*** 4–5 mm long, concave, somewhat fleshy, main veins 5; side lobes 1–1.5 × 0.8 mm, oblong, rounded, erect; mid-lobe 1.6–2 × 3 mm, obovate-flabellate when expanded, rounded-obtuse, attenuate at base; disc with two short basal keels extending almost to the base of the side lobes. ***Column*** 2.8–3 mm long, gently curved; foot distinct; apical hood ovate, truncate, entire; rostellum triangular-ovate, prominent; stelidia 0.8 mm long, borne a little way below stigmatic cavity, triangular-linear, subobtuse or acute, somewhat falcate - upcurved, reaching to level with stigma; anther-cap 0.8 × 0.8 mm, cordate, cucullate. ***Fruit*** globose. Plate 13C & D.

HABITAT AND ECOLOGY: Upper montane ridge forest; oak-laurel forest; mossy forest. Alt. 1700 to 2200 m. Flowering observed in January, July, August, September and October.

DISTRIBUTION IN BORNEO: BRUNEI: Bukit Retak. KALIMANTAN TIMUR: Mt. Batu Harun. SABAH: Crocker Range. SARAWAK: Batu Lawi area; Mt. Mulu National Park; Mt. Murud area; Mt. Pagon Periuk.

GENERAL DISTRIBUTION: Endemic to Borneo.

DERIVATION OF NAME: Named after Mt. Mulu in Sarawak, the type locality.

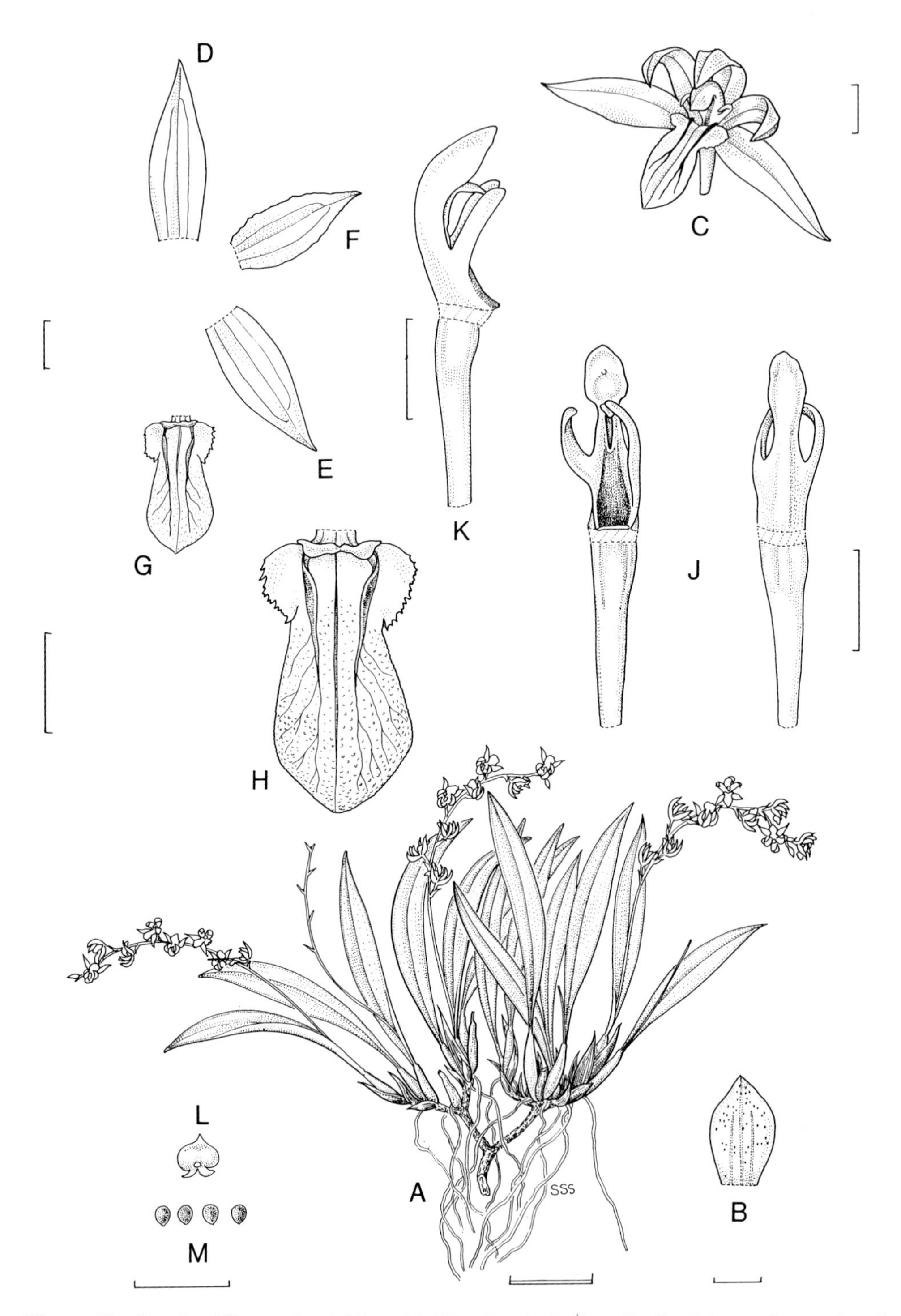

Figure 59. Dendrochilum ochrolabium J.J. Wood. - A: habit. - B: floral bract, flattened. - C: flower, oblique view. - D: dorsal sepal. - E: lateral sepal. - F: petal. - G & H: lip, flattened. - J: pedicel with ovary and column, front and back views. - K: pedicel with ovary and column, side view. - L: anther-cap, back view. - M: pollinia. All drawn from *de Vogel* 8351 (holotype) by Susanna Stuart-Smith. Scale: single bar = 1 mm; double bar = 1 cm.

59. DENDROCHILUM OCHROLABIUM J. J. Wood

Dendrochilum ochrolabium *J.J. Wood* in Wood and Cribb, Checklist Orchids Borneo: 187, fig. 24 J & K (1994). Type: Borneo, Sabah, Sipitang District, ridge east of Maga River, *c.*1.5 km south of confluence with Pa Sia River, *de Vogel* 8351 (holotype L, isotype K).

Tufted ***epiphyte. Pseudobulbs*** 0.5–1.2 × 0.2–0.3 cm, ovoid to elliptic, or oblong-cylindrical. ***Leaves*** very minutely black punctate-ramentaceous, particularly on lower surface; blade 1.2–2.6(–3) × 0.3–0.4 cm, narrowly elliptic, obtuse and usually minutely mucronate, margin minutely papillose, appearing raised and thickened in dried material; petiole 0.1–0.3 cm long. ***Inflorescence*** 5- to 8-flowered; flowers borne 2.3 mm apart; peduncle 1.2–2 cm long, filiform; rachis 1.5–2.5 cm long, quadrangular, sparsely ramentaceous; floral bracts 2–2.5 × 1.1–1.2 mm, oblong-elliptic, acute, concave, clasping pedicel with ovary, sparsely ramentaceous. ***Flowers*** slightly scented; sepals and petals pale green or very pale greenish; lip ochre, middle part brownish, back green with two longitudinal brown stripes or orange with pale green side lobes; column pale green. ***Pedicel*** with ***ovary*** 1.9–2 mm long, outer wall fleshy and translucent. ***Dorsal sepal*** 3.9–4 × 1.1 mm, narrowly ovate-elliptic, acute. ***Lateral sepals*** 3.9–4 × 1.2–1.3 mm, ovate-elliptic, acute. ***Petals*** 3 × 1.1 mm, ovate-elliptic, acute. ***Lip*** 2.8 mm long, 1.2–1.3 mm wide across side lobes, 1.4–1.5 mm wide across mid-lobe, stipitate to column-foot by a tiny claw, minutely papillose; side lobes auriculate, irregularly toothed; disc with two somewhat papillose raised keels joined at the base to a narrow transverse flange, each terminating a little above base of mid-lobe; mid-lobe obovate, obtuse, convex, minutely papillose. ***Column*** *c.* 1.8 mm long, with a foot, curved; stelidia *c.* 0.8 mm long, ligulate, acute, shorter than apical hood, arising from middle; apical hood ovate, very obscurely 3-lobed; anther-cap minute, cucullate, extinctoriform. Plate 14A.

HABITAT AND ECOLOGY: The type locality is in open, low, dry stunted forest 5 to 10 metres high on a sandstone ridge, with a dense undergrowth of terrestrial orchids and other herbs. It is also recorded from very low and open podsol forest. Alt. 1300 to 1500 m. Flowering observed in October and December.

DISTRIBUTION IN BORNEO: SABAH: Sipitang District. SARAWAK: Ulu Sungai Entulu; Mt. Penrissen.

GENERAL DISTRIBUTION: Endemic to Borneo.

NOTE: *D. ochrolabium* is related to *D. dulitense* Carr from Sarawak but differs in having slightly broader, narrowly elliptic, minutely black punctate-ramentaceous leaves, pale green flowers with an ochre lip, pedicel with ovary with a translucent outer wall, slightly smaller sepals and petals lacking basal papillae, a distinctly smaller, narrower lip without an elevated median vein, and ligulate, acute stelidia.

DERIVATION OF NAME: The specific epithet is derived from the Latin *ochraceus,* ochre-yellow or yellowish-brown, and *labium*, lip, referring to the colour of the lip.

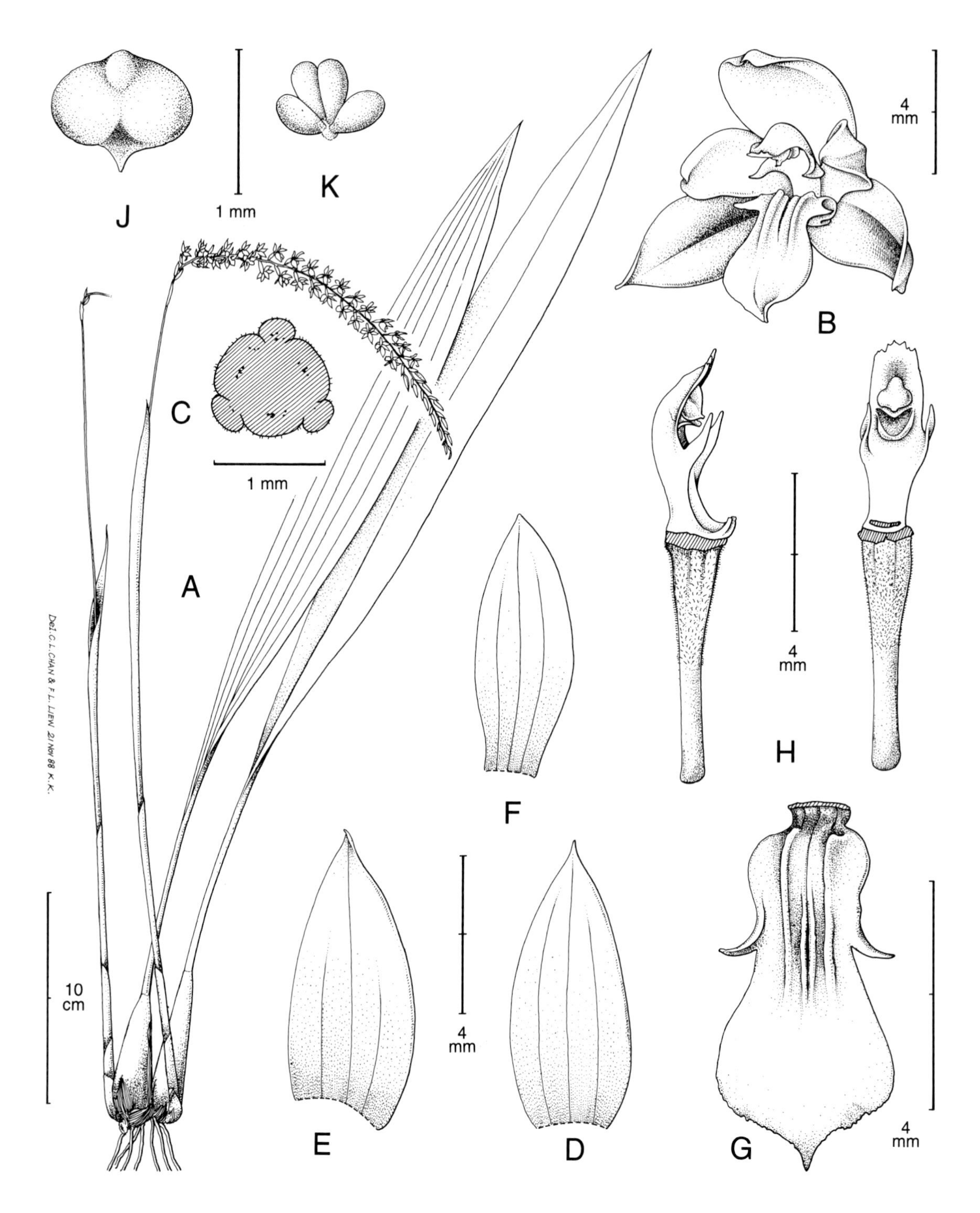

Figure 60. Dendrochilum oxylobum Schltr. - A: habit. - B: flower, oblique view. - C: ovary, transverse section. - D: dorsal sepal - E: lateral sepal. - F: petal. - G: lip, flattened. - H: pedicel with ovary and column, front and side views. - J: anther-cap, back view. - K: pollinia. All drawn from *Chan & Lohok* s.n., cultivated at Tenom Orchid Centre, by C.L. Chan and Lucy F.L. Liew.

60. DENDROCHILUM OXYLOBUM Schltr.

Dendrochilum oxylobum *Schltr.* in Repert. Spec. Nov. Regni Veg. 9:431 (1911). Type: Borneo, Sarawak, Kuching, *Beccari* 1125 (holotype FI).

Dendrochilum viridifuscum J.J. Sm. in Bull. Jard. Bot. Buitenzorg, ser.2:11 (1917). Type: Borneo, Kalimantan, Kota Waringin, *van Nouhuys* s.n., cult. Bogor no.16 (holotype BO).

Erect ***epiphyte*** or ***terrestrial. Rhizome*** abbreviated, *c.* 3 cm long, tough, branching, covered in sheaths. ***Roots*** 1–1.5 mm in diameter, flexuous, smooth. ***Pseudobulbs*** 3.3–7 × 0.4–1.6 cm, shortly cylindrical or oblong-ovoid, approximate, shiny pale green, aggregated on rhizome, when immature concealed by 4–5 imbricate, tubular, acute, brown, minutely furfuraceous-punctate sheaths, the uppermost up to 14 cm long which encloses the petiole and part of the peduncle. ***Leaves*** large, thin-textured, erect or suberect, suffused brown when young; blade 12–42 × 1.6–4.6 cm, lanceolate, acute, sometimes slightly constricted below apex, mid–vein sulcate on adaxial surface, prominent and raised on abaxial surface; petiole 1.5–15 cm long, sulcate. ***Inflorescence*** synanthous, subdensely many-flowered; flowers borne (3.5–)4–5 mm apart; peduncle 16–30 cm long, terete, erect or porrect, pale brownish-green, glabrous or very sparsely brown furfuraceous; rachis 8–26 cm long, quadrangular, subalate, recurved to pendulous, with a solitary sterile bract 5–11 mm long at base, pale green, suffused pale brown, sparsely furfuraceous-punctate; floral bracts 5–8 × 4.7–5 mm, ovate-elliptic to suborbicular, shortly apiculate, convex, flesh to brownish. ***Flowers*** 8–9.3 mm across, sweetly scented; sepals and petals pale yellow suffused with brown to cinnamon or orange-brown; lip dark brown, often lemon at base, margin of mid-lobe sometimes paler brown; column white, often flushed brownish dorsally, yellowish at base, foot yellow or chestnut. ***Pedicel*** with ***ovary*** 4–6 mm long, narrowly clavate, glabrous or furfuraceous-punctate. ***Sepals*** and ***petals*** 3-veined. ***Dorsal sepal*** 8–9 × 3 mm, narrowly elliptic or ovate-elliptic, acute to acuminate. ***Lateral sepals*** 8.4–1 × 3 mm, obliquely oblong-elliptic, apex acute to acuminate, slightly carinate. ***Petals*** 7.5–8 × 2.2–2.5 mm, ovate-oblong or narrowly elliptic, acute, sometimes slightly erose. ***Lip*** 6.5–8 mm long, *c.* 2 mm wide at base, 3-lobed, stipitate to column-foot by a short claw; free portion of side lobes 1.1–1.5 mm long, triangular-subulate, acute, erect or porrect, lower margins sometimes erose; mid-lobe 4–5 × 2.5–3.5 mm, broadly rhomboid or broadly elliptic, cuneate at base, apex acutely triangular-apiculate, distal margins erose; disc with two low parallel keels originating near base of lip, becoming dilated at the middle, then gradually diminishing and finally terminating above base of mid-lobe, never joined to form a U at the base. ***Column*** 4–5 mm long; foot 0.7–0.8 mm, concave; apical hood 2 × 1.3–1.4 mm, oblong-quadrangular, 3- or 5-dentate, or irregularly toothed; stelidia *c.* 2.6–2.7 mm long, subulate or ligulate-falcate, acute to acuminate, median, arising just below stigmatic cavity, extending just beyond rostellum; rostellum broadly triangular, acute, convex; anther-cap ovate, cucullate, apex recurved, acute. Plate 14B & C.

HABITAT AND ECOLOGY: Lower montane forest on sandstone; podsol forest with

Dacrydium, Rhododendron malayanum, Tristania, etc. Alt. 400 to 900 m. Flowering observed in May and November.

DISTRIBUTION IN BORNEO: KALIMANTAN: Kota Waringin. KALIMANTAN TIMUR: Balikpapan area. SABAH: Crocker Range, Kimanis road; Nabawan area. SARAWAK: Kuching area.

GENERAL DISTRIBUTION: Endemic to Borneo.

NOTE: *D. oxylobum* belongs to section *Platyclinis.*

DERIVATION OF NAME: The specific epithet is derived from the Greek *oxy*, sharp, and the Latin *lobus*, a lobe, in reference to the shape of the lip.

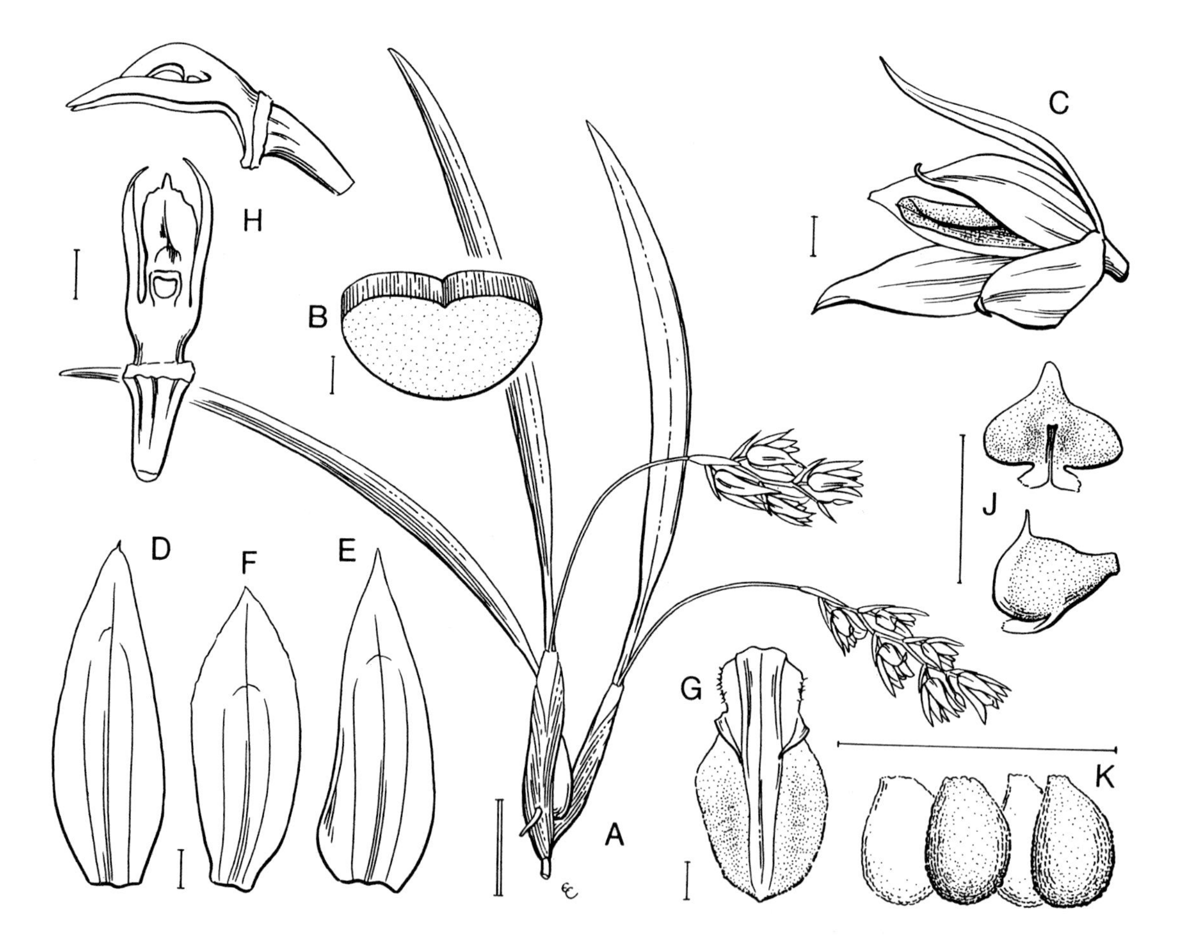

Figure 61. Dendrochilum pachyphyllum J.J. Wood & A. Lamb. - A: habit. - B: leaf, transverse section. - C: flower, side view. - D: dorsal sepal. - E: lateral sepal. - F: petal. - G: lip, flattened. - H: pedicel with ovary and column, front and side views. - J: anther-cap, back and side views. - K: pollinia. All drawn from *Lamb* AL 674/86 (holotype) by Eleanor Catherine. Scale: single bar = 1 mm; double bar = 1 cm.

61. DENDROCHILUM PACHYPHYLLUM J.J. Wood & A. Lamb

Dendrochilum pachyphyllum *J.J. Wood & A. Lamb* in Wood and Cribb, Checklist Orchids Borneo: 189, fig. 21, plate 9E (1994). Type: Borneo, Sabah, Crocker Range, Mt. Alab, Sinsuron Road, *Lamb* AL 674/86 (holotype K).

Tufted clump-forming ***epiphyte,*** sometimes ***lithophytic. Pseudobulbs*** 0.7–1.2 × 0.3–0.5 to 1.8 × 0.2–0.3 cm, oblong-elliptic to narrowly fusiform, clothed in acute, brown cataphylls when young. ***Leaves*** narrowly linear-elliptic, acute, thick and fleshy, curved, sulcate above; blade 2–7 × 0.3–0.6 cm; petiole 0.6–1.3 cm long, narrow, sulcate. ***Inflorescences*** 3- to 5-flowered, pendulous, shorter than leaves; peduncle 1–7 cm long, filiform, yellowish–green; rachis 0.6–1.8 cm long, thicker than peduncle, fractiflex, salmon-pink; floral bracts 4–7 mm long, ovate-acuminate, brown. ***Flowers*** *c.* 1 cm across; pedicel with ovary brownish pink; sepals translucent pink to salmon-pink or pale brownish, mid-vein darker red to brownish pink; petals translucent pink to salmon-pink; lip green at base, mid-lobe white or dull yellow, central area red to brownish red, side lobes white; column pink; anther-cap white. ***Pedicel*** with ***ovary*** 1.2 mm long. ***Sepals*** acute, somewhat carinate. ***Dorsal sepal*** 7–9 × 2–3 mm, ovate-elliptic, curved. ***Lateral sepals*** 7–9 × 2.5–3 mm, ovate-elliptic. ***Petals*** 6–8 × 2–2.8 mm, elliptic, acute, minutely erose towards apex. ***Lip*** 5.5–7 mm long, 2 mm wide at base, 3–3.5 mm wide across mid-lobe, 3-lobed, spathulate, 3-veined; mid-lobe 4.5–5 mm long, oblong-elliptic, obtuse, rather fleshy at centre, shallowly concave, margin thin, erose; side lobes minutely toothed, free portion 1–1.5 mm long, linear-triangular, acute; disc with a raised keel 1 mm wide extending from base of lip to upper portion of mid-lobe, its base retuse, its margins raised. ***Column*** 4 mm long, with a foot, its apex irregularly 3– to 4–toothed, the central tooth longer than the outer; stelidia 4 × 0.5–0.6 mm, linear–ligulate, subacute, borne from near column base, exceeding apex; anther–cap ovate–cordate, minutely papillose. Plate 14D & 15.

HABITAT AND ECOLOGY: Mixed hill-dipterocarp forest; lower montane mossy forest; mossy sandstone rocks and shale banks along roadside cuttings; very low and open podsol forest. Alt. 1300 to 2000 m. Flowering observed in October, November and December.

DISTRIBUTION IN BORNEO: SABAH: Crocker Range, Mt. Alab; Sipitang District. SARAWAK: Ulu Sungai Limbang.

GENERAL DISTRIBUTION: Endemic to Borneo.

NOTES: *D. pachyphyllum* belongs to section *Platyclinis* and is distinguished from all other species by its short, acute, thick and fleshy leaves, pendulous, filiform peduncles and brownish salmon-pink flowers.

DERIVATION OF NAME: The specific epithet is derived from the Greek *pachy,* thick or stout, and *phyllum,* leaf, referring to the fleshy leaves.

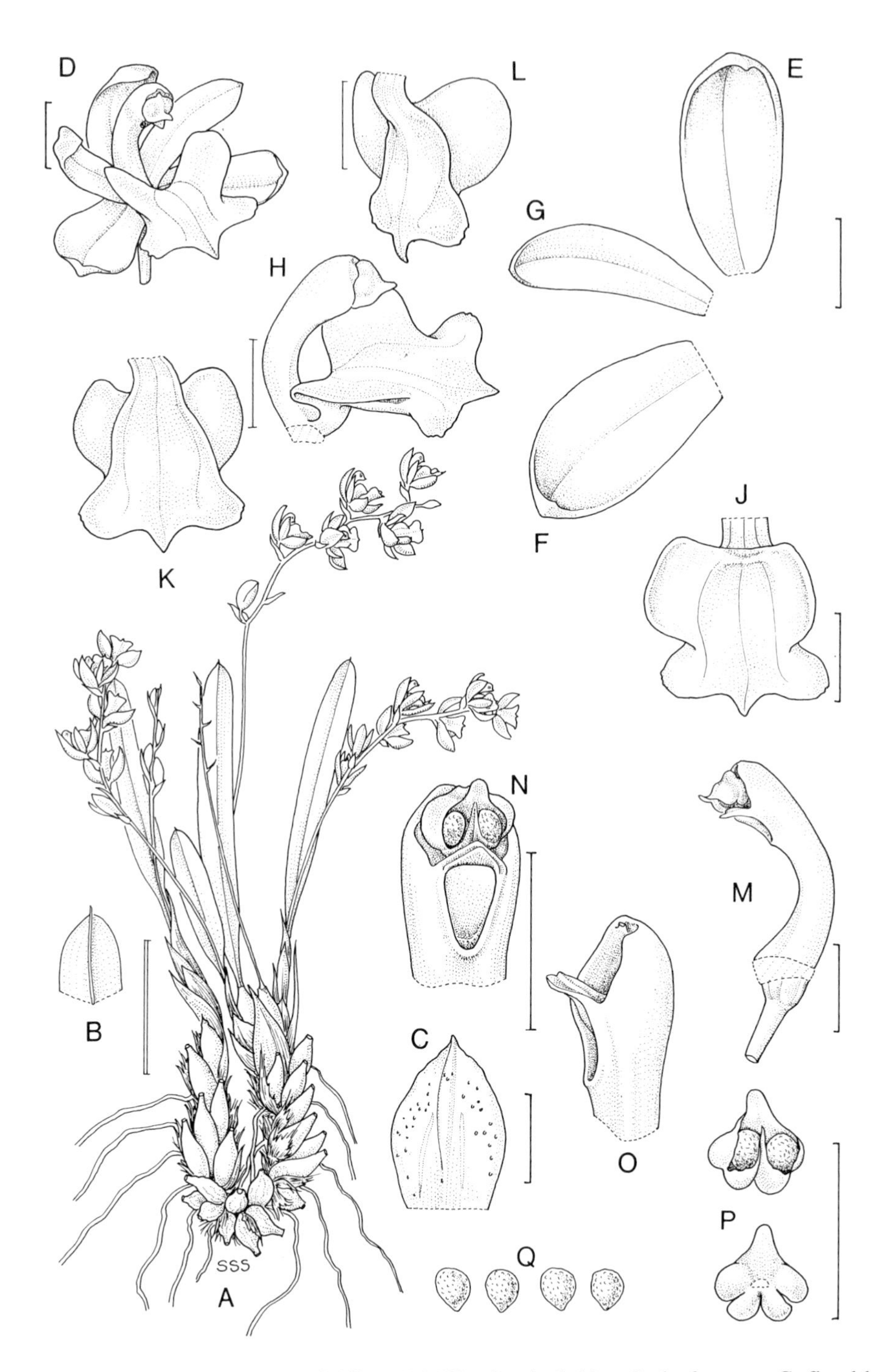

Figure 62. Dendrochilum pandurichilum J.J. Wood. - A: habit. - B: leaf apex. - C: floral bract, flattened. - D: flower, oblique view. - E: dorsal sepal. - F: lateral sepal. - G: petal. - H: lip and column, oblique view. - J: lip, flattened. - K & L: lip, back views. - M: pedicel with ovary and column, side view. - N: column with anther-cap, front view. - O: column with anther-cap removed, side view. - P: anther-cap with pollinia, front and back views. - Q: pollinia. A (habit) drawn from *Comber* 102 (holotype) and B-Q from *de Vogel* s.n., cultivated at Hortus Botanicus, Leiden, no. 911260A by Susanna Stuart-Smith. Scale: single bar = 1 mm; double bar = 1 cm.

62. DENDROCHILUM PANDURICHILUM J.J. Wood

Dendrochilum pandurichilum *J.J .Wood* in Wood & Cribb, Checklist Orchids Borneo: 190, fig. 23 A–C (1994). Type: Borneo, Sabah, Sipitang District, Mt. Lumaku, *J.B. Comber* 102 (holotype K).

Clump–forming ***epiphyte. Pseudobulbs*** 5–7 × 2–3 mm, ovoid–elliptic, yellowish, enclosed in pale brown cataphylls when young. ***Leaves*** 0.8–3.5 × 0.2–0.4 cm, narrowly ligulate, obtuse and mucronate; petiole 1–3 mm long. ***Inflorescence*** laxly 3– to 8 –flowered, opening from top down; peduncle 1–3.5 cm long, filiform, naked; rachis 0.6–1.8 cm long; floral bracts 2–3 mm long, subulate, acute. ***Flowers*** pale orange. ***Pedicel*** with ***ovary*** 1.8 mm long, clavate. ***Dorsal sepal*** 3 × 1–1.1 mm, ovate–elliptic, subacute, concave, cucullate, median vein prominent and dorsally carinate. ***Lateral sepals*** 2.9 × 1.5–1.6 mm, triangular-ovate, subacute, somewhat concave. ***Petals*** 2.9–3 × *c.*0.8–0.9 mm, linear, acute, slightly falcate. ***Lip*** 2.1–2.8 mm long, 2.4–2.5 mm wide across keels, pandurate, side lobes absent, mid-lobe trilobulate, 2–2.1 mm wide when flattened, outer lobules obtuse, median lobule tooth-like, subacute; disc with prominent oblong, rounded wing-like keels each 0.8 × 1 mm, looking, at first sight, like side lobes. ***Column*** 3 mm long (when straightened), narrow, arcuate; hood entire; with a foot; stelidia absent; anther-cap cucullate.

HABITAT AND ECOLOGY: The habitat of the type locality on Mount Lumaku in S.W. Sabah is lower montane forest, while in Sarawak it is recorded from riparian forest. It is curious, therefore, that it has also been collected in lowland "kerangas" forest at only 10 metres above sea level in Brunei. Here the forest grows on pure white sand, is about 30 metres in height, contains numerous large *Agathis* and has an undergrowth of slender pole trees. De Vogel has collected this species in ridge forest up to 20 metres high on sandstone with an understorey of climbing bamboo, rattan palms, etc. Alt. around sea level to 1900 m. Flowering observed in January, October and December.

DISTRIBUTION IN BORNEO: BRUNEI: Badas Forest Reserve. SABAH: Sipitang District. SARAWAK: Mt. Batu Lawi area.

GENERAL DISTRIBUTION: Endemic to Borneo.

NOTES: *D. pandurichilum* is the type species of the monotypic section *Falsiloba* of subgenus *Acoridium.* This is distinct in having a prominently clawed lip without side lobes but proximally with two prominent, semi-erect, wing-like keels reminiscent of side lobes.

DERIVATION OF NAME: The specific epithet is derived from the Latin *panduratus,* fiddle-shaped, and the Greek *chilos,* a lip, referring to the distinctive and elegant lip shape.

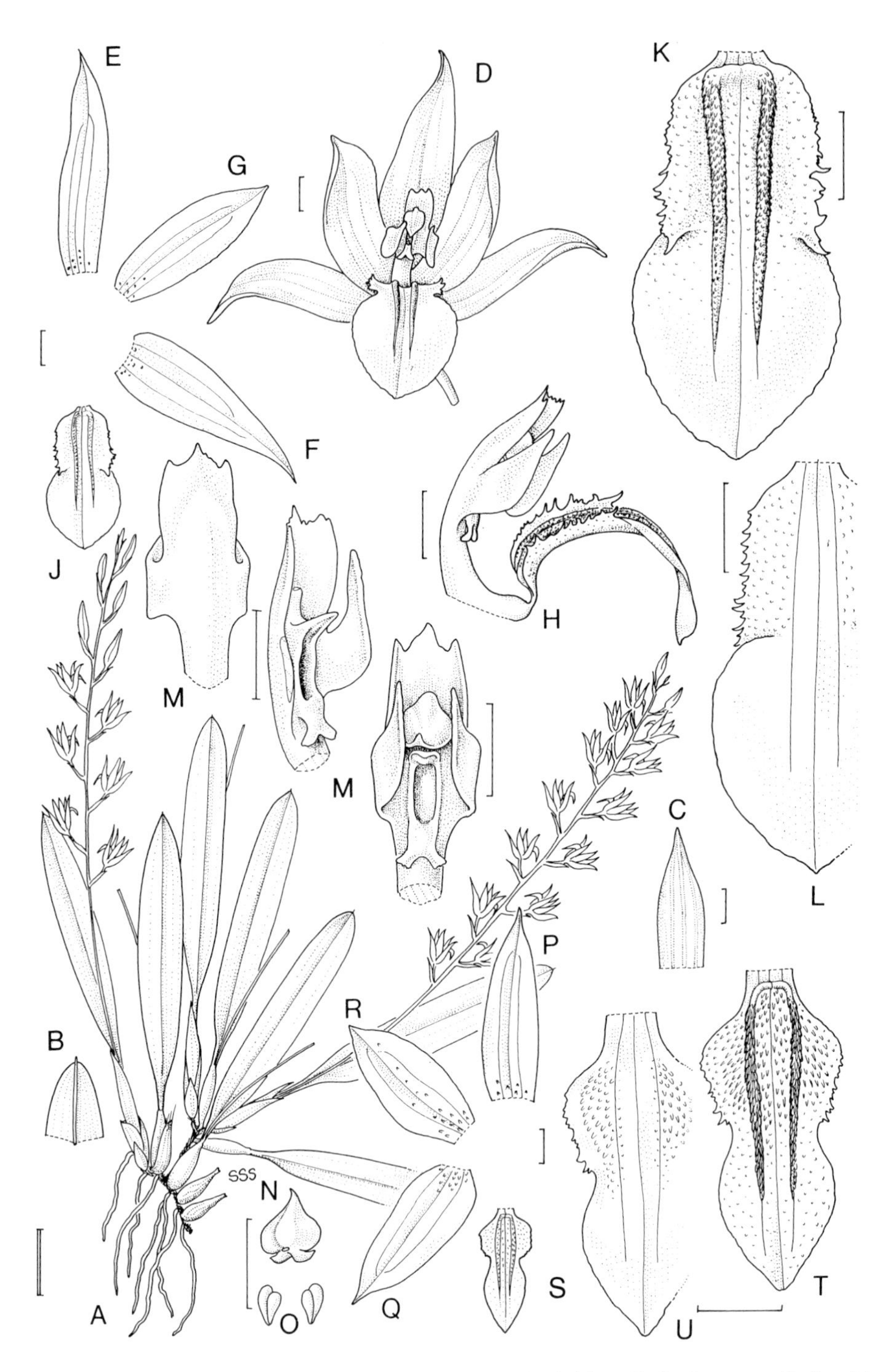

Figure 63. Dendrochilum papillilabium J.J. Wood. - A: habit. - B: leaf apex. - C: floral bract, flattened. - D: flower, front view. - E: dorsal sepal. - F: lateral sepal. - G: petal. - H: column and lip, side view. - J: lip, flattened. - K: lip, front view. - L: lip, back view. - M: column, back, oblique and front views. - N: anther-cap, back view. - O: pollinia. - P: dorsal sepal. - Q: lateral sepal. - R: petal. - S: lip, front view. - T: lip, front view. - U: lip, back view. A–O drawn from *Vermeulen & Duistermaat* 666 (holotype) and P–U from *Aban* SAN 95230 by Susanna Stuart-Smith. Scale: single bar = 1 mm; double bar = 1 cm.

63. DENDROCHILUM PAPILLILABIUM J.J. Wood

Dendrochilum papillilabium *J.J. Wood* in Wood & Cribb, Checklist Orchids Borneo: 191, fig. 24 L & M (1994). Type: Borneo, Sabah, Crocker Range, Keningau to Kimanis road, *Vermeulen & Duistermaat* 666 (holotype L).

Tufted ***epiphyte*** producing many long, branching roots. ***Pseudobulbs*** 0.5–1 × 0.4–0.5 cm, ovoid to elliptic, enclosed by several acute to acuminate greyish-brown cataphylls up to 2 cm long when young. ***Leaves*** coriaceous; blade 1.5–5.5 × 0.6–1 cm, oblong-elliptic, obtuse and mucronate to acute; petiole 0.2–1 cm long, sulcate. ***Inflorescence*** many-flowered, the uppermost often opening first; flowers borne 2–5 mm apart; peduncle 1.7–4 cm long; rachis 5–7 cm long; floral bracts 3–4(–5) mm long, ovate-elliptic, acute. ***Flowers*** described as light green, pale greenish, lip brighter green. ***Pedicel*** with ***ovary*** 2.2 mm long, clavate. ***Sepals*** and ***petals*** with a few brown ramentaceous scales at the base. ***Dorsal sepal*** 6 × 2 mm, ovate-elliptic, acute, concave, slightly carinate. ***Lateral sepals*** 5.5 × 2–2.1 mm, slightly obliquely ovate-elliptic, acute, slightly carinate. ***Petals*** 5 × 2.2 mm, oblong to oblong-elliptic, acute, slightly concave. ***Lip*** 3.5–3.8 mm long, 1.5 mm wide across side lobes, 1.1–1.2 mm wide across mid-lobe, decurved; side lobes shallowly rounded, margin irregularly toothed, upper surface papillose-hairy except along margins; mid-lobe elliptic, obtuse, minutely papillose; disc with two papillose-hairy fleshy ridges joined near base of lip and terminating on lower part of mid-lobe. ***Column*** 3.5 mm long; foot 0.6–07 mm long, curved; stelidia 2 mm long, falcate, acute, almost equal or subequal to apical hood, arising at the middle; apical hood irregularly toothed; rostellum prominent; lower margin of stigmatic cavity developed into a distinct bilobed flange; anther-cap 0.9 × 0.9 mm, cucullate, with a triangular-acute 'tail'.

HABITAT AND ECOLOGY: Lower montane ridge forest on sandstone; mossy forest. Alt. 1500 to 1600 m. Flowering observed in March and December.

DISTRIBUTION IN BORNEO: SABAH: Crocker Range; Bukit Monkobo.

GENERAL DISTRIBUTION: Endemic to Borneo.

NOTES: *D. papillilabium* is closely related to *D. tenompokense* Carr, but can be distinguished by the papillose lip and distinctly bilobed flange-like lower margin of the stigmatic cavity. These characters also distinguish it from broad-leaved forms of *D. linearifolium* Hook.f. (syn. *Platyclinis pulchella* Ridl.) from Peninsular Malaysia, which it superficially resembles.

DERIVATION OF NAME: The specific epithet is derived from the Latin *papillatus*, having papillae, and *labium*, lip, referring to the papillose lip.

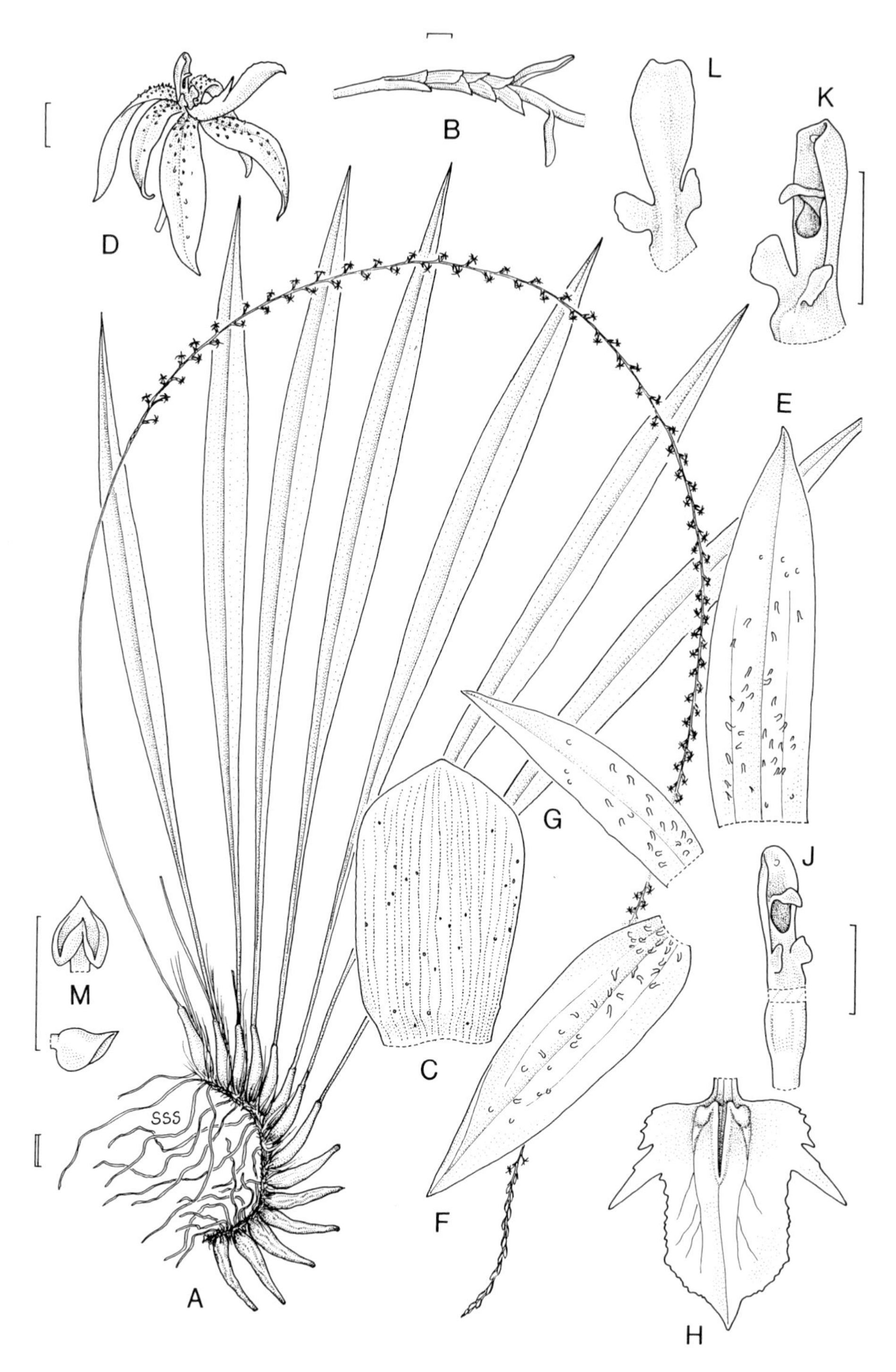

Figure 64. Dendrochilum papillitepalum J.J. Wood. - A: habit. - B: non-floriferous bracts at junction of peduncle and rachis. - C: floral bract, flattened. - D: flower, side view. - E: dorsal sepal. - F: lateral sepal. - G: petal. - H: lip, flattened. - J: pedicel with ovary and column, oblique view. - K: column, oblique view. - L: column, back view. - M: anther-cap, front and back views. All drawn from *Awa & Lee* S. 47676 (holotype) by Susanna Stuart-Smith. Scale: single bar = 1 mm; double bar = 1 cm.

64. DENDROCHILUM PAPILLITEPALUM J.J. Wood

Dendrochilum papillitepalum *J.J. Wood* in Orchid Rev. 103 (1201): 8, fig. 4 (1995). Type: Borneo, Sarawak, Ulu Sungai Sipayan, Mt. Pagon, Limbang, *Awa & Lee* S. 47676 (holotype K, isotypes KEP, SAR).

Epiphyte. Rhizome 6 cm long, 2 mm in diameter, clothed in fibrous sheaths. ***Roots*** produced in a large mass, wiry, 0.6–0.8 mm in diameter, very minutely papillose. ***Cataphylls*** ?3, up to 6 cm long, brown, speckled darker brown, becoming fibrous. ***Pseudobulbs*** crowded on rhizome, 2.5–3 × 0.6–0.7 cm, fusiform, rugose in dried material. ***Leaf blade*** 23.5–25 × 1.3–1.6 cm, narrowly elliptic-ligulate, acute, gradually attenuated below, main nerves 4–5; petiole 4.6–5.5 cm long, narrow, sulcate. ***Inflorescence*** many-flowered, pendulous, subdense; flowers borne 2–2.5 mm apart; peduncle 21 cm long, terete, filiform; rachis quadrangular, 42 cm long, with a group of 7 imbricate, obtuse, apiculate sterile bracts 2 mm long, sparsely ramentaceous; floral bracts 3 × 1.6 mm, oblong-ovate. ***Flowers*** pale green, column reddish. ***Pedicel*** with ***ovary*** 2–3 mm long, narrowly clavate. ***Dorsal*** and ***lateral sepals*** 4–4.1 × 1.1 mm, narrowly ovate-elliptic, acute, papillate on adaxial surface, particularly towards base, veins 3. ***Petals*** 3.3 × 0.8–0.9 mm, narrowly elliptic, acute, papillate on adaxial surface, particularly towards base, veins 1–3. ***Lip*** 2.5–2.6 mm long, 1.5 mm wide across side lobes, 1.1–1.3 mm wide across mid-lobe, stipitate to column-foot, glabrous; side lobes rounded, auriculate, margin serrate to lacerate, apex acuminate; mid-lobe oblong-ovate to ovate-elliptic, acute to shortly acuminate, margin minutely erose, veins 3; disc with an M-shaped callus composed of two small swollen basal bosses linked by two low, raised median ridges terminating on lower half of mid-lobe. ***Column*** 1.6 mm long, *c.* 0.3–0.4 mm wide at apex; foot short; apical hood entire, obtuse, somewhat truncate; rostellum triangular, acute; stelidia arising just above base, spathulate, rounded to truncate, reaching to base of stigmatic cavity; anther-cap cucullate, acute, surface uneven.

HABITAT AND ECOLOGY: Riparian forest, growing one metre above ground level. Alt. 530 m. Flowering observed in August.

DISTRIBUTION IN BORNEO: SARAWAK: Ulu Sungai Sipayan.

GENERAL DISTRIBUTION: Endemic to Borneo.

NOTES: A curious species closely related to *D. gracile* (Hook. f.) J.J. Sm. and *D. lyriforme* J.J. Sm. but distinguished by the smaller flowers with papillate sepals and petals, a character not found in any other Bornean species. The lip is shorter and the short distinctive stelidia are obtuse and spathulate, and only reach to as far as the base of the stigmatic cavity. It is further distinguished from *D. gracile* by the group of imbricate sterile bracts at the base of the rachis, and from *D. lyriforme* by the narrower leaves.

DERIVATION OF NAME: The specific epithet is derived from the Latin *papillatus,* have papillae, i.e. nipple-like protuberances, and *tepalum,* a division of the perianth, i.e. either sepal or petal.

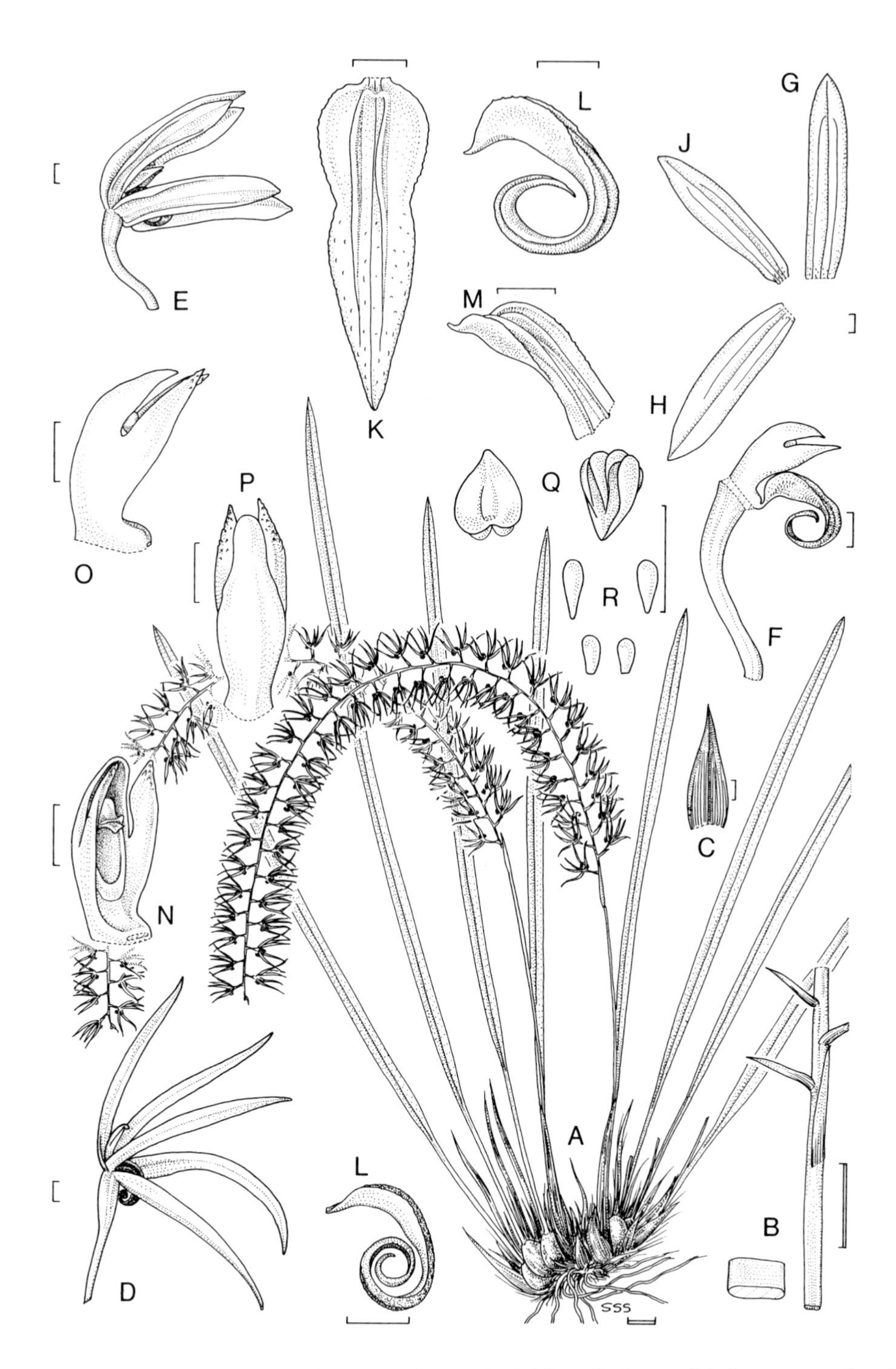

Figure 65. Dendrochilum planiscapum Carr. - A: habit. - B: base of inflorescence showing flattened peduncle and with a transverse section through the peduncle. - C: floral bract, flattened. - D & E: flowers, side view. - F: pedicel, lip and column, side view. - G: dorsal sepal. - H: lateral sepal. - J: petal. - K: lip, flattened. - L: lip, natural position. - M: base of lip. - N: column, oblique view. - O: column, side view. - P: lip, back view. - Q: anther-cap, front and back views. - R: pollinia. A, D & L drawn from *Meijer* 48111, and B & C, E–K and M–R from *Giles & Woolliams* s.n. by Susanna Stuart-Smith. Scale: single bar = 1 mm; double bar = 1 cm.

65. DENDROCHILUM PLANISCAPUM Carr

Dendrochilum planiscapum *Carr* in Gard. Bull. Straits Settlem. 8: 228 (1935). Type: Borneo, Sabah, Mt. Kinabalu, Tenompok, 1440 m, August 1933, *Carr* 3663, SFN 28020 (holotype SING, isotypes AMES, K).

Tufted ***epiphyte*** to 30 cm high. ***Pseudobulbs*** 1.5–2.7 × 1.2 cm, ovoid, wrinkled. ***Leaf blade*** 15–30 × 0.6–0.85 cm, linear, apex conduplicate, acute, grooved above, keeled beneath, rigid; petiole 6–7 cm long, sulcate. ***Inflorescence*** emerging from apex of pseudobulb with almost fully expanded leaf, stout, many-flowered; peduncle 11–16 cm long, laterally compressed, naked, with an apical bract-like acuminate sheath *c.* 1.5 cm long; rachis 18–24 cm long, quadrangular; floral bracts 0.5–0.9 cm long, subulate, involute. ***Flowers*** opening from bottom of spike; sepals and petals bright yellow-green or citron, the sepals suffused red at base; lip brown; column pale yellow, the hood whitish, the stelidia whitish suffused pale salmon. ***Pedicel*** with ***ovary*** 5–7 mm long, narrow. ***Sepals*** somewhat carinate. ***Dorsal sepal*** *c.* 8.5–1.1 cm long, *c.* 1.8 mm wide, narrowly oblong-ligulate, acute. ***Lateral sepals*** *c.* 0.7–1 cm long, *c.* 1.8 mm wide, narrowly linear-elliptic to ligulate, acute, margins revolute. ***Petals*** *c.* 6.8–7 cm long, *c.* 1.7 mm wide, oblong to oblong-elliptic, acute, sometimes falcate, margin strongly revolute. ***Lip*** *c.* 4.8 × 2 mm, subentire, strongly recurved apically; side lobes very obscure, scarcely rounded; mid-lobe narrowly triangular-ovate or triangular-elliptic, acute or obtuse; disc with two keels fading near apex as raised veins, median vein prominent on mid-lobe. ***Column*** *c.* 2.8 mm high; apical hood triangular, 2- to 3-toothed; stelidia arising on either side of stigma, broadly triangular, acuminate, longer than apical hood. Plate 16A.

HABITAT AND ECOLOGY: Lower and upper montane forest; mossy forest; recorded as epiphytic on *Castanopsis*. Alt. 1300 to 2400 m. Flowering observed in February, August and September.

DISTRIBUTION IN BORNEO: SABAH: Mt. Alab; Mt. Kinabalu.

GENERAL DISTRIBUTION: Endemic to Borneo.

NOTE: A distinctive species belonging to section *Platyclinis* instantly recognised by the flattened peduncle and strongly recurved lip apex.

DERIVATION OF NAME: The specific epithet is derived from the Latin *planus*, even or flat, and *scapus*, leafless floral axis or peduncle, in reference to the flattened peduncle.

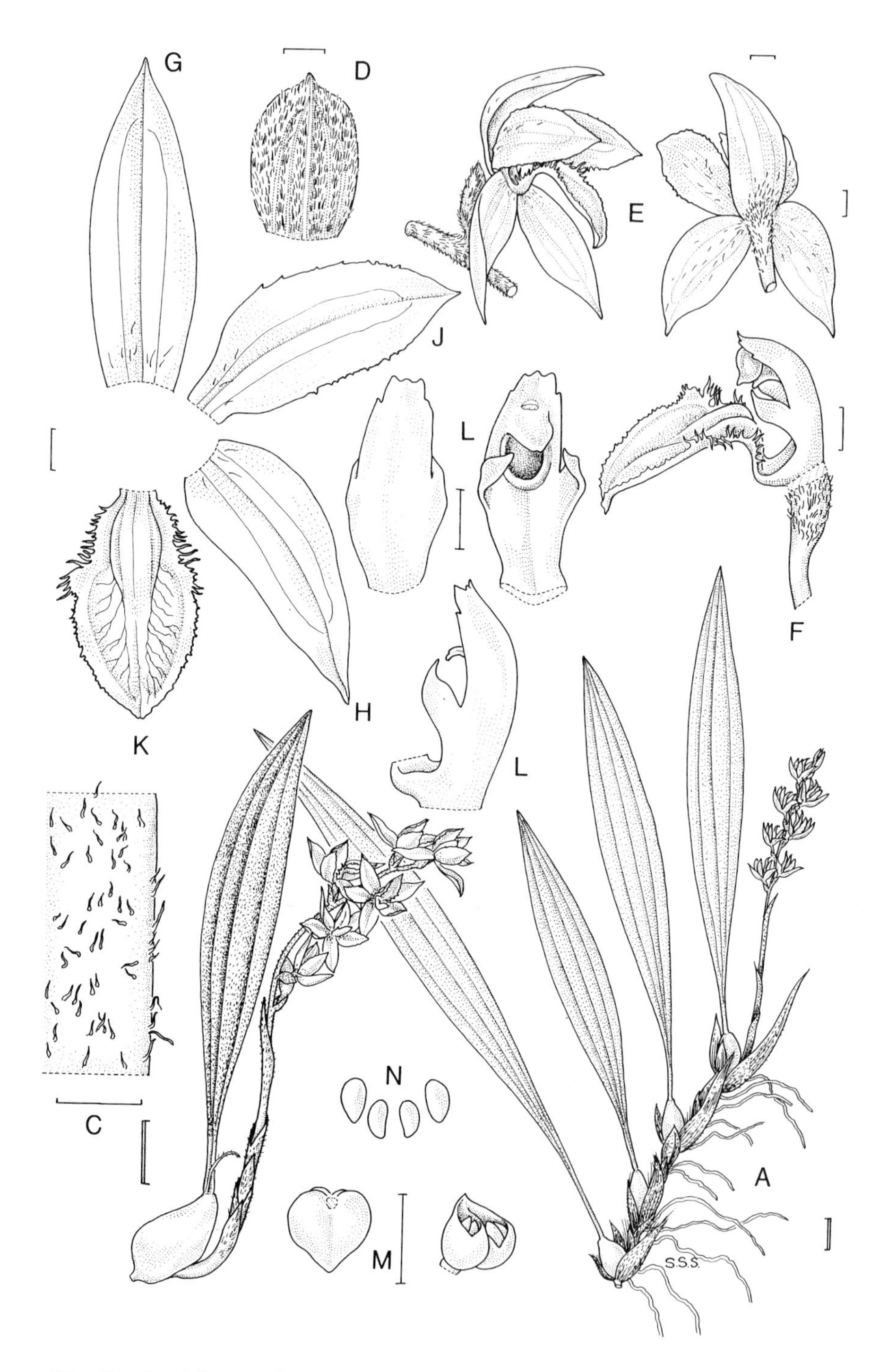

Figure 66. Dendrochilum pubescens L.O. Williams. - A & B: habit. - C: close-up of leaf indumentum. - D: floral bract, flattened. - E: flowers, side and back views. - F: pedicel with ovary, lip and column, side view. - G: dorsal sepal. - H: lateral sepal. - J: petal. - K: lip, flattened.- L: column, back, front and side views. - M: anther-cap, back and oblique views. - N: pollinia. A (habit) drawn from *Johns* 7341 and B–N from *Thomas* 197 by Susanna Stuart-Smith. Scale: single bar = 1 mm; double bar = 1 cm.

66. DENDROCHILUM PUBESCENS L.O. Williams

Dendrochilum pubescens *L.O. Williams* in Bot. Mus. Leafl. Harv. Univ. 6: 58 (1938). Type: Borneo, Sarawak, Gunung Temabok, Upper Baram Valley, *Moulton* 6763 (holotype AMES, isotype SING).

Epiphyte. Rhizome 10–42 cm long, tough, to 0.5 cm in diameter, branching, clothed in greyish to pale brown, brown furfuraceous-pubescent sheaths, which become fibrous with age. **Roots** 1–1.2 mm in diameter, flexuous, smooth. ***Pseudobulbs*** 1–2.5 × 0.6–1.2 cm, ovate, often somewhat flattened, smooth, deeply furrowed when dry, borne 1–2.5 cm apart on rhizome, enclosed in cataphylls when young. ***Leaves*** shortly brown- or black- furfuraceous-pubescent, slightly less so on adaxial surface, thin-textured; blade (6.5–)7–16.5(–25) × 0.6–2(–3.5) cm, oblong elliptic or narrowly elliptic, acute to acuminate, somewhat plicate, with 3 prominent and usually 4 less prominent veins; petiole 1–3.5 cm long, sulcate, furfuraceous-pubescent. ***Inflorescence*** synanthous, laxly to subdensely (2–)6– to 12-flowered; flowers borne 4–10 mm apart; peduncle 5–12 cm long, terete, densely brown or black finely setose-pubescent, enclosed at first by the developing leaf; rachis 4–5 cm long, quadrangular, densely brown or black finely setose-pubescent, porrect to pendulous; floral bracts 4–5 mm long, oblong-ovate, apiculate, densely brown or black setose-pubescent on abaxial surface. ***Flowers*** *c.* 1 cm across; translucent green, yellow, orange or brownish ochre, the lip green flushed with ochre or orange-brown, the column pale green or white. ***Pedicel*** with ***ovary*** 2.6–4 mm long, clavate, densely brown or black setose-pubescent, less so on base of pedicel. ***Sepals*** and ***petals*** sparsely brown or black setose-pubescent on abaxial surface, petals with very few hairs, 3-veined. ***Dorsal sepal*** 7–8 × 2.8–3.5 mm, ovate-elliptic or narrowly elliptic, subacute or acute, somewhat fleshy, shallowly concave, with a few basal hairs on adaxial surface. ***Lateral sepals*** 6.5–7.8 × 3 mm, ovate-elliptic or narrowly elliptic, dorsally acutely carinate at apex. ***Petals*** 6–7.5 × 2.2–3.2 mm, oblong-elliptic, subacute or acute, sometimes shortly clawed, sometimes oblique, erose to irregularly serrulate, with a few basal hairs on adaxial surface. ***Lip*** 4.5–6 × 2.5–3(–4.1) mm, oblong-ovate or oblong-elliptic, subacute or acute, shortly clawed, curved, fleshy, margins thinner, erose to serrulate, lower half often irregularly lacerate; disc with two fleshy parallel keels, not united at base. ***Column*** 3–4 mm long, curved; foot 0.6–1 mm long; apical hood oblong, entire to irregularly obscurely lobed; stelidia 1 mm long, median, just reaching rostellum, ligulate, acute, incurved; rostellum prominent, broadly triangular; stigmatic cavity with a swollen lower margin; anther-cap 0.8–0.9 × 1 mm, ovate, cucullate, acute. Plate 16B & C.

HABITAT AND ECOLOGY: Mixed dipterocarp forest to 40 metres high, with little undergrowth; mixed dipterocarp to "kerangas" transitional forest; lower montane ridge-top forest; recorded as an epiphyte on the trunks of canopy trees near a waterfall. Alt. sea level to 900 m. Flowering observed in January, April, November and December.

DISTRIBUTION IN BORNEO: BRUNEI: Batu Melintang; Belait Melilas, Sungai Ingei; Temburong River Valley. SARAWAK: Kapit District, Hose Mountains; Mt. Temabok.

GENERAL DISTRIBUTION: Endemic to Borneo.

NOTES: A member of section *Platyclinis* which is unlikely to be confused with any of the other species native to Borneo on account of the densely brown- or black-pubescent leaves, sheaths, inflorescence and sepals. The related *D. vestitum* J.J. Sm., from Bunguran Island in the North Natuna Archipelago north-west of Borneo, has smaller leaves, more numerous smaller flowers, a bidentate column hood and longer stelidia.

DERIVATION OF NAME: The specific epithet *pubescens* is Latin and refers to the hairy nature of this species.

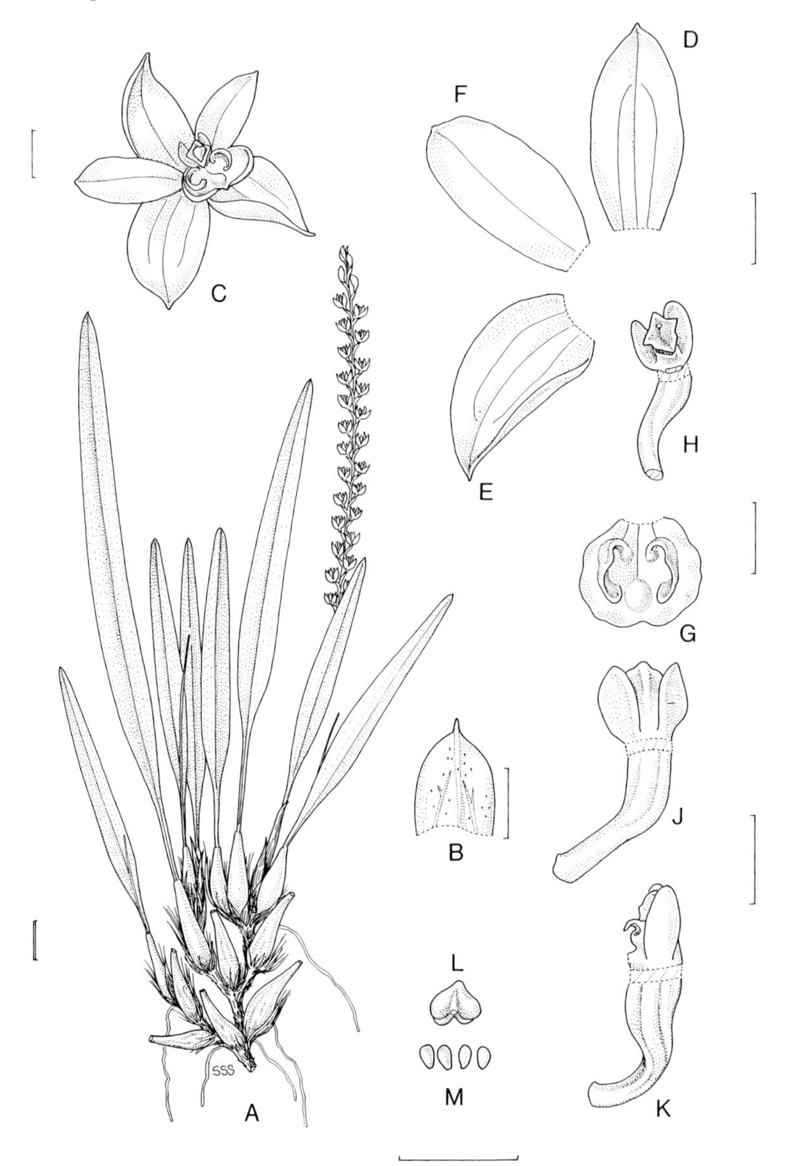

Figure 67. Dendrochilum scriptum Carr. - A: habit. - B: floral bract, flattened. - C: flower, oblique view. - D: dorsal sepal. - E: lateral sepal. - F: petal.- G: lip, flattened. - H: pedicel with ovary and column, front view. - J: pedicel with ovary and column, back view. - K: pedicel with ovary and column, side view - L: anther-cap, back view. - M: pollinia. All drawn from *Carr* 3597 (holotype) by Susanna Stuart-Smith. Scale: single bar = 1 mm; double bar = 1 cm.

67. DENDROCHILUM SCRIPTUM Carr

Dendrochilum scriptum *Carr* in Gard. Bull. Straits Settlem. 8: 234 (1935). Type: Borneo, Sabah, Mt. Kinabalu, above Kamborangah, *Carr* 3597 (holotype SING).

Epiphyte or ***lithophyte. Rhizome*** *c.* 4 cm long, shortly creeping, branched, densely covered with dry sheaths, which become fibrous with age. ***Pseudobulbs*** 1.5–2.5 × 0.4–0.7 cm, narrowly ovoid, minutely wrinkled, borne up to 1 cm apart on rhizome, usually much less, forming an acute angle with rhizome, red. ***Leaves*** rigid; blade 5.5–13 × 0.7–1.1 cm, narrowly ligulate-elliptic, shortly acute, obtuse or minutely cuspidate, main veins 5 or 6; petiole 1.6–3 cm long, sulcate. ***Inflorescence*** synanthous, erect, subdensely many-flowered; flowers borne 2–2.5 mm apart; peduncle to 6 cm long, terete; rachis to 8.5 cm long, quadrangular; floral bracts 2–3.5 mm long, ovate, obtuse. ***Flowers*** *c.* 6 mm across, non–resupinate, with a rather musty odour of over-ripe fruit, sepals and petals salmon-pink, ochre, orange or yellow, sepals often tipped brownish salmon with a salmon median line outside; lip salmon-pink or reddish brown with paler keels, margins ochre, or orange; column dark reddish brown; anther-cap cream with a dark salmon-pink median streak. ***Pedicel*** with ***ovary*** 1.8–2 mm long, narrowly clavate. ***Sepals*** 3-veined. ***Dorsal sepal*** 3 × 2 mm, oblong-elliptic, obtuse or subacute. ***Lateral sepals*** 3 × 2.5–3 mm, falcate-ovate, obtuse or acute. ***Petals*** 3 × 1.7 mm, oblong, obtuse, 1- to 3-nerved. ***Lip*** *c.* 1.8 × 2.7 mm, transversely oblong, shallowly retuse; disc with a large roughly M-shaped keel in the lower two thirds provided below the apex with a low fleshy keeled cushion. ***Column*** 1 mm long; foot absent; apical hood very shortly triangular, obtuse; stelidia basal, as long as apex of hood, broadly elliptic, obtuse; stigmatic cavity with a swollen lower margin; anther-cap ovate, cucullate. Plate 16D & 17A.

HABITAT AND ECOLOGY: A twig epiphyte in upper montane forest and scrub on extreme ultramafic substrate, preferring an open canopy; also found abundantly as a riverine species on mossy limbs overhanging rivers and sometimes on boulders in and adjacent to rivers. Alt. 2600 m and above, not observed above 3200 m. Flowering observed in March (in full bloom) and July.

DISTRIBUTION IN BORNEO: SABAH: Mt. Kinabalu only.

GENERAL DISTRIBUTION: Endemic to Borneo.

NOTES: This species belonging to section *Eurybrachium* was, until recently, only known from the type collection. Barkman (pers. comm.) reports that it is a "weed" in certain places on Mt. Kinabalu where one specimen seen growing on a boulder in a river had over 25 flower spikes. There are two colour forms, the commoner being salmon-pink or orange, the rarer having yellow sepals and petals and an orange lip.

DERIVATION OF NAME: The specific epithet *scriptum* is Latin, meaning written matter, and refers to the letter M-shaped keel on the lip.

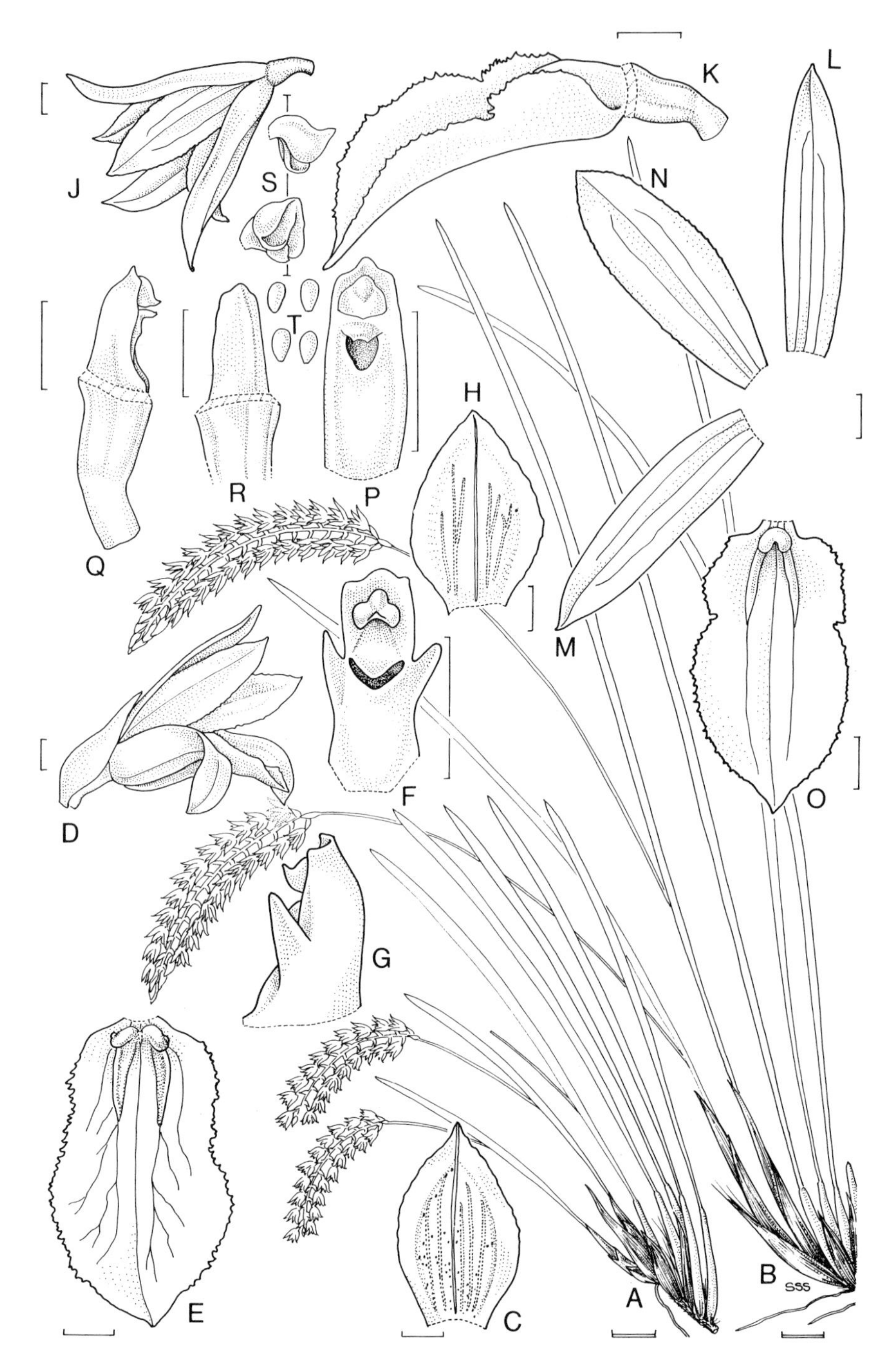

Figure 68. Dendrochilum stachyodes (Ridl.) J.J. Sm. - A & B: habits. - C: floral bract, flattened. - D: flower with floral bract, side view. - E: lip, flattened. - F: column with anther-cap, front view. - G: column with anther-cap, side view. - H: floral bract, flattened. - J: flower, side view. - K: pedicel with ovary, lip and column, side view. - L: dorsal sepal. - M: lateral sepal. - N: petal. - O: lip, flattened. - P: column with anther-cap, front view. - Q: pedicel with ovary, column and anther-cap, side view. - R: upper portion of ovary and column, back view. - S: anther-cap, side and oblique views. - T: pollinia. A (habit) drawn from *Gardner* 48, B–G from *J. & M.S. Clemens* 33177 and H–T from *Wood* 605 by Susanna Stuart-Smith. Scale: single bar = 1 mm; double bar = 1 cm.

68. DENDROCHILUM STACHYODES (Ridl.) J.J. Sm.

Dendrochilum stachyodes *(Ridl.) J.J.Sm.* in Recueil Trav. Bot. Néerl. 1: 77 (1904). Type: Borneo, Sabah, Mt. Kinabalu, 3300 m, *Haviland* 1097 (holotype BM, isotypes K, SING).

Platyclinis stachyodes Ridl. in Trans. Linn. Soc. London, Bot., ser. 2 (4): 234 (1894).

Acoridium stachyodes (Ridl.) Rolfe, Orchid Rev. 12: 220 (1904).

A clump-forming ***lithophyte.*** 10–15 cm high. ***Pseudobulbs*** (1–)2–3 × 0.4–0.5(–0.6) cm, tufted, fusiform, covered by several brown overlapping cataphylls which become fibrous and disintegrate with age. ***Leaf blade*** 6–8(–12) × 0.4–0.5(–0.6) cm, linear-ligulate, obtuse to acute, often falcate, coriaceous, striolate, narrowly petiolate; petiole 1–2(–2.5) cm long. ***Inflorescence*** apical, curving, densely many-flowered, racemose; peduncle 6–9(–10) cm long, terete, filiform, wiry, naked; rachis 2–5(–6) cm long, quadrangular, sulcate; floral bracts 4 × 3 mm, ovate, acute, glumaceous, 5- to 7-veined. ***Flowers*** opening from the top of the spike; creamy white to straw-coloured, the column and ovary pinkish. ***Ovary*** 1–1.5 mm long, sessile. ***Sepals*** linear-oblong, acute. ***Dorsal sepal*** 6 × 1 mm. ***Lateral sepals*** 6 × 1–1.5 mm. ***Petals*** 5.5–6 × 2 mm, narrowly oblong-elliptic, acute, margin minutely erose. ***Lip*** 5 × 3 mm, obscurely 3-lobed, narrowly elliptic to oblong-elliptic, acute, margin erose-denticulate, 3-veined, basal callus cuneate, 2-lobed; side lobes subovate, obtuse; mid-lobe much longer. ***Column*** 0.9–1 mm long, usually without stelidia; stelidia, when present, borne opposite stigmatic cavity, triangular, subacute; apical hood obscurely tridentate; rostellum semiovate, fleshy; anther-cap cucullate. Plate 17B & 18A.

HABITAT AND ECOLOGY: Granitic rock crevices or on ultramafic substrate; upper montane forest; growing in the open or between scrub patches. Alt. 2400 to 3700 m. Flowering observed in January, February, June, July, August, October, November and December.

DISTRIBUTION IN BORNEO: SABAH: Mt. Kinabalu only.

GENERAL DISTRIBUTION: Endemic to Borneo.

NOTES: *D. stachyodes*, together with *Bulbophyllum coriaceum* Ridl., *Coelogyne papillosa* Ridl. and *Eria grandis* Ridl., attains a greater altitude than any other Bornean orchid. It forms extensive colonies along rock crevices, usually in the open but sometimes under *Leptospermum*, and is obviously able to withstand adverse climatic conditions involving great temperature extremes. The mass flowering of *D. stachyodes* among the windswept granite rock faces is one of the delights of upper Kinabalu. Such displays have, however, suffered in recent years from periodic droughts resulting from the El Niño/Southern Oscillation phenomenon, that of 1983 being particularly severe.

Clemens 33177 (AMES, BM, E, HBG) differs from other collections in having longer, narrowly linear leaves and small stelidia on the column.

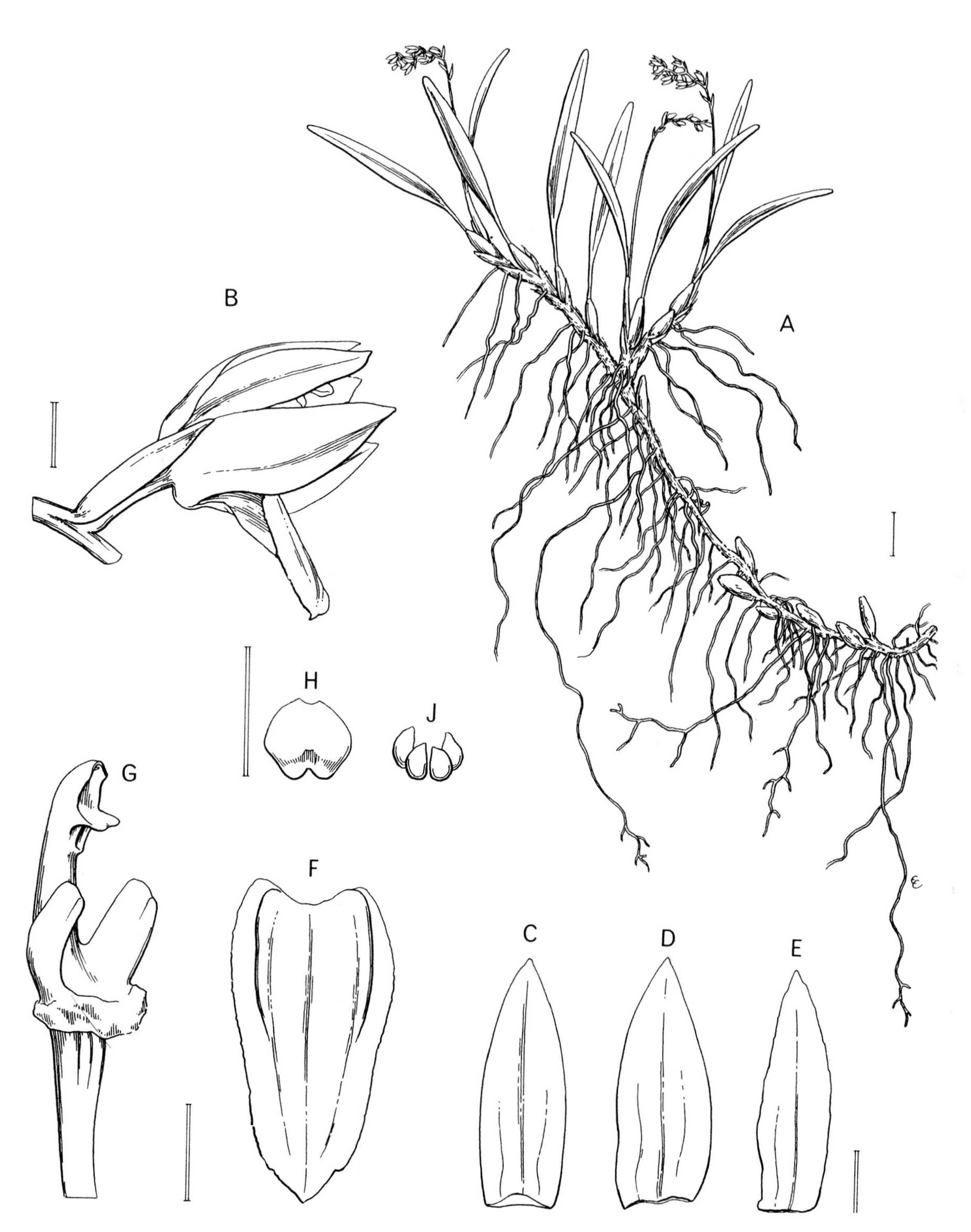

Figure 69. Dendrochilum suratii J.J. Wood. - A: habit. - B: flower, side view. - C: dorsal sepal. - D: lateral sepal. - E: petal. - F: lip, flattened.- G: pedicel with ovary and column, oblique view. - H: anther-cap, back view. - J: pollinia. All drawn from *Wood* 905 (holotype) by Eleanor Catherine. Scale: single bar = 1 mm; double bar = 1 cm.

DERIVATION OF NAME: The specific epithet is derived from the Greek *stachys*, a spike, referring to the inflorescence. Ridley described the inflorescences as "reminding one of an ear of wheat, whence the specific name."

69. DENDROCHILUM SURATII J.J. Wood

Dendrochilum suratii *J.J. Wood* in Lindleyana 7(2): 77, fig. 3 (1991). Type: Borneo, Sabah, Tambunan District, Mt. Trus Madi, *Surat* in *Wood* 905 (holotype K).

Trailing ***epiphyte. Rhizome*** up to 15 cm or more long, bearing numerous long roots, enclosed in pale brown, darker brown speckled, fibrous sheaths. ***Pseudobulbs*** 6–8 mm apart, 1–1.2 × 0.4–0.6 cm, ovate, plum-purple. ***Leaves*** linear-elliptic to ligulate, obtuse, slightly constricted 0.8–1.3 cm below apex, attenuate into a sulcate petiole; blade 3–5.2 × 0.3–0.8 cm; petiole 0.5–0.7 cm long. ***Inflorescence*** 5- to 8-flowered; peduncle 2.7–4 cm long, erect, filiform; rachis 1.6–2.2 cm long, curved, quadrangular, sparsely ramentaceous; floral bracts 4 mm long, 1.5 mm wide when flattened, ovate-elliptic, apex ± truncate with a small central tooth, involute, sparsely ramentaceous. ***Flowers*** resupinate, opening from top of inflorescence; yellow, flushed plum-purple. ***Pedicel*** with ***ovary*** 2 mm long, clavate. ***Sepals*** ovate-elliptic, acute. ***Dorsal sepal*** 4 × 1.1 mm. ***Lateral sepals*** 4 × 1.5 mm. ***Petals*** 4 × 1 mm, narrowly elliptic, acute. ***Lip*** 3.5–3.6 mm long, 1.9–2 mm wide near base, ovate, acute, margin minutely erose, curved, apex deflexed; disc with three prominent veins, the outer two developed into low basal keels and becoming obscure distally, the central a prominent line. ***Column*** 2.6 × 0.6 mm, gently curved; stelidia 1 mm long, basal, oblong, obtuse; anther-cap ovate, cucullate.

HABITAT AND ECOLOGY: Upper montane ericaceous forest, growing on exposed branches and twigs. Alt. 2400 to 2500 m. Flowering observed in June.

DISTRIBUTION IN BORNEO: SABAH: Mt. Trus Madi only.

GENERAL DISTRIBUTION: Endemic to Borneo.

NOTES: This species is abundant at high altitudes on Mt. Trus Madi where it is often found in the company of *Coelogyne plicatissima* Ames & C. Schweinf. Only a few plants were seen in flower during June and full flowering probably takes places later in the year.

DERIVATION OF NAME: Named after Andi Surat, employed at the Agricultural Park (formerly Agricultural Research Station), Tenom, Sabah, who has accompanied me on several field trips in Sabah including an ascent of Mt. Trus Madi in 1988.

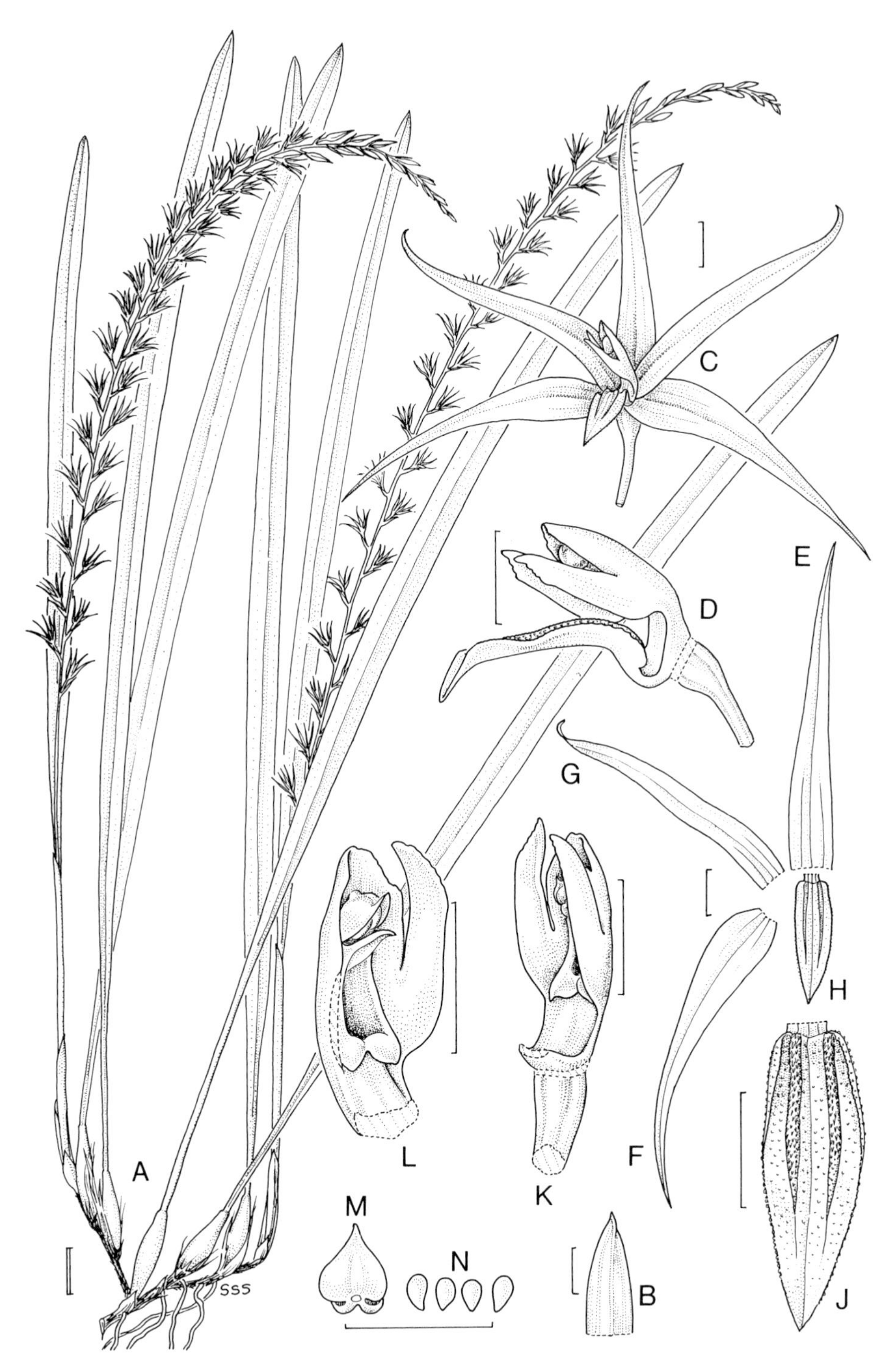

Figure 70. Dendrochilum tenuitepalum J.J. Wood. - A: habit. - B: floral bract, flattened. - C: flower, front view. - D: pedicel with ovary, lip and column, side view. - E: dorsal sepal. - F: lateral sepal.- G: petal. - H & J: lip, flattened. - K: pedicel with ovary and column, oblique view. - L: column with stelidium removed, oblique view. - M: anther-cap, back view. N: pollinia. All drawn from *Vermeulen & Duistermaat* 1008 (holotype) by Susanna Stuart-Smith. Scale: single bar = 1 mm; double bar = 1 cm.

70. DENDROCHILUM TENUITEPALUM J.J. Wood

Dendrochilum tenuitepalum *J.J. Wood* in Wood & Cribb, Checklist Orchids Borneo: 195, fig. 24 N & O (1994). Type: Borneo, Sabah, Sipitang District, ridge between Sungai Maga headwaters and Sungai Malabid headwaters, *Vermeulen & Duistermaat* 1008 (holotype L, isotype K).

Creeping, clump-forming ***epiphyte. Rhizome*** tough, producing numerous wiry roots. ***Pseudobulbs*** 1–1.8 × 0.4 – 0.6 cm, ovate-elliptic, gently curved, borne 0.5 cm apart on rhizome, enclosed in cataphylls when young. ***Leaves*** linear-ligulate, obtuse and mucronate or apiculate, conduplicate at base; blade 10.5–21.5 × 0.4–0.6 cm; petiole 2.5–5 cm long, sulcate. ***Inflorescence*** many-flowered; flowers borne 3–4 mm apart; peduncle 6–9 cm long, lower half enclosed in 2–3 acute to acuminate cataphylls 1–4 cm long; rachis 14–18 cm long, gently curving; floral bracts 2–5 mm long, ovate-elliptic, acuminate. ***Flowers*** unscented; very pale greenish. ***Pedicel*** with ***ovary*** 2.5–2.6 mm long, narrowly clavate. ***Sepals*** and ***petals*** narrowly lanceolate, acuminate. ***Dorsal sepal*** 6–6.5 mm long, 0.8–0.9 mm wide at base. ***Lateral sepals*** 7–7.1 mm long, 1–1.1 mm wide at base. ***Petals*** 7–7.1 mm long, 1 mm wide at base. ***Lip*** 2.8 mm long, 1 mm wide near base, narrowly ovate-elliptic, acute, entire, minutely papillose, margin irregular in places, erect at base; disc with two low papillose keels extending from base and terminating at or just beyond middle of lip, not linked by a transverse basal ridge. ***Column*** 2 mm long, with a foot; hood minutely serrulate; stelidia *c.* 1.8 mm long, a little longer than hood, ligulate, obtuse to subacute, minutely serrulate at apex, inserted a little above base; anther-cap cucullate, extinctoriform.

HABITAT AND ECOLOGY: Open, low mossy forest. Alt. 1600 m. Flowering observed in December.

DISTRIBUTION IN BORNEO: SABAH: Sipitang District.

GENERAL DISTRIBUTION: Endemic to Borneo.

NOTES: *D. tenuitepalum* is distinguished from the closely related *D. mucronatum* J.J. Sm. from Sarawak by its longer and slightly narrower sepals and petals, non-erose petals and entire, narrowly ovate-elliptic, minutely papillose lip. It is known only from the type collection.

DERIVATION OF NAME: The specific epithet is derived from the Latin *tenuis,* slender, thin and *tepalum*, a division of the perianth, either sepal or petal, referring to the long, narrow sepals and petals.

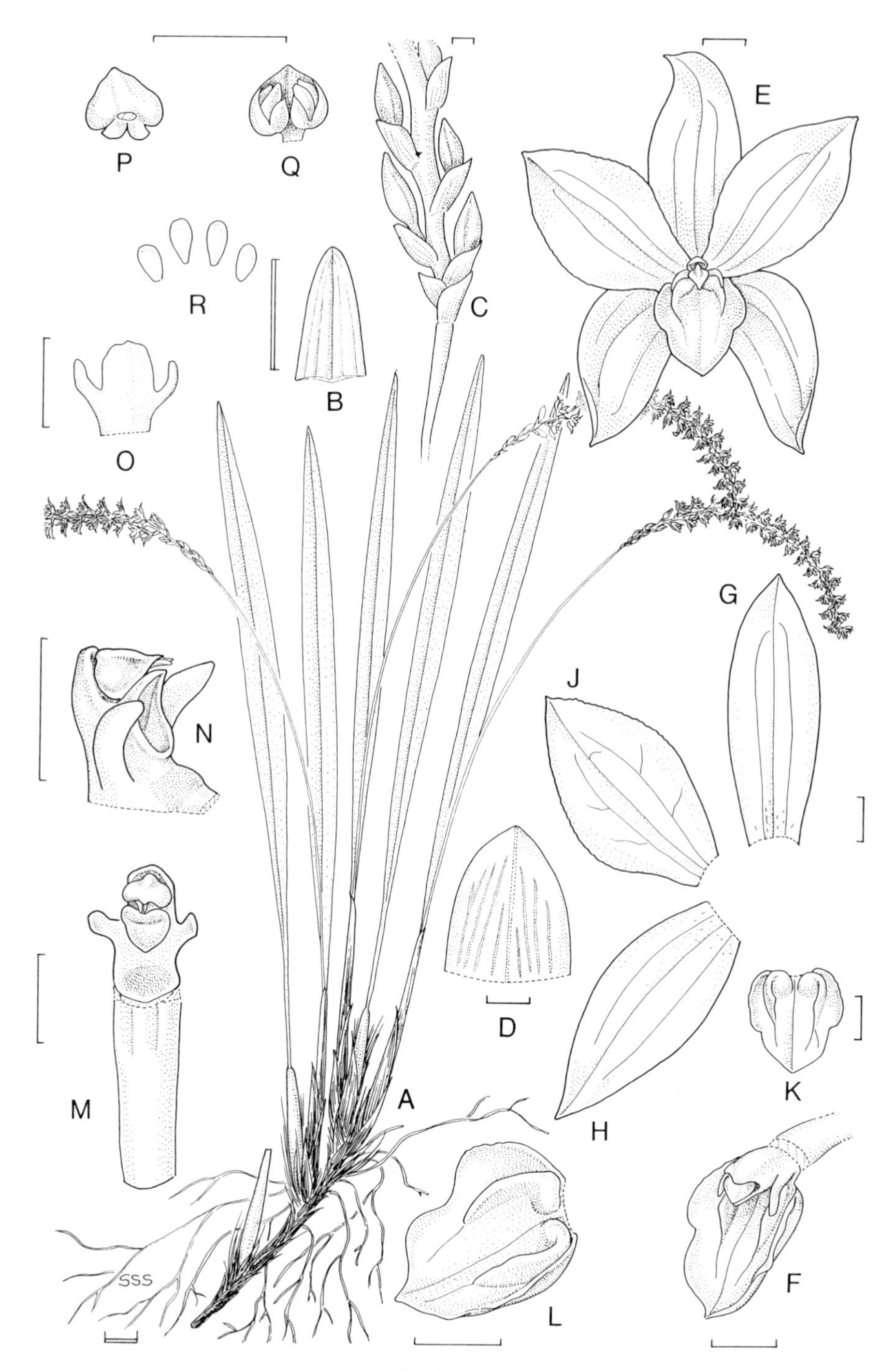

Figure 71. Dendrochilum trusmadiense J.J. Wood. - A: habit. - B: leaf apex. - C: base of inflorescence. - D: floral bract. - E: flower, front view. - F: pedicel with ovary, lip and column, oblique view.- G: dorsal sepal. - H: lateral sepal. - J: petal. - K: lip, flattened. - L: oblique view. - M: pedicel with ovary, column and anther-cap, front view. - N: column and anther-cap, side view. - O: column, back view. - P: anther-cap, back view. - Q: anther-cap with pollinia, front view. - R: pollinia. All drawn from *Wood* 886 (holotype) by Susanna Stuart-Smith. Scale: single bar = 1 mm; double bar = 1 cm.

71. DENDROCHILUM TRUSMADIENSE J.J. Wood

Dendrochilum trusmadiense *J.J. Wood* in Lindleyana 5(2): 93, fig. 8 (1990). Type: Borneo, Sabah, Tambunan District, Mt. Trus Madi, *Wood* 886 (holotype K, isotypes L, UKMS).

Clump-forming ***epiphyte. Rhizome*** up to 10 cm long. ***Pseudobulbs*** 1.5–3 × 0.6–0.8 cm, caespitose, fusiform or cylindrical, rugose, yellowish when dry, enclosed in three or four fibrous cataphylls. ***Leaves*** oblong-elliptic to narrowly elliptic, obtuse, attenuate below; blade (6–)9–20 × 0.8–2 cm; petiole 1–3(–4) cm long, sulcate. ***Inflorescence*** erect to curved, emerging from apex of pseudobulbs with a fully expanded leaf, flowers opening from the top of the spike; peduncle 10–20 cm long, slender, wiry; rachis 7–13 cm long, quadrangular; floral bracts 3 × 2–2.5 mm, ovate, acute, prominently veined, pale salmon-brown. ***Flowers*** resupinate; sweetly scented; pedicel with ovary tan; sepals and petals spreading, translucent pale lemon-yellow; lip chocolate-brown; column whitish. ***Pedicel*** with ***ovary*** 2.5–3 mm long clavate, curved. ***Sepals*** and ***petals*** minutely sparsely hirsute at base. ***Dorsal sepal*** 5.5 × 2 mm, oblong-elliptic, acute. ***Lateral sepals*** 6 × 2.5 mm, ovate-elliptic, acute, slightly carinate at apex. ***Petals*** 5 × 2.8–3 mm, elliptic, mucronate. ***Lip*** 2 × 2 mm, obscurely shallowly 3-lobed, ovate, obtuse, somewhat concave; side lobes obscurely crenulate; margins of mid-lobe incurved; disc bicallose, calli thick and fleshy at base, extending as raised keels to halfway along the lip. ***Column*** 0.8 mm long; stelidia 0.8 mm long, basal, oblong, obtuse, slightly curved; anther-cap ovate, cucullate.

HABITAT AND ECOLOGY: Twig epiphyte in upper montane ericaceous ridge-top forest. Alt. 1900 to 2000 m. Flowering observed in June.

DISTRIBUTION IN BORNEO: SABAH: Mt. Trus Madi only.

GENERAL DISTRIBUTION: Endemic to Borneo.

NOTES: This beautiful species, so far only recorded from Mt. Trus Madi, bears a striking resemblance to *D. kamborangense* Ames (section *Platyclinis*) from Mt. Kinabalu. The lip of *D. trusmadiense*, however, is quite different and the column structure typical of section *Eurybrachium*.

DERIVATION OF NAME: The specific epithet is named after the type locality, Mt. Trus Madi (2642 m), the second highest peak in Borneo.

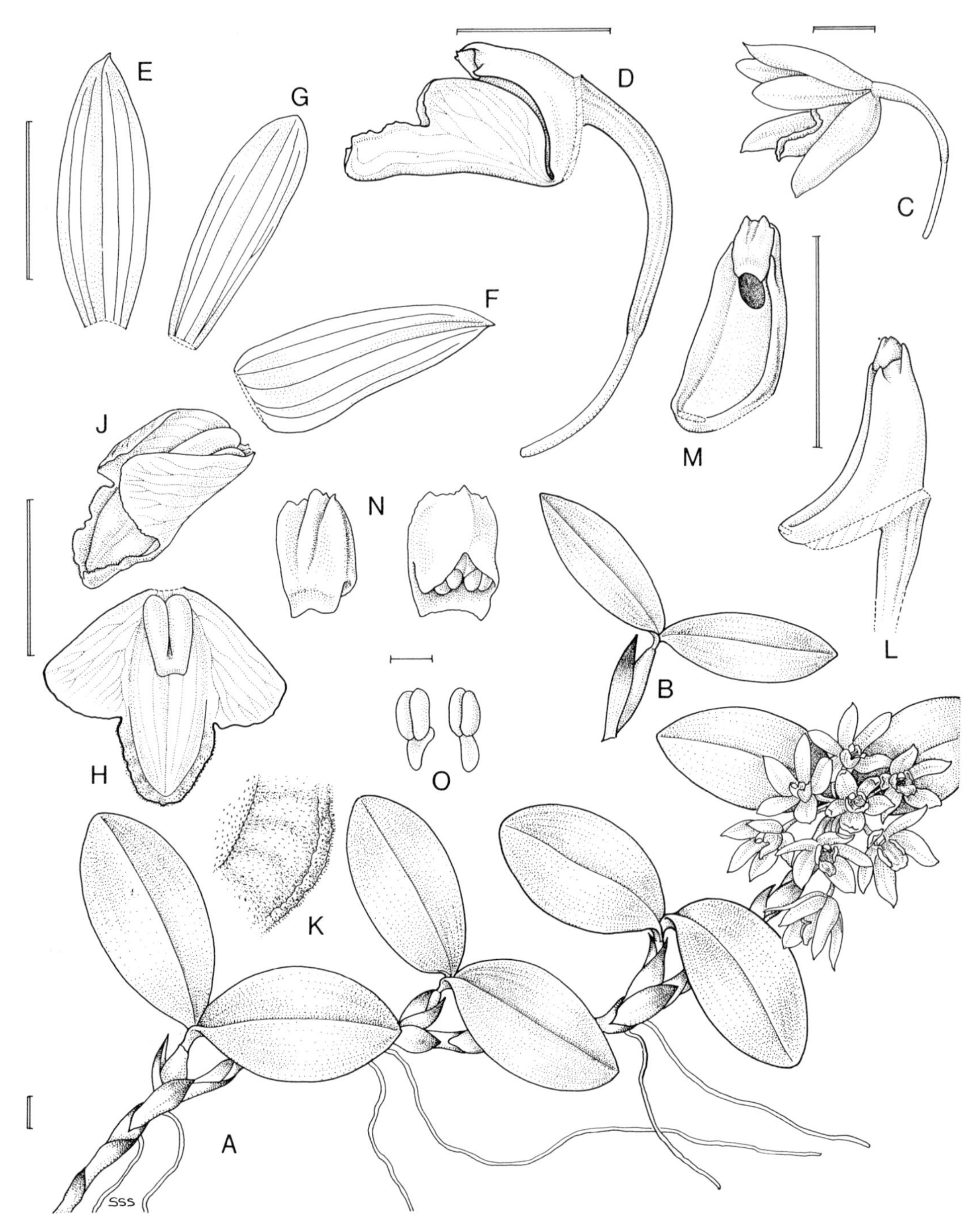

Figure 72. Epigeneium speculum (J.J. Sm.) Summerh. - A & B: habits. - C: flower, side view. - D: pedicel with ovary, lip and column with anther-cap, side view. - E: dorsal sepal. - F: lateral sepal. - G: petal. - H: lip, flattened. - J: lip, natural position. - K: close-up detail of margin of lip mid-lobe. - L: upper portion of ovary and column with anther-cap, side views. - M: column with anther-cap, oblique view. - N: anther-cap, back view and front view showing pollinia. - O. pollinaria. A (habit) drawn from *Synge* S. 449, B (habit) from *Lamb* AL 447/85 and C–O from *Wood* 597 by Susanna Stuart-Smith. Scale: single bar = 1 mm; double bar = 1 cm.

72. EPIGENEIUM SPECULUM (J.J. Sm.) Summerh.

Epigeneium speculum *(J.J. Sm.) Summerh.* in Kew Bull. 12: 264 (1957). Type: Borneo, Kalimantan, Bukit Kasian, *Nieuwenhuis* 274 (holotype L).

Dendrobium speculum J.J. Sm. in Bull. Dép. Agric. Indes Néerl. 5: 34 (1907).

Sarcopodium speculum (J.J. Sm.) Carr in Gard. Bull. Straits Settlem. 8: 109 (1935).

Katherinea specula (J.J. Sm.) A.D.Hawkes in Lloydia 19: 97 (1956).

Creeping ***terrestrial*** or ***epiphyte. Rhizome*** long creeping, up to 40 cm or more long, 3–5 mm in diameter, dark olive-green, almost entirely enclosed by imbricate, obtuse chestnut-brown sheaths 1–2 cm long. ***Roots*** 1 mm in diameter, rugulose to minutely papillose. ***Pseudobulbs*** 1.1–3 × 1–1.2 cm, 2-leaved, ovate or oblong-elliptic, obtusely 4-angled, dark olive-green, enclosed in an obtuse 3 × 1–1.2 cm cataphylls when immature. ***Leaves*** 6–9 × 2.3–4.5 cm, oblong, oblong-elliptic, or elliptic, obtuse, shallowly retuse, tough and leathery, adaxial surface minutely rugulose, narrowed at base to a petiole 0.3–1 cm long. ***Inflorescences*** borne one or two at a time between the leaves, 3- to 4-flowered; peduncle and rachis 5–11 mm long; floral bracts 3–4 mm long, ovate, acute. ***Flowers*** 1.5–2 cm across; sweetly scented; sepals lemon-cream, flushed pink on reverse; petals cream; lip cream with fine purple-red stripes on the side lobes, underside of the mid-lobe flushed yellow; column-foot stained purple-red. ***Pedicel*** with ***ovary*** 2.5–3 cm long, slender. ***Dorsal sepal*** 1.5–2 × 0.6–0.7 cm, oblong-elliptic, subacute. ***Lateral sepals*** 1.6–2 × 0.6–0.7 cm, obliquely ovate-elliptic, acute. ***Mentum*** obtuse. ***Petals*** 1.5–1.7 × 0.4–0.55 cm, ligulate or narrowly oblanceolate, subacute. ***Lip*** 1.2–1.3 cm long, 2–3 mm wide at base, 1.4–1.6 cm wide across side lobes when flattened; side lobes 8–9 × 7–8 mm, oblong-ovate, obtuse, erect; mid-lobe 5–7 mm long, 7 mm wide at base, 3.5–5 mm wide at apex, oblong-ovate, shallowly retuse, thick and fleshy, margin thin-textured, slightly uneven and undulate; disc with a swollen oblong, shiny basal callus 5–6 × 3 mm, sulcate at base. ***Column*** 6–7 × 4–6 mm, oblong; foot 5–7 mm long, gently incurved; anther-cap 2 × 2.1 mm, oblong; pollinia four. Plate 18B.

HABITAT AND ECOLOGY: Terrestrial in mossy clumps in wet podsol forest composed of *Dacrydium, Eugenia, Garcinia*, etc. with a *Rhododendron longiflorum* and *R. malayanum* understorey, and a field layer of *Calamus*, ferns, *Bromheadia finlaysoniana, Dendrochilum simplex, Nepenthes ampullaria*, etc.; "transitional moss forest, in moderate sun". Alt. 400 to 900 m. Flowering observed in September and October.

DISTRIBUTION IN BORNEO: KALIMANTAN: Bukit Kasian. SABAH: Nabawan area. SARAWAK: Mt. Dulit.

GENERAL DISTRIBUTION: Endemic to Borneo.

DERIVATION OF NAME: The generic name is derived from the Greek *epi*, upon, and *geneion*, chin, in reference to the lateral sepals and column-foot which form a chin or mentum. The specific epithet is derived from the Greek *speculum*, a mirror, in reference to the shiny callus on the lip.

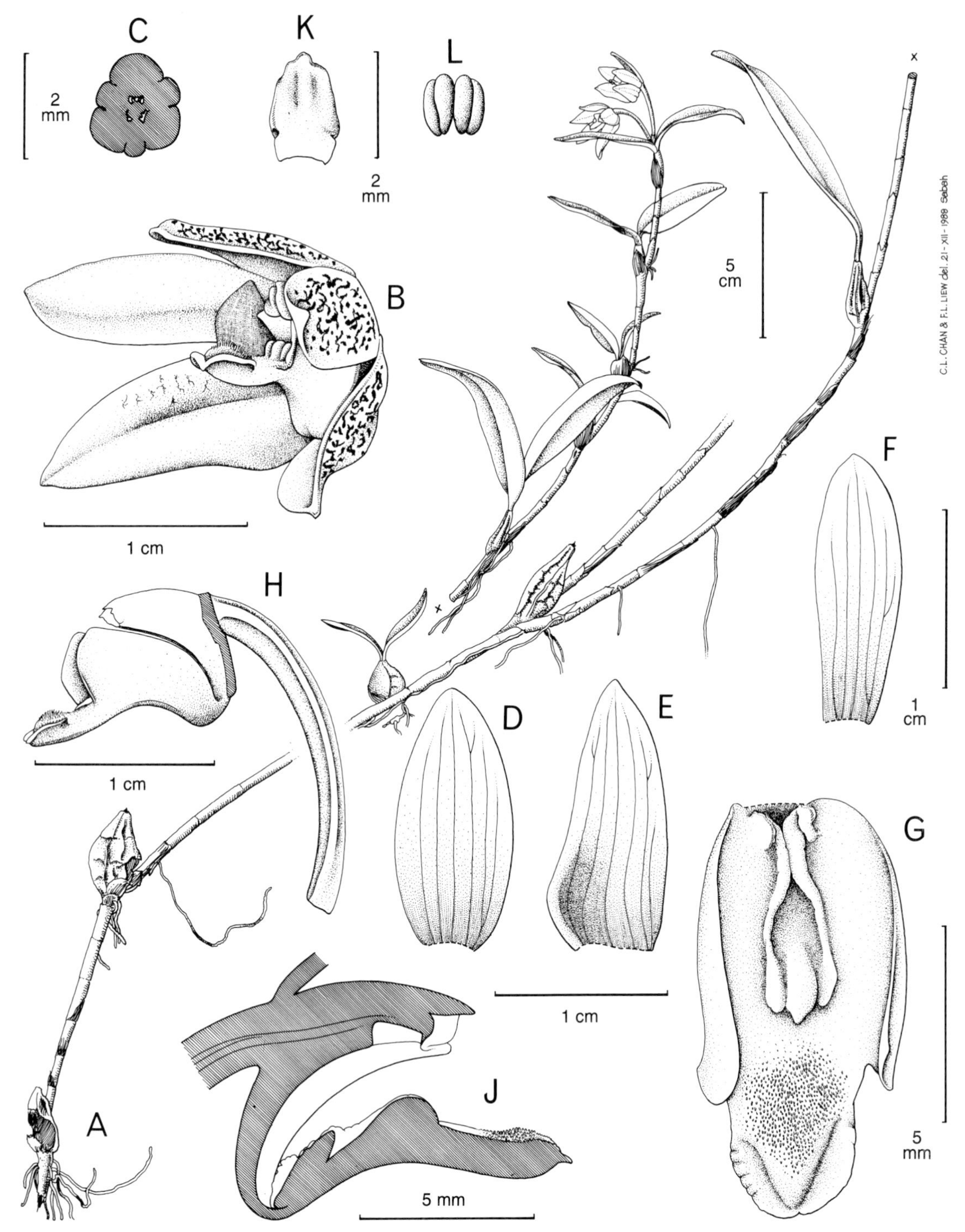

Figure 73. Epigeneium tricallosum (Ames & C. Schweinf.) J.J. Wood. - A: habit. - B: flower, oblique view. - C: ovary, transverse section. - D: dorsal sepal. - E: lateral sepal. - F: petal.- G: lip, front view. - H: pedicel with ovary, lip and column, side view. - J: pedicel with ovary, lip and column, longitudinal section. - K: anther-cap, back view. - L: pollinia. All drawn from *Chan & Gunsalam* s.n. by C.L. Chan and Lucy F.L. Liew.

73. EPIGENEIUM TRICALLOSUM (Ames & C. Schweinf.) J.J. Wood

Epigeneium tricallosum (*Ames & C. Schweinf.*) *J.J .Wood* in Lindleyana 5(2): 99 (1990). Type: Borneo, Sabah, Mt. Kinabalu, *J. Clemens* s.n. (holotype AMES).

Dendrobium tricallosum Ames & C. Schweinf., Orchidaceae 6: 114 (1920).

Creeping ***epiphyte,*** or sometimes scrambling ***terrestrial. Rhizome*** up to 60 cm or more long, branching, 2–3 mm in diameter, mustard-yellow, shiny, internodes 7–18 mm long, covered with scarious cinnamon-brown imbricate, acute cataphylls when immature which become fibrous with age. ***Roots*** 1–1.5 mm in diameter, verruculose-papillose. ***Pseudobulbs*** 0.6–2.6 × 0.4–0.8 cm, ovoid, irregularly 4-angled, 2-leaved, shiny yellow, coarsely rugose, reclining on rhizome, borne 1–10 cm apart, enclosed by scarious cataphylls when immature. ***Leaves*** spreading; blade 1.5–9 × 0.6 × 1.5 cm, oblong-elliptic to ligulate-lanceolate, obtuse, minutely obliquely bilobed, rigid, thick, leathery, mid-vein prominently sulcate above, carinate beneath, margins somewhat revolute; petiole 2–5 mm long, sulcate, complicate. ***Inflorescences*** borne between the leaves, 2- to 3-flowered, sessile; rachis 2–3 mm long, fractiflex; floral bracts 2 mm long, ovate, acute. ***Flowers*** 2.5 cm across; pedicel with ovary pinkish; sepals and petals cream, blotched and stained lilac-pink on reverse; lip cream, the side lobes stained lilac-pink at base, the mid-lobe with a dark purple, or brownish red basal patch; column creamy white. ***Pedicel*** with ***ovary*** 1.5–2 cm long, slender, finely grooved. ***Dorsal sepal*** 0.9–1.5 × 0.46–0.7 cm, broadly oblong or oblong-elliptic, rounded at the slightly oblique apex which is dorsally slightly carinate, concave. ***Lateral sepals*** 1–1.6 × 0.44–0.55 cm, obliquely ovate, rounded at the subacute apex which is dorsally conspicuously carinate. ***Mentum*** obtuse. ***Petals*** 1–1.6 × 0.37 – 0.5 cm, oblong-ligulate or oblong-elliptic, apex obtuse, fleshy. ***Lip*** 9–1.2 cm long, *c.* 1.5 mm wide at base, 0.75–0.9 cm wide across side lobes when flattened; side lobes 8–9 × 4 mm, broadly rounded, erect, thin-textured; mid-lobe 3–4.9 × 3–4 mm, oblong, obtuse, sometimes slightly retuse, verrucose or papillose and thickened, margin thin–textured; disc with two parallel keels thickened at base, extending to a little below junction of side lobes and mid-lobe where they terminate in two pyriform calli, between which is a short median callus, sometimes the three swellings are ill-defined. ***Column*** 5–6 mm long; foot 5 mm long; gently incurved; anther-cap 2 × 1.8 mm, ovate, cucullate, obtuse. Plate 18C.

HABITAT AND ECOLOGY: Lower montane forest; ridge-top forest with *Dacrydium*, *Leptospermum*, *Phyllocladus*, *Podocarpus* and *Tristania*, associated with *Chelonistele* spp., *Dilochia rigida* and *Gahnia* spp. Alt. 800 to 1800 m. Flowering observed from April to June and from October to December.

DISTRIBUTION IN BORNEO: BRUNEI: Mt. Pagon. SABAH: Mt. Alab; Mt. Kinabalu.

GENERAL DISTRIBUTION: Endemic to Borneo.

DERIVATION OF NAME: The specific epithet is derived from the Latin *tri*, three, and *callus*, a hardened thickening, referring to the three terminal swellings on the disc of the lip.

74. ERIA CARICIFOLIA J.J. Wood var. CARICIFOLIA

Eria caricifolia *J.J. Wood* in Lindleyana 5(2): 93, fig. 9 (1990). Type: Borneo, Sarawak, Mt. Temabok, Upper Baram Valley, *Moulton* 6673 (holotype K, isotype SING).

Terrestrial or ***epiphyte. Pseudobulbs*** absent, replaced by a stem 4–5 cm long, concealed by leaf sheaths. ***Cataphylls*** brown, unspotted. ***Leaves*** 3–5, distichous, sedge-like; blade 35–60 × 0.4–0.5 cm, linear, ligulate, acute or obliquely acute; sheath 4.5–13.5 cm long, with papery, brown margins. ***Inflorescence*** lax to subdense; peduncle and rachis puberulus; peduncle emerging from two or three basal sheaths, 9–10 cm long; rachis 9–13 cm long; floral bracts 1.5–2 mm long, ovate, acute, glabrous. ***Flowers*** non-resupinate;

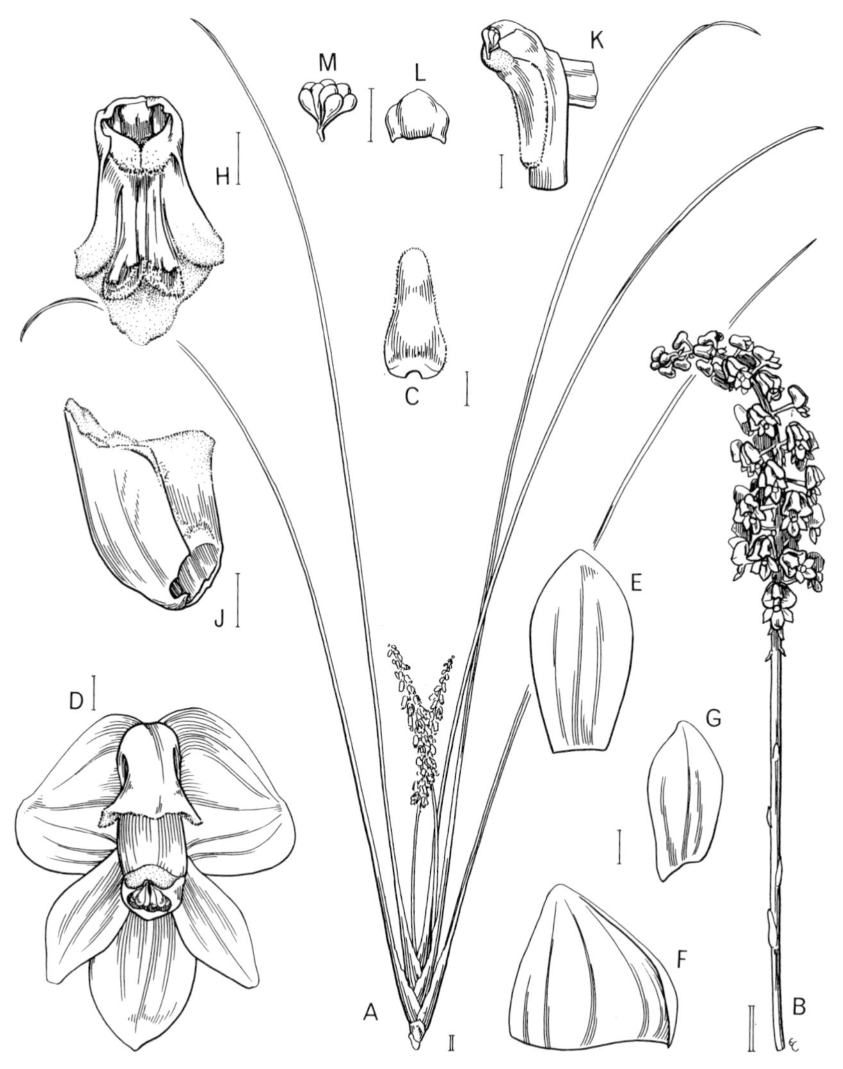

Figure 74. Eria caricifolia J.J. Wood var. **caricifolia** . - A: habit. - B: inflorescence. - C: floral bract. - D: flower, front view. - E: dorsal sepal. - F: lateral sepal.- G: petal. - H: lip, front view. - J: lip, side view. - K: column with anther-cap removed showing pollinia, oblique view. - L: anther-cap, back view. - M: pollinia. All drawn from *Moulton* 6673 (holotype) by Eleanor Catherine. Scale: single bar = 1 mm; double bar = 1 cm.

colour not recorded. ***Pedicel*** with ***ovary*** 0.6–1 cm long, puberulus. ***Dorsal sepal*** 4–5 × 3 mm, oblong, obtuse, glabrous. ***Lateral sepals*** 5 × 5.5 mm, broadly ovate, oblique, obtuse, glabrous. ***Mentum*** 4–5 mm long, obtuse. ***Petals*** 4 × 2 mm, oblong-elliptic, obtuse to subacute. ***Lip*** 3-lobed, concave, cymbiform, saccate at base, minutely papillose, 5 mm long, 4.5–5 mm wide when flattened; disc with a prominent, glabrous median nectary, attenuated distally just below apex of mid-lobe; side lobes 1.8 mm long, narrowly triangular, acute, each with a papillose-hairy, spathulate, flange-like basal callus extending on to mid-lobe; mid-lobe 0.5 mm long, triangular, subacute. ***Column*** 1.8 mm long; foot 4–5 mm long; anther-cap 0.8 × 1 mm, ovate, cucullate; pollinia eight.

HABITAT AND ECOLOGY: Lower montane forest. Alt. 1200 m. Flowering observed in November.

DISTRIBUTION IN BORNEO: SARAWAK: Mt. Temabok.

GENERAL DISTRIBUTION: Endemic to Borneo.

NOTES: This distinctive plant has proportionally longer and narrower leaves than the closely related *E. cymbidifolia* Ridl. from Sumatra, Sabah and Sarawak. The inflorescence also appears to be somewhat laxer, although this may not be a constant character.

DERIVATION OF NAME: The generic name is derived from the Greek *erion,* wool, referring to the woolly hairs found on the inflorescence and flowers of many species. The specific epithet is derived from the Latin *carex*, a sedge, and *-folius*, -leaved.

var. **glabra** *J.J. Wood* in Lindleyana 5(2): 95 (1990). Type: Borneo, Sabah, Sipitang District, slope north of Sungai Malabid, west of Long Pa Sia to Long Samado trail, *Vermeulen & Duistermaat* 1000 (holotype K).

Terrestrial. Cataphylls pale brown on straw–yellow, with darker brown spotting. ***Leaves*** 2–8; blade 28–30 × 0.3–0.4 cm; sheath 6–8 cm long. ***Inflorescence*** lax, peduncle and rachis glabrous; peduncle 4–6 cm long; rachis 4–5 cm long. ***Flowers*** unscented; sepals white; petals suffused violet; lip purple at base, apex white. ***Pedicel*** with ***ovary*** and ***sepals*** glabrous.

HABITAT AND ECOLOGY: Lower montane *Agathis-Lithocarpus* forest. Alt. 1300 to 1500 m. Flowering observed in December.

DISTRIBUTION IN BORNEO: SABAH: Sipitang District.

GENERAL DISTRIBUTION: Endemic to Borneo.

NOTES: This variant collected from near the Sarawak border is distinguished by the spotted cataphylls, shorter inflorescence with glabrous peduncle, rachis and pedicel with ovary. The presence or absence of spotted cataphylls in the type variety is unknown because only part of a cataphyll remains on the type specimen.

DERIVATION OF NAME: The specific epithet is derived from the Latin *glaber*, without hair, and refers to the glabrous inflorescence and flowers.

Figure 75. Eria grandis Ridl. - A: habit. - B: leaf. - C: inflorescence. - D: flower, side view. - E: pedicel with ovary, lip and column, side view. - F: dorsal sepal. - G: lateral sepal. - H: petal. - I: lip, front view. - J: lip, longitudinal section. - K: ovary and column, front view. - L: column, side view. - M: anther-cap, back and inner views. - N: pollinia. - O: stellate hair from ovary, close-up. - P: fruit. A–C drawn from *Lamb* SAN 91583, D–O from *Beaman* 8307 and P from *Chew et al.* 733 by Susanna Stuart-Smith. Scale: single bar = 1 mm; double bar (A & B) = 5 cm; double bar = 1 cm.

75. ERIA GRANDIS Ridl.

Eria grandis *Ridl.* in Stapf in Trans. Linn. Soc. Lond., Bot. 4: 237 (1894). Type: Borneo, Sabah, Mt. Kinabalu, 3000–3700 m, *Haviland* 1157 (holotype SING, isotype K).

Robust ***terrestrial*** 60 cm to 1 m tall. ***Roots*** branching, pubescent. ***Stem*** up to 1.3 cm in diameter, covered below with 6–9 imbricate, ovate to lanceolate, acute to acuminate, stiff cataphylls 2–20 cm long, cataphyll surface usually purple-spotted, margins pale, chartaceous, becoming ragged. ***Leaves*** 3–5; blade 18–48 × 1.2–3 cm, ensiform, subcoriaceous, apex asymmetrical to unequally bifid, subacute to acuminate, articulated to an imbricate sheath *c.* 1.5–12 cm long. ***Inflorescences*** borne in axils of leaves, subtended by an upper sheath, erect, densely many-flowered; flowers borne 5–10 mm apart; peduncle and rachis densely whitish-pubescent; peduncle 8–15 cm long, dark purple-pink, with several obtuse remote sterile bracts 1–1.8 cm long; rachis 12–19 cm long, dark purple-pink; floral bracts 4–8 × 2–2.5 mm, oblong-ovate or ovate, obtuse or subacute, chartaceous, reflexed. ***Flowers*** not opening very wide; pedicel with ovary dark purple; sepals and petals pale lilac-pink or whitish, stained deep purple-red at base inside, or dark pink outside, whitish flushed pale pink inside; lip dark purple-red, the margin and apex whitish to pale pink, the calli deep purple-red, shiny; column dark purple, the stigma whitish; anther-cap blackish purple. ***Pedicel*** with ***ovary*** 1–1.5 cm long, narrowly clavate, densely whitish-pubescent. ***Sepals*** slightly verrucose and whitish-pubescent on exterior. ***Dorsal sepal*** 1 × 0.4–0.45 cm, oblong-elliptic, obtuse, cucullate, concave. ***Lateral sepals*** 1–1.1 × 0.6–0.65 cm (flattened), obliquely ovate-elliptic, obtuse or acute. ***Mentum*** rounded. ***Petals*** 0.7–0.75 × 0.35–0.4 cm, narrowly elliptic, obtuse, concave, glabrous. ***Lip*** 0.6–0.7 cm long, 0.5 cm deep, 0.3–0.35 cm wide (unflattened), saccate, cymbiform, adnate to column-foot, sulcate along median line on undersurface, raised inside, glabrous; side lobes broadly rounded, erect; mid-lobe reduced to a fleshy, triangular, acute beak; with a broad, fleshy swollen callus on each side of upper part of interior wall, extending from near base to base of mid-lobe, broadest and most swollen at front. ***Column*** 2.5 × 2.5 mm, raised and fleshy at base, with a few short hairs on dorsal surface and sides of stelidia; foot 3 mm long; stelidia broad, rounded; anther-cap 1 × 1 mm, globose, cucullate, slightly papillose. Plate 18D & E.

HABITAT AND ECOLOGY: Upper montane forest and scrub, often on ultramafic substrate; frequently forming dense swards beneath *Leptospermum*. Alt. 2100 to 3700 m. Flowering observed in January, February, July, August, October, November and December.

DISTRIBUTION IN BORNEO: SABAH: Mt. Kinabalu only.

GENERAL DISTRIBUTION: Endemic to Borneo.

NOTES: This robust species is confined to areas above 2100 m on Mt. Kinabalu, where it often forms continuous stands. Kitayama (1991: 42) noted that it is a dominant species in the upper subalpine forest. It is found at a higher elevation than any other orchid species in Borneo.

DERIVATION OF NAME: The specific epithet *grandis* is Latin for large, great or tall, and refers to the robust nature of the plant.

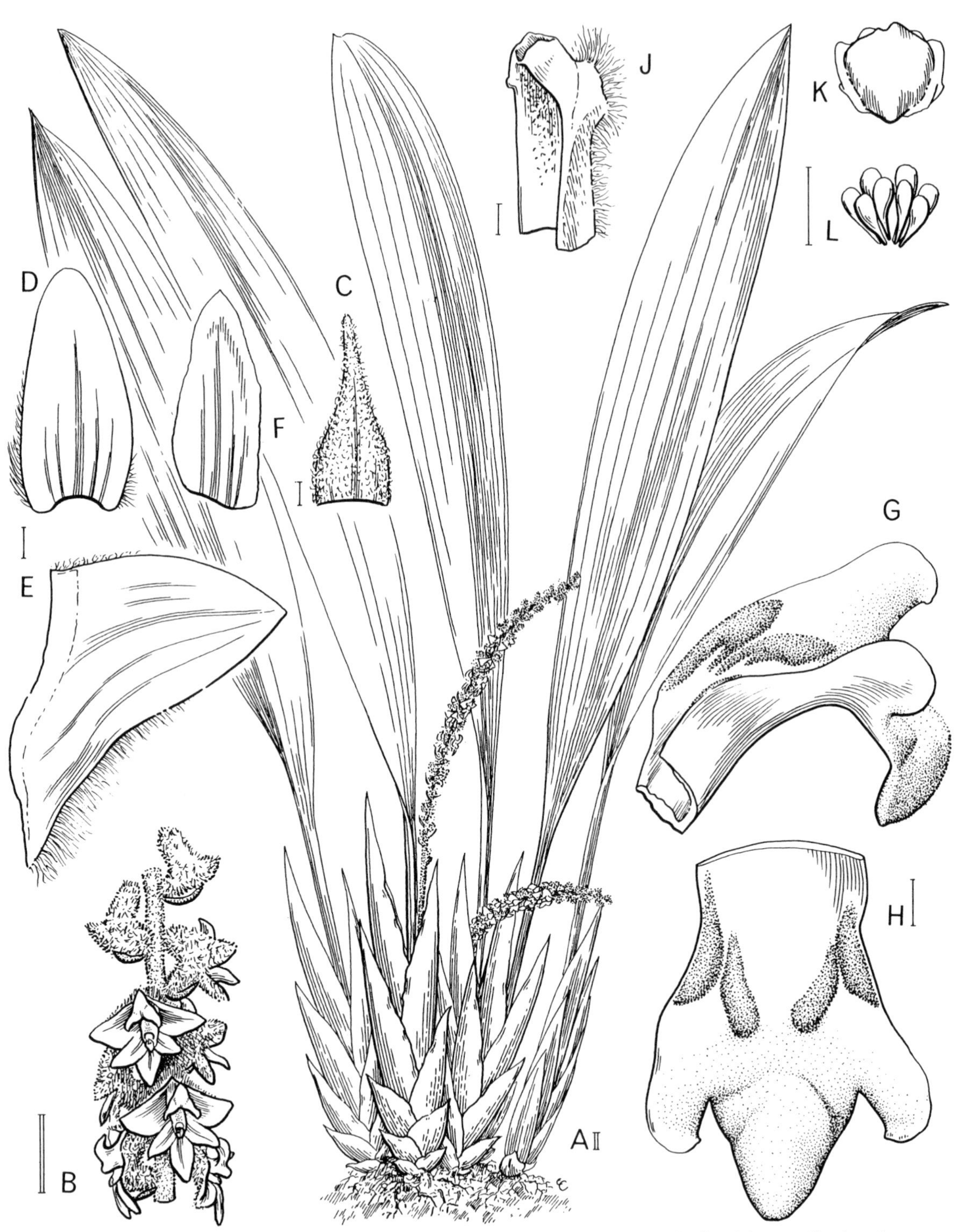

Figure 76. Eria lanuginosa J.J. Wood. - A: habit. - B: inflorescence. - C: floral bract. - D: dorsal sepal. - E: lateral sepal. - F: petal.- G: lip, side view. - H: lip, flattened. - J: column, oblique view. - K: anther-cap, back view. - L: pollinia. All drawn from *Wood* 725 (holotype) by Eleanor Catherine. Scale: single bar = 1 mm; double bar = 1 cm.

76. ERIA LANUGINOSA J.J. Wood

Eria lanuginosa *J.J. Wood* in Lindleyana 5(2): 95, fig. 10 (1990). Type: Borneo, Sabah, Sipitang District, banks of Pa Sia River near Long Pa Sia, *Wood* 725 (holotype K).

Robust, clump-forming ***epiphyte. Pseudobulbs*** 1-leaved, thick and fleshy, 7–8 cm long, covered by 6–8 imbricate, acute, pale brown cataphylls 2–15 cm long, cataphyll margins often paler fawn-coloured ***Leaves*** 30–40 × 6 cm, narrowly elliptic to oblong-elliptic, ensiform, acute, erect to spreading; petiole 10–15 cm long. ***Inflorescence*** erect or decurved, densely many-flowered, peduncle and rachis purplish olive-green, covered in a dense buff to whitish lanuginose indumentum; peduncle 6 cm long, concealed by outer sheath, arising from several imbricate basal sheaths, bearing 5–6 triangular-ovate, acuminate sterile bracts each 7–8 mm long; rachis 18–20 cm long; floral bracts 6–10 mm long, triangular-ovate, acute to acuminate, unevenly lanuginose, stiff. ***Flowers*** non–resupinate; 1.5–1.8 cm across; sepals and petals white, the mentum flushed lilac-pink; lip white flushed lilac pink at base and on the side lobes, the mid-lobe cream ageing to yellow; column white, flushed dark lilac-pink at apex; ovary and outer surface of sepals covered in a thick, furry, buff to whitish lanuginose indumentum. ***Ovary*** 5–6 mm long, sessile. ***Dorsal sepal*** 8 × 4 mm, oblong or oblong-ovate, obtuse. ***Lateral sepals*** 11 × 8–9 mm, obliquely triangular-ovate, obtuse to subacute, united to column-foot to form an obtuse ***mentum*** 6 mm long. ***Petals*** 7 × 3 mm, oblong-elliptic, obtuse. ***Lip*** strongly recurved, 1 cm long, 0.4 cm broad at base, 3-lobed; side lobes 2 × 1 mm, oblong, obtuse, erect; mid-lobe 4–4.5 × 3 mm, triangular-ovate, obtuse, thickened, papillose; disc with two central, two-lobed, slightly raised papillose calli. ***Column*** 3 mm long, oblong, with shallow rounded wings, outer surface covered in a buff to whitish lanuginose indumentum; foot 5 mm long, inner surface hirsute; anther-cap 1 × 1.5 mm, cucullate. Plate 19A & B.

HABITAT AND ECOLOGY: Riverside trees; riparian forest. Alt. 900 to 1000 m. Flowering observed in September and October.

DISTRIBUTION IN BORNEO: SABAH: Sipitang District. SARAWAK: Sungai Menalio.

GENERAL DISTRIBUTION: Endemic to Borneo.

NOTES: *E. lanuginosa* belongs to section *Hymeneria* Lindl. and is closely related to *E. hyacinthoides* (Blume) Lindl. distributed in Peninsular Malaysia, Singapore, Java, Sumatra and Bali. It differs in having pseudobulbs with one rather than two leaves, a denser lanuginose indumentum on the peduncle, rachis and flowers, and a lip lacking the two prominent crests at the base of the mid-lobe found in *E. hyacinthoides*.

DERIVATION OF NAME: The specific epithet is derived from the Latin *lanuginosus*, meaning woolly or downy, and refers to the densely woolly covering on the inflorescence.

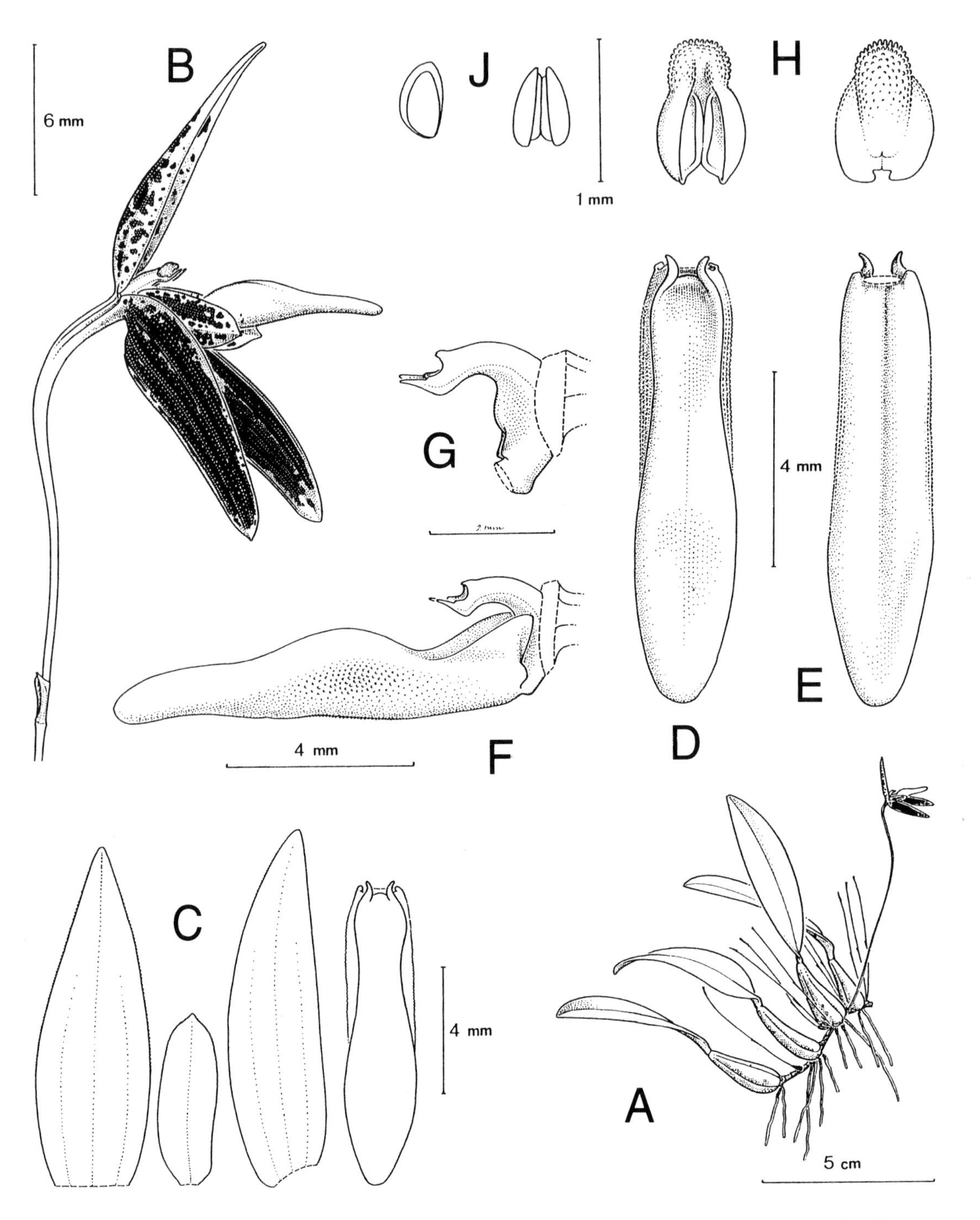

Figure 77. Hapalochilus lohokii (J.J. Verm. & A. Lamb) Garay, Hamer & Siegerist. - A: habit. - B: flower, side view. - C: flower analysis, from left to right: dorsal sepal, petal, lateral sepal, lip. - D: lip, front view. - E: lip, back view. - F: lip and column, side view.- G: anther-cap, front and back views. - H: pollinia, left: one pair, right: two pairs. All drawn from *Lamb* AL 459/86 (holotype) by Jaap Vermeulen.

77. HAPALOCHILUS LOHOKII (J.J. Verm. & A. Lamb) Garay, Hamer & Siegerist

Hapalochilus lohokii (*J.J. Verm. & A. Lamb*) *Garay, Hamer and Siegerist* in Nord. J. Bot. 14(6): 643 (1994). Type: Borneo, Sabah, *Lamb* AL569/86 (holotype L).

Bulbophyllum lohokii J.J. Verm & A. Lamb in Blumea 38 : 339, fig. 4 (1994).

Clump-forming ***epiphyte. Rhizome*** 1.5–2 mm in diameter, creeping. ***Pseudobulbs*** 0.8–2 × 0.5–0.9 cm, borne 0.7–1.2 cm apart, ovoid, hardly flattened, green, stained purple. ***Leaf blade*** 2–6 × 0.8–1.4 cm, elliptic, acute; petiole 0.5–2 mm long. ***Inflorescences*** usually borne together on a short sympodium, 3–5 cm long, 1-flowered; peduncle 2.5–4.5 cm long, purple; sterile bracts 5 or 6, 2–3 mm long; floral bracts 1.8–2.5 mm long, tubular, acute to acuminate. ***Flowers*** widely opening; pedicel with ovary purple; sepals and petals yellowish or translucent white, with dark purple veins and blotches; lip yellow, sometimes with a few tiny purple spots below, the base sometimes purplish-brown; column white. ***Pedicel*** with ***ovary*** 1.4–1.8 cm long, with the node *c.* 0.8 mm from floral bract. ***Dorsal sepal*** 10.5–13 × 3–4 mm, ovate, acute to acuminate, thin, margins finely papillose, surface glabrous. ***Lateral sepals*** 10–13 × 3.2–3.5 mm, obliquely ovate, acute to acuminate, margins slightly papillose. ***Petals*** 4–5.8 × 1.2–2 mm, elliptic, acuminate, thin, margins finely papillose near apex, surface glabrous. ***Lip*** firmly fused to column-foot, immobile, 3-lobed; side lobes inserted near base of lip, pointing backwards, triangular, obtuse, thin, glabrous; mid-lobe 8–10.5 × 1.6–2.4 mm, ± straight, but very soft, adaxially concave near base, convex towards apex, surface glabrous, abaxially without a median ridge and surface papillose halfway towards margins. ***Column*** 2–2.3 mm long; stelidia 0.5–0.7 mm long, subulate, acute, with an obscure tooth along upper margin; foot with two small lateral wings close to attachment of lip; stigma protruding at its base, slit-like; anther-cap with an abaxial ridge. Plate 19C & D.

HABITAT AND ECOLOGY: Upper montane forest; usually on small branches near the forest floor; on moss covered *Dacrydium* in open, sunny situations. Alt. 1700 to 2000 m. Flowering observed in March and from June to October.

DISTRIBUTION IN BORNEO: SABAH: Crocker Range (Mt. Alab; near Tambunan, etc.). Mt. Kinabalu.

GENERAL DISTRIBUTION: Endemic to Borneo.

NOTE: *Hapalochilus*, formerly a section of *Bulbophyllum*, is primarily New Guinean, *H. lohokii* being the first species to be recorded from Borneo.

DERIVATION OF NAME: The generic name is derived from the Greek *hapalo*, soft, and *chilos*, lip, and refers to the soft texture of the lip. The specific epithet honours Harry Lohok, Superintendant of Poring Orchid Centre near Ranau in Sabah.

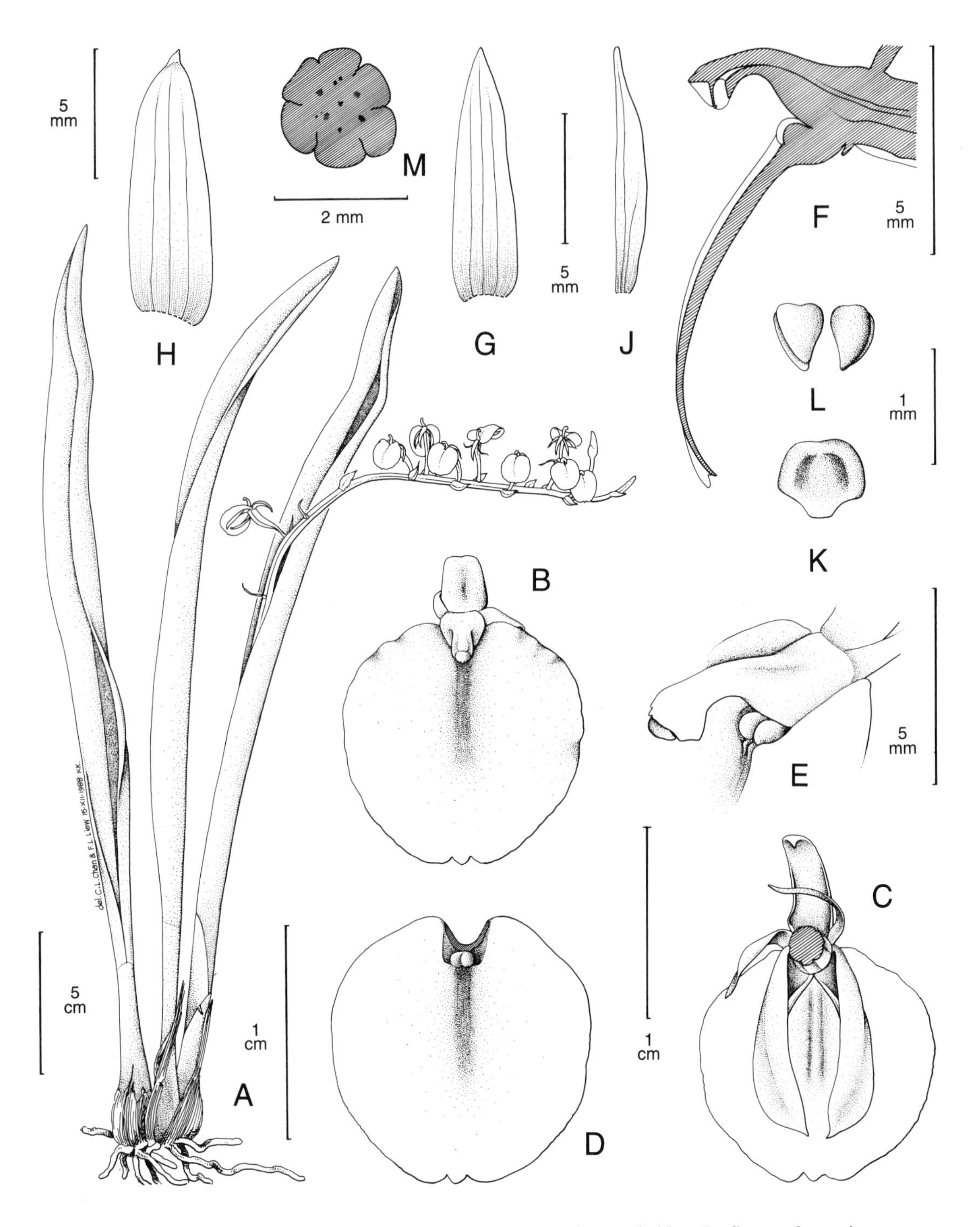

Figure 78. Liparis aurantiorbiculata J.J. Wood & A. Lamb. - A: habit. - B: flower, front view. - C: flower, back view. - D: lip, flattened. - E: column and base of lip showing calli. - F: column and lip, longitudinal section.- G: dorsal sepal. - H: lateral sepal. - J: petal. - K: anther-cap, back view. - L: pollinia. - M: ovary, transverse section. All drawn from *Chan & Gunsalam* s.n. by C.L. Chan and Lucy F.L. Liew.

78. LIPARIS AURANTIORBICULATA J.J. Wood & A.Lamb

Liparis aurantiorbiculata *J.J. Wood* in Wood, Beaman & Beaman, The Plants of Mt. Kinabalu 2, Orchids: 243, fig. 41, plate 56D (1993). Type: Borneo, Sabah, Mt. Kinabalu, Pinosuk Plateau, *Lamb* AL22/82 (holotype K).

Erect ***epiphyte. Rhizome*** thick and tough, concealed by sheaths. ***Pseudobulbs*** 6–11 cm long, to 0.5 cm wide above, 1 cm wide at base, attenuate above, swollen below, enclosed by 2–3 green to pale brown acuminate cataphylls 6–11 cm long, unifoliate. ***Leaves*** (28–)36–46 × 1.7–2.2 cm, jointed, ensiform, acute, conduplicate toward base, erect to spreading. ***Inflorescence*** a stiffly erect, laxly 12- to 15-flowered raceme, as long as or shorter than the leaves; peduncle 9–12 cm long, terete; rachis 11–16 cm long, 4-winged; floral bracts 0.7–1.2 × 0.3–0.4 cm, ovate, acute to acuminate. ***Flowers*** porrect to spreading; orange, with a yellowish-green column. ***Pedicel*** with ***ovary*** 1.6–2 cm long, clavate. ***Dorsal sepal*** and ***petals*** strongly reflexed. ***Dorsal sepal*** 1–1.1 × 0.2–0.25 cm, narrowly triangular-ovate or narrowly oblong-ovate, acute. ***Lateral sepals*** 1.1 × 0.3 cm, oblong, twisted, apex carinate, subulate, hidden behind lip. ***Petals*** 1 × 0.2–0.3 cm, linear, acute, sometimes obliquely linear. ***Lip*** 1.1–1.2 × 1.1–1.2 cm, broadly ovate to orbicular, apex shallowly retuse, with a triangular, acute tooth in the sinus, clasping column at base, flat, margin minutely erose, upper and lower surface minutely papillose, particularly in apical half, disc with a small, fleshy bilobed basal callus. ***Column*** 4.5 mm long, thickened and 2 mm wide at base, narrow and curved in middle, apex with short, rather truncate wings; anther-cap 0.5 × 0.6–0.7 mm, cucullate; pollinia four, in two groups. Plate 19E & F.

HABITAT AND ECOLOGY: Lower montane forest. Alt. 1200 to 1700 m. Flowering observed in November and December.

DISTRIBUTION IN BORNEO: SABAH: Mt. Kinabalu.

GENERAL DISTRIBUTION: Endemic to Borneo.

NOTES: *L. aurantiorbiculata* belongs to section *Coriifoliae* and is closely related to *L. grandis*, so far only known from Mt. Kinabalu, but may be distinguished by its much longer pseudobulbs, narrower leaves, shorter clavate pedicel with ovary, broader triangular-ovate to oblong-ovate sepals and broader, broadly ovate to orbicular lip which has a shallowly retuse, never abruptly mucronate, apex and a small fleshy bilobed basal callus.

DERIVATION OF NAME: The generic name is derived from the Greek *liparos,* fat, greasy or shining, refering to the smooth shiny leaves of many species. The specific epithet is derived from the Latin *aurantiacus*, orange, and *orbicularis*, circular, in reference to the lip.

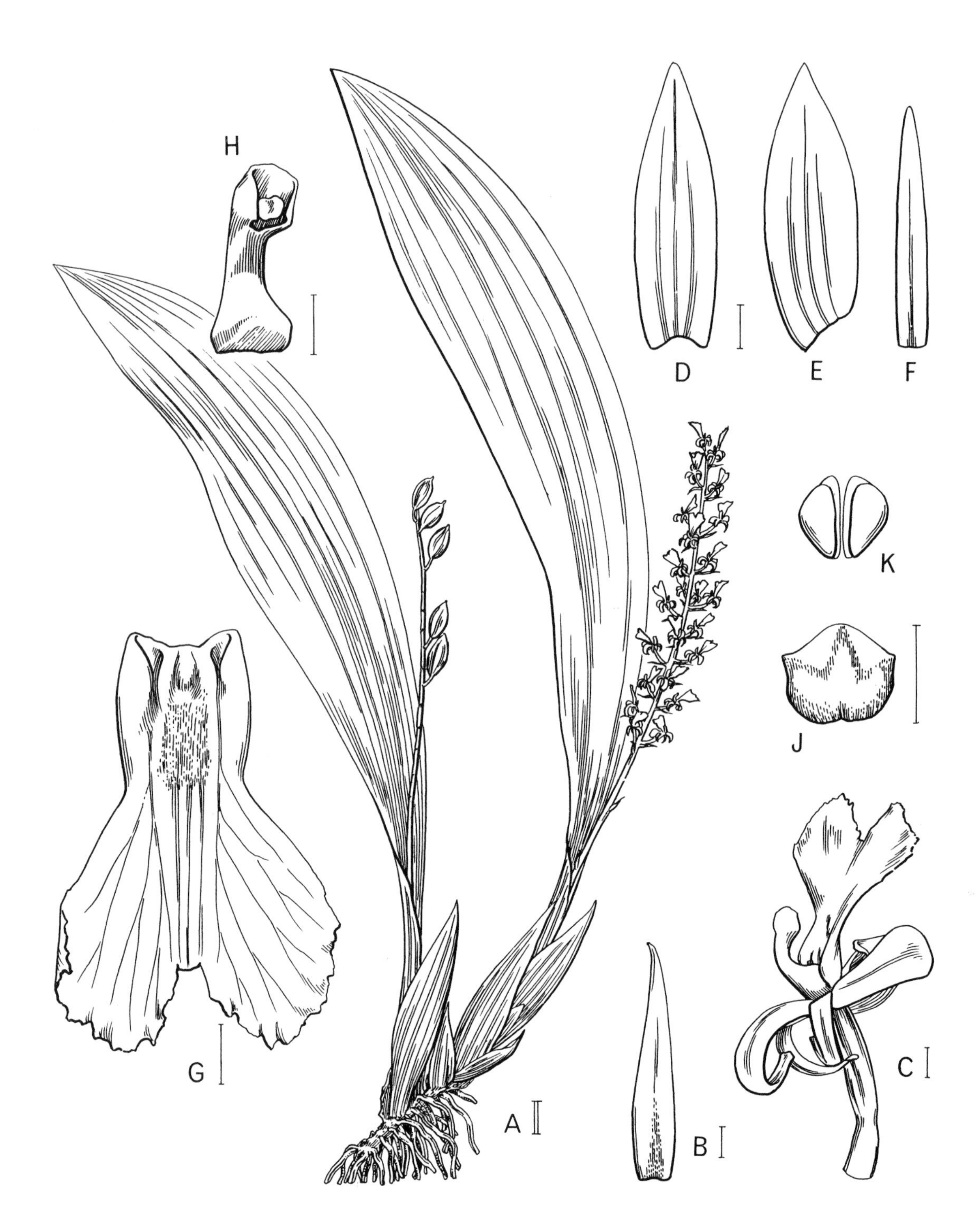

Figure 79. Liparis kinabaluensis J.J. Wood. - A: habit. - B: floral bract. - C: flower, side view. - D: dorsal sepal. - E: lateral sepal. - F: petal.- G: lip, front view. - H: column, front view. - J: anther-cap, back view. - K: pollinia. A (habit) drawn from *J. & M.S. Clemens* 30141, 50099 & 51462 and B-K from *Brentnall* 124 by Eleanor Catherine. Scale: single bar = 1 mm; double bar = 1 cm.

79. LIPARIS KINABALUENSIS J.J. Wood

Liparis kinabaluensis *J.J. Wood* in Lindleyana 5(2): 84, fig. 2 (1990). Type: Borneo, Sabah, Mt. Kinabalu, Penibukan Ridge, east of Dahobang River, 1200 m, 2 November 1933, *J. & M.S. Clemens* 50099 (holotype K).

Epiphyte, sometimes ***lithophytic. Stem*** to 5 cm or more long, creeping. ***Pseudobulbs*** 0.8–1 × 1–1.2 cm, 1-leaved, ovoid, enclosed in papery brown ovate, acute, carinate, imbricate cataphylls 2–6 cm long. ***Leaves*** smooth, coriaceous, fresh green; blade 16 × 4–32 × 6 cm, oblong-elliptic, apex acute, subulate-apiculate, conduplicate towards base; sheaths 1 cm long, enclosed by 3–4 imbricate, ovate, acute cataphylls 3–9 cm long. ***Inflorescence*** many-flowered, subdense; peduncle and rachis obscurely sulcate, pale green; peduncle 5–8 cm long, emerging from an acute, carinate basal sheath 2 cm long, bearing 3–5 acuminate sterile bracts 7–8 mm long; rachis 4.5–10 cm long; floral bracts 7–9 mm long, triangular-ovate, acuminate, thin-textured, whitish-green. ***Flowers*** faintly scented of cucumber; non-resupinate or with lip held in horizontal plane on same inflorescence; cream, the lip with an orange flush at the base. ***Pedicel*** with ***ovary*** 7–8 mm long, curved distally. ***Dorsal sepal*** 6 × 2 mm, oblong-ovate, obtuse, recurved. ***Lateral sepals*** 5.5–6 × 2.5 mm, oblong-ovate, acute, recurved. ***Lip*** flat, not geniculate-deflexed, flabellate, apex bilobed, 7 mm long, 2.5 mm wide at base, 5 mm wide across lobules, base rather fleshy, with semi-erect sides and a fleshy quadrate concave callus, apical lobules rounded, erose. ***Column*** 3 mm long, 1 mm wide at base, apex with obscure rounded wings; anther-cap ovate, cucullate. Plate 20A & B.

HABITAT AND ECOLOGY: Hill forest on ultramafic substrate; lower montane forest; rocky places. Alt. 1000 to 1800 m. Flowering observed in February, October and November.

DISTRIBUTION IN BORNEO: SABAH: Mt. Kinabalu.

GENERAL DISTRIBUTION: Endemic to Borneo.

NOTE: *L. kinabaluensis* is closely related to *L. monophylla* Ames from Leyte Island in the Philippines but is distinguished by its much shorter, ovoid pseudobulbs and larger flowers with a flabellate, bilobed lip.

DERIVATION OF NAME: The specific epithet refers to Mt. Kinabalu, the type locality.

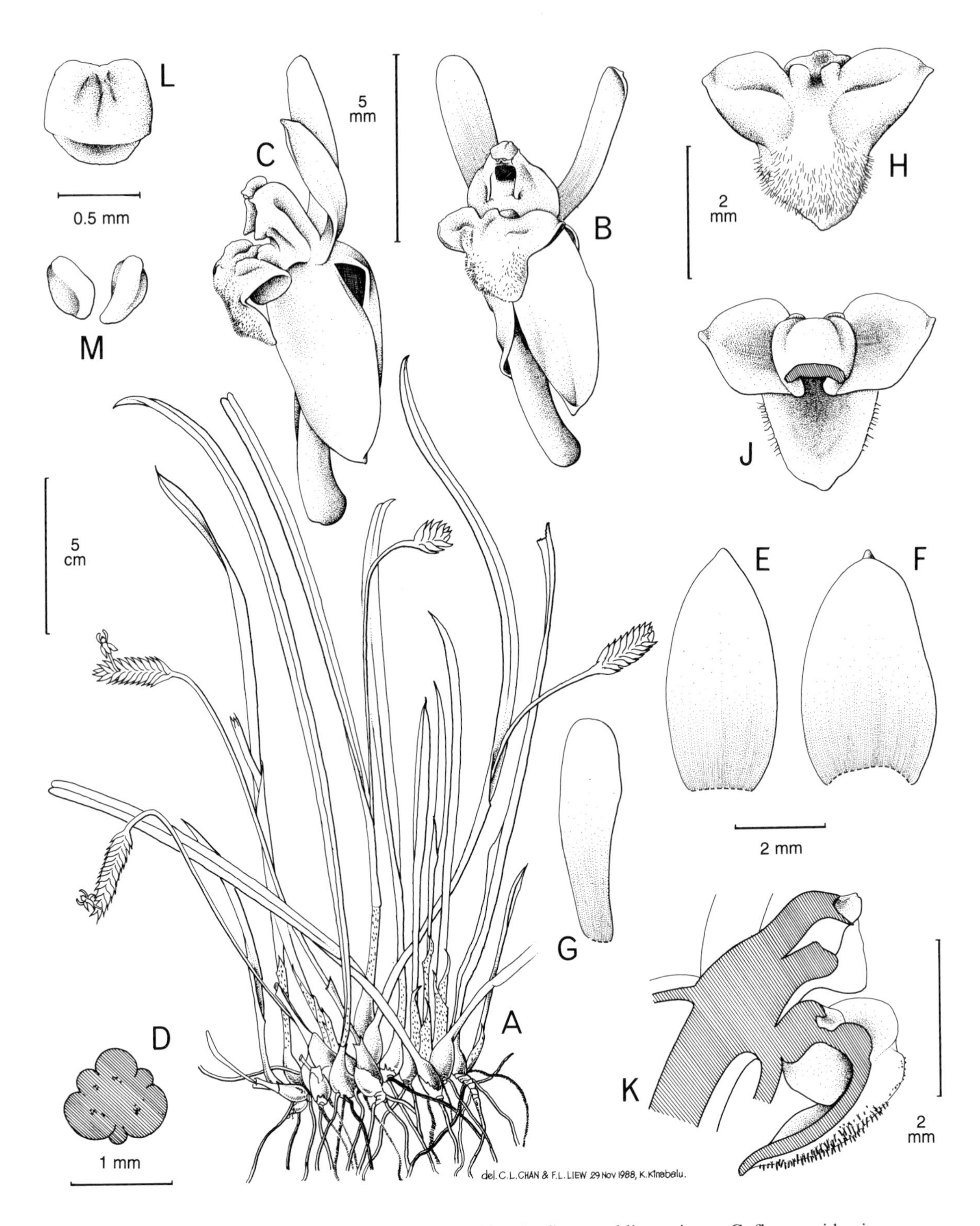

Figure 80. Liparis lobongensis Ames. - A: habit. - B: flower, oblique view. - C: flower, side view. - D: ovary, transverse section. - E: dorsal sepal. - F: lateral sepal.- G: petal. - H: lip, front view. - J: lip, back view. - K: pedicel with ovary, column and lip, longitudinal section. - L. anther-cap, back view. - M. pollinia. All drawn from *Chan* 121/89 by C.L. Chan and Lucy F.L. Liew.

80. LIPARIS LOBONGENSIS Ames

Liparis lobongensis *Ames,* Orchidaceae 6: 92 (1920). Type: Borneo, Sabah, Mt. Kinabalu, Lubang (Lobong), October 1915, *J. Clemens* 219 (holotype AMES).

Epiphyte. Roots wiry, elongate, forming a dense mass. ***Rhizome*** creeping, *c.* 2 mm in diameter. ***Cataphylls*** 3–7, 0.5–4.8 cm. long, imbricate, ovate-elliptic, acuminate, speckled blackish brown. ***Pseudobulbs*** (0.8–)1.5–5 × 0.3–0.4 cm, approximate, pyriform below, long-attenuated distally, rugose, concealed by cataphylls. ***Leaves*** (10–)18–36 × 0.5–0.7 cm, articulated, linear, ligulate, grass-like, acute, apiculate, attenuated below, thin-textured. ***Inflorescence*** with flowers opening one or rarely two at a time in succession; peduncle shorter than leaf, 8–24 cm long, erect, conspicuously two-winged, particularly so distally, naked; rachis 1.5–4.5 cm long, elongating during anthesis; floral bracts 6–9 mm long, distichous, lanceolate acute, rigid, persistent, obliquely ascending. ***Flowers*** unscented, greenish, yellow or cream lip brownish red on reverse, column reddish. ***Pedicel*** with ***ovary*** 4–5 mm long. ***Dorsal sepal*** 5.5 × 2.3–2.4 mm, oblong-elliptic, acute, reflexed. ***Lateral sepals*** 5.5 x 2.5 mm, oblong-ovate, obtuse and mucronate, margins often revolute, reflexed. **Petals** 6 × 1–1.25 mm, linear-oblanceolate, obtuse, attenuated below, erect. ***Lip*** 5 mm long, 3-lobed; side lobes auriculate, thickened with a convex bullate callus; mid-lobe 3 × 2 mm, sharply deflexed, oblong-ovate, lingulate, truncate or rounded, margins and distal half papillose, middle of upper half smooth, depressed; disc with a 2- to 3-lobed callus near column. ***Column*** 2 mm long, *c.* 2 mm wide at base, stout, with two triangular wings below stigma; anther-cap quadrate, smooth. Plate 20C & D.

HABITAT AND ECOLOGY: Lower montane forest, sometimes on ultramafic substrate. Alt. 800–2300 m. Flowering observed in February, May, September, October and November.

DISTRIBUTION IN BORNEO: SABAH: Mt. Kinabalu.

GENERAL DISTRIBUTION: Endemic to Borneo.

NOTES: *L. lobongensis* belongs to section *Distichae* which is particularly diverse in New Guinea and distinguished by having rigid distichous floral bracts and flowers appearing in succession over a long period. *L. gibbosa* Finet is the most widespread species, ranging from Myanmar (Burma) to Samoa and frequently found in Borneo. Another Bornean endemic, *L. lingulata* Ames, is closely related to *L. lobongensis* but distinguished by its peduncle which is always longer than the leaf, and the salmon flowers with a smaller triangular-lingulate lip.

DERIVATION OF NAME: The specific epithet refers to the type locality.

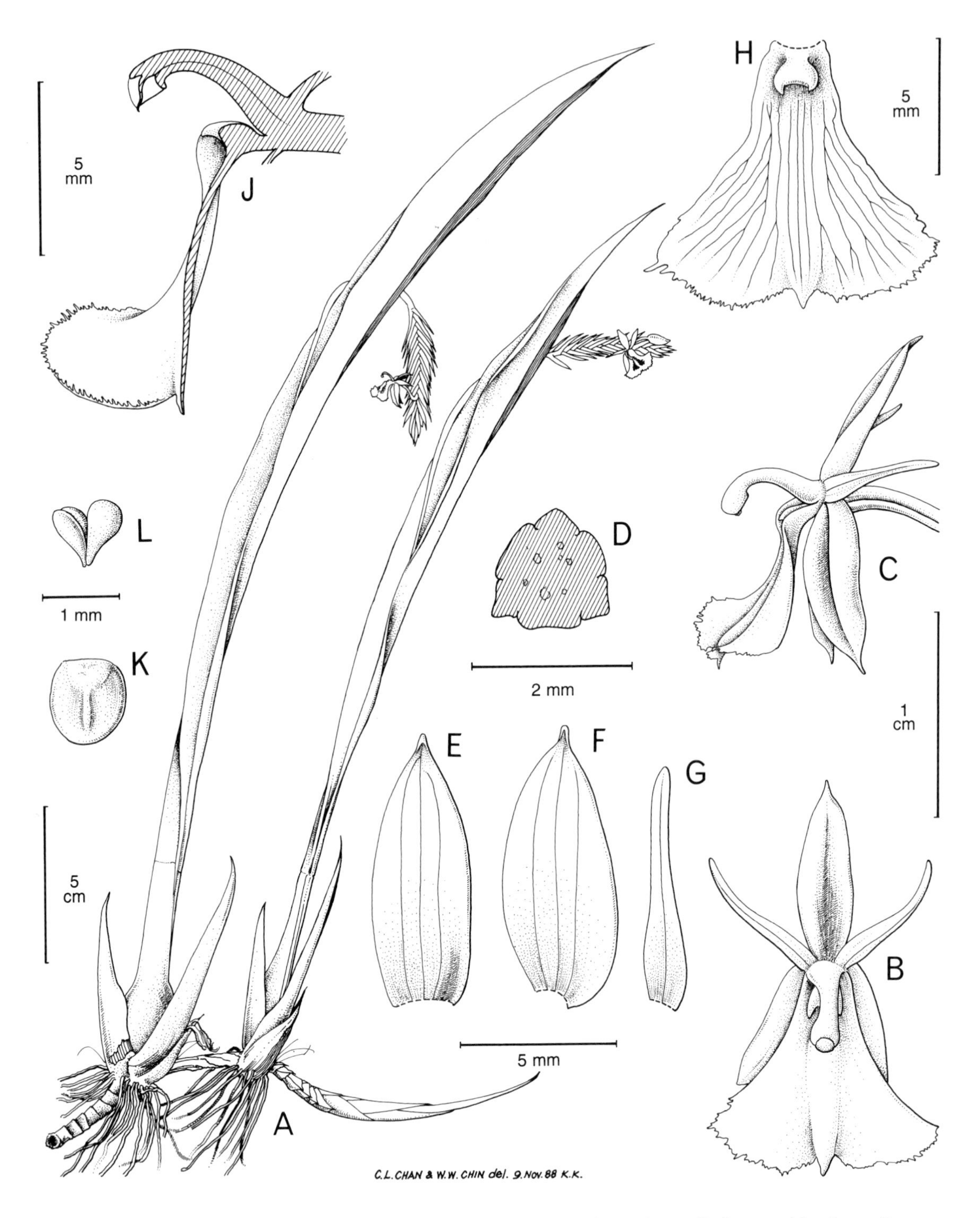

Figure 81. Liparis pandurata Ames. - A: habit. - B: flower, front view. - C: flower, side view. - D: ovary, transverse section. - E: dorsal sepal. - F: lateral sepal.- G: petal. - H: lip, front view. - J: pedicel with ovary, lip and column, longitudinal section. - K: anther-cap, back view. - L: pollinia. All drawn from *Chan* s.n. by C.L. Chan and Chin Wan Wai.

81. LIPARIS PANDURATA Ames

Liparis pandurata *Ames*, Orchidaceae 6 : 94 (1920). Type: Borneo, Sabah, Mt. Kinabalu, Lubang (Lobong), *J. Clemens* 117 (holotype AMES).

Epiphyte. Rhizome 18 cm or more long, creeping, branching, covered in numerous imbricate, acuminate sheaths 1–1.5 cm long which become fibrous with age. **Pseudobulbs** 3.5–8.5 cm long, 0.8–1.2 cm wide near base, borne 2.5–8.5 cm apart, elongate, pyriform, attenuated above into a slender complanate neck, 1-leaved, concealed by 3–4 diverging ensiform cataphylls up to 8 cm long. ***Leaf blade*** 18–38 × 1.5 – 3 cm, ligulate, acute, attenuate below, deciduous; petiole 1–3 cm long, conduplicate. ***Inflorescence*** shorter than leaf; peduncle 12–25 cm long, up to 4 mm wide below rachis, conspicuously 2-winged, naked; rachis 2.5–9.5 cm long; floral bracts 5–7 mm long, distichous, falcate, acute, conduplicate, rigid, obliquely ascending, closed at apex. ***Flowers*** opening in succession, usually one at a time; unscented; described as dark peach, dark salmon, orange, brownish orange or dull greenish salmon, the lip brighter salmon-pink, darker at the centre; column salmon-pink, apex green. ***Pedicel*** with ***ovary*** 1–1.2 cm long, narrow. ***Sepals*** reflexed, pendent. ***Dorsal sepal*** 8–9 × 3 mm, oblong, apex acute and dorsally slightly carinate. ***Lateral sepals*** 9 × 3.5 mm, obliquely oblong, apex apiculate and dorsally carinate. ***Petals*** 6.5–8 mm long, 1.9 mm wide at base, narrower above, linear-ligulate, acute, reflexed. ***Lip*** 8–10 × 12–13 mm, 3.5 mm wide at base, pandurate, retuse, apiculate, margin irregularly serrate-denticulate, ascending for 1.5 mm, then sharply reflexed; disc with a bicornute basal callus. ***Column*** 7 mm long, 2 mm wide at base, narrow above, then broader at apex, strongly arcuate above middle; anther-cap cucullate. Plate 20E.

HABITAT AND ECOLOGY: Oak–laurel forest; lower montane forest; swamp forest. Alt. 1200 to 2600 m. Flowering observed from February to July, and September and December.

DISTRIBUTION IN BORNEO: SABAH: Mt. Kinabalu; Sipitang district, Ulu Long Pa Sia.

GENERAL DISTRIBUTION: Endemic to Borneo.

DERIVATION OF NAME: The specific epithet is derived from the Latin *panduratus*, fiddle-shaped, in reference to the lip.

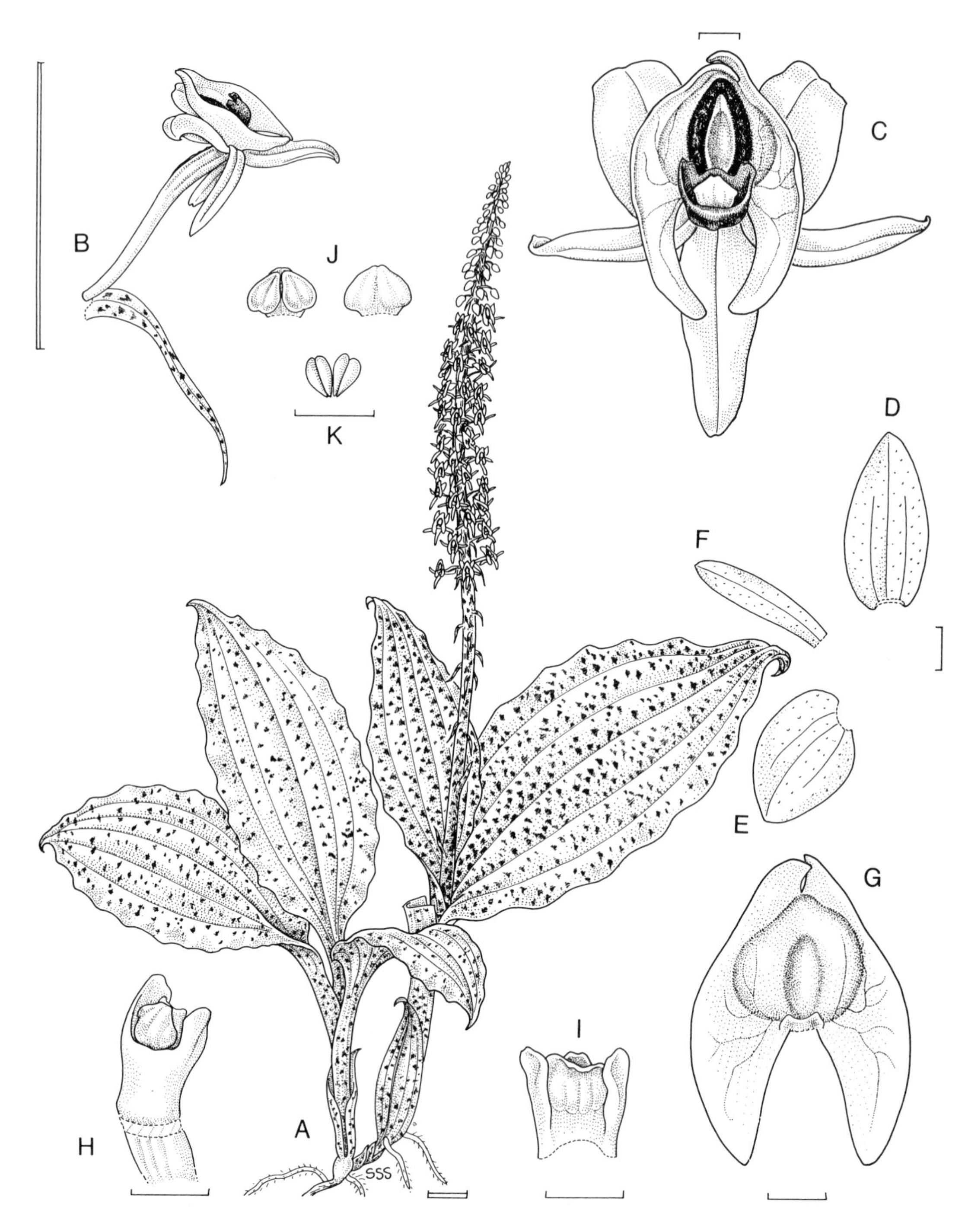

Figure 82. Malaxis punctata J.J. Wood. - A: habit. - B: floral bract and flower, side view. - C: flower, front view. - D: dorsal sepal. - E: lateral sepal. - F: petal. - G: lip, front view. - H: upper portion of ovary, column and anther-cap, oblique view. - I: column, front view. - J: anther-cap, inner and back views. - K: pollinia. All drawn from *Sands* 4011 (holotype) by Susanna Stuart-Smith. Scale: single bar = 1 mm; double bar = 1 cm.

82. MALAXIS PUNCTATA J.J. Wood

Malaxis punctata *J.J. Wood* in Kew Mag. 4(2): 76, plate 78 (1987). Type: Borneo, Sabah, Mt. Kinabalu, between Kiau and Dahobang River, *Sands* 4011 (holotype K).

Erect, glabrous ***terrestrial*** 12–14 cm high. ***Stems*** partially thickened into 1- or 2-noded, weak, fleshy, pale green ***pseudobulbs***, each 1.5–2.5 cm long, 1–1.2 cm broad, entirely enclosed by sheathing leaf-bases when immature, with roots appearing from the nodes, and with 1–3 acute, pale green, purple-spotted, imbricate basal sheaths 1.8–4 cm long. ***Cataphylls*** leafy, 4–7 cm long (including petiole), blade 1–1.6 cm broad, ovate-elliptic, acute, slightly undulate, veins prominent, particularly on the petiole, entirely purple-spotted, spots gradually fading on upper surface. ***Leaves*** 3, 1 large, 2 smaller, 5–12 × 2.3–5 cm, ovate-elliptic, acute to acuminate, asymmetrical at the base, passing into a sheathing petiole 2.5–4 cm long, thin-textured, plicate, margin undulate, veins prominent below, pale green and prominently spotted purple when young, the mature leaves darker green, paler below, spots less distinct above, but remaining prominent below, rarely few in number. ***Inflorescence*** terminal, a dense many-flowered raceme up to 17 cm long; peduncle 3–3.5 cm long, angular, pale green, flecked purple; rachis angular, paler green or purple, unmarked; floral bracts 3–5 mm long, linear-triangular to subulate, acuminate, pale green or purple, becoming reflexed and adpressed to the rachis. ***Flowers*** non-resupinate, lip borne uppermost, the lowermost remote; scented of fresh cucumber; pale green fading to straw, or purple, the lip with a brown horseshoe-shaped border to the fovea; column dark green. ***Pedicel*** with ***ovary*** slender, up to 5 mm long, curved and drooping in bud, becoming horizontal. ***Dorsal sepal*** 3.5–4 × 1 mm, oblong-elliptic, obtuse, revolute, erect. ***Lateral sepals*** 3 × 1.8 mm, ovate-elliptic, obtuse, slightly revolute, spreading. ***Petals*** 3 × 0.5 mm, linear, obtuse, strongly revolute, spreading. ***Lip*** 4–6 × 3–3.5 mm, sagittate, auricles ± obtuse, apical portion bilobulate, lobes short, obtuse and overlapping, fovea narrowly elliptic. ***Column*** 1 mm long, with porrect, oblong, obtuse wings; anther-cap minute, ovate, obtuse; pollinia four, in two groups. Plate 21A–D.

HABITAT AND ECOLOGY: Mixed hill-dipterocarp forest often with a palm understorey; mossy forest; oak-chestnut forest; lower montane forest, often on ultramafic substrate. Alt. 800 to 1800 m. Flowering observed in May, June, July, September, October and December.

DISTRIBUTION IN BORNEO: SABAH: Crocker Range, Sinsuron road; Mt. Kinabalu; Sipitang District.

GENERAL DISTRIBUTION: Endemic to Borneo.

NOTES: *Malaxis* are shade-loving plants and consequently are typical inhabitants of the humid forest floor environment where they grow in the deep accumulations of leaf litter. Most species are restricted to lowland and mid-zones, becoming rarer at higher elevations. *M. punctata* is closely related to *M. calophylla* (Rchb.f.) Kuntze, an attractive species native to Sikkim, Myanmar (Burma) and Peninsular Malaysia to Thailand, Cambodia and Borneo. The leaves of *M. calophylla,* however, are distinctly variegated, having a broad, dark bronze-green central band surrounded by a paler greenish-cream margin. Spotting is restricted to this paler green margin and to the undersurface.

DERIVATION OF NAME: The generic name *malaxis* is Greek and means softening, referring to the soft and tender texture of the stems and leaves. The specific epithet is derived from the Latin *punctatus*, dotted or spotted, in reference to the leaves.

Figure 83. Neuwiedia zollingeri Rchb.f. var. **javanica** (J.J. Sm.) de Vogel. - A: habit. - B: flower, side view. - C: lateral sepal. - D: close-up of glandular and non-glandular hairs on lateral sepal. - E: ovary, transverse section. - F: close-up of glandular hairs on ovary. - G: pedicel with ovary, stamens and style, side view. - H: distal portion of style showing stigma, side and oblique views. All drawn from *Beaman* 7416 by Susanna Stuart-Smith. Scale: single bar = 1 mm; double bar = 1 cm.

N.B. Fresh flowers in liquid preservative were unavailable during the preparation of this figure. Details of the lip, etc. were unable to be discerned from the poorly preserved herbarium material examined.

83. NEUWIEDIA ZOLLINGERI Rchb.f. var. JAVANICA (J.J. Sm.) de Vogel

Neuwiedia zollingeri *Rchb.f.* in Bonplandia 5: 58 (1857). Type: Java, *Zollinger* 2808 (holotype W).

var. **javanica** (*J.J. Sm.*) *de Vogel* in Blumea 17(2): 329 (1969). Type: Java, Djampang Tengah, Artana, *J.J. Smith* s.n. (lectotype L, isolectotype BO).

Neuwiedia javanica J.J. Sm. in Bull. Jard. Bot. Buitenzorg, ser. 2, 14 : 5 (1914).

Terrestrial. Stem 10–15 cm long, enclosed by disintegrating leaf sheaths below, producing numerous often woolly aerial roots up to 4 mm in diameter. ***Leaves*** spreading; blade 15–35 × 1–7.5 cm, broadly elliptic to linear, acuminate, papery-textured, plicate, glabrous; petiole 5–15 cm long, sheathing at base. ***Inflorescence*** usually entirely hairy, sometimes bracts only hairy, hairs of two types: long and glandular mainly on the sepals, glandless usually on upper leaves, rachis, bracts and pedicel with ovary; peduncle 8–15 cm long, bearing usually *c.* 3 ovate, acute to acuminate sterile bracts 2–4 cm long, rachis 6–17 cm long, densely 30- to 60-flowered, flowers borne 3–5 mm apart; floral bracts 1–2 × 0.2–0.8 cm, lanceolate to linear, acuminate. ***Flowers*** up to 3 cm long, fragrant; yellow. ***Pedicel*** with ***ovary*** 6–11 cm long, ovary ellipsoid. ***Dorsal sepal*** 18–20 × 3.5–4.5 mm, linear or linear-lanceolate, mid-vein projecting to form a 1–1.5(–2.3) mm long apical cusp. ***Lateral sepals*** 18.5–20 × 4 mm, cusp 1.4–2.2 mm long, obovate-lanceolate. ***Petals*** 16–19 × 5 mm, obovate-oblong to oblanceolate, cusp 0.5–1(–1.2) mm long. ***Lip*** 17–19 × 5–6 mm, obovate-oblong to oblanceolate, cusp 1–1.2 mm long. ***Column*** 4–6 × 0.5–1 mm; lateral anthers 5–7.5 mm long; median anther 3.5–6 mm long, the long loculus at base 0.2–0.4 mm longer; free part of style 5–6 mm long. ***Fruit*** a fleshy, white, berry-like capsule, 8–10 × 6–9 mm, with a beak 2–3 mm long. Plate 22A & B.

HABITAT AND ECOLOGY: Hill forest on ultramafic substrate, with *Gymnostoma sumatrana*, *Agathis*, etc. Alt. 600 to 1000 m. Flowering observed in January, February, March and November.

DISTRIBUTION IN BORNEO: SABAH: Mt. Kinabalu; Telupid area.

GENERAL DISTRIBUTION: Sumatra, Java, Bali and Borneo.

NOTES: *Neuwiedia*, containing eight species, belongs, together with *Apostasia*, to the subfamily Apostasioideae which are usually considered to be the most primitive orchids. Unlike the majority of orchids, which only have one fertile stamen, *Neuwiedia* has three. The pollen grains are powdery and granular and never aggregated into pollinia, which is otherwise universal, except in some slipper orchids. The typical variety is native to Java and differs in being totally glabrous and having a contracted raceme and shorter lateral anthers.

DERIVATION OF NAME: The generic name is dedicated to Prince Maximilian von Neuwied (1782–1867). The specific epithet honours the Swiss botanist and explorer Heinrich Zollinger (1818–1859). The varietal epithet refers to the island of Java.

Figure 84. Paphiopedilum kolopakingii Fowlie. - A: habit. - B: leaf apex. - C: part of inflorescence. - D: dorsal sepal. - E: synsepal. - F: petal.- G: lip, longitudinal section. - H: column, side view. - J: column, oblique view. - K: anthers. A–C drawn from a cultivated specimen at RBG Kew and D–K from *Fowlie* s.n. by Judi Stone. Scale: single bar divided into mm; double bar divided into cm.

84. PAPHIOPEDILUM KOLOPAKINGII Fowlie

Paphiopedilum kolopakingii *Fowlie* in Orchid Digest 48: 41 (1984). Type: Borneo, Kalimantan, cult. *Kolopaking* LKW82K1 (holotype UCLA).

P. topperi Braem & Mohr in Schlechteriana 1(2): 15 (1988) and Orchid Digest 53(4): 155–156 (1989). Type: Borneo, Kalimantan, cult. *Topper* in Herb. Schlechter Inst. 23/10.1987/01 (holotype Schlechter Inst. Lahnau).

Large ***terrestrial*** with a short stem and clustered growths on a short rhizome. ***Roots*** 5–8 mm in diameter, fleshy, pubescent. ***Leaves*** 8–10 arranged in a fan, 40–80 × 6–12 cm, coriaceous, linear-ligulate, rounded at apex. ***Inflorescence*** 40–70 cm long, laxly 6- to 14-flowered; peduncle and rachis terete, suberect, purple, densely pubescent; floral bracts 4–6 cm long, elliptic-lanceolate, acuminate, ochre striped with purple. ***Flowers*** 7–10 cm across, opening simultaneously; sepals creamy-pale yellow with longitudinal purple stripes; petals and lip yellow with ochre venation; staminode yellow. ***Pedicel*** with ***ovary*** 5.5–6.5 cm long, glabrous. ***Dorsal sepal*** 3.5–6.5 × 2–3.5 cm, elliptic-ovate, acuminate. ***Synsepal*** 3.5–4.7 × 2–2.6 cm, similar in shape to dorsal sepal. ***Petals*** 5–7 × 0.6–0.8 cm, linear-tapering, rounded at tip, deflexed at 45 degrees to the horizontal, half twisted and finely glandular-pubescent in apical part. ***Lip*** 4.1–6 × 0.9–2 cm, deeply slipper-shaped, with acuminate auricles inserted within. ***Column*** 6–7 mm long; staminode 9–15 × 6–10 mm, subquadrate-ovate, slightly retuse or obtuse at apex, pubescent on upper and side margins. Plate 22C & D.

HABITAT AND ECOLOGY: Steep sloping ground in hill and lower montane forest; growing between stones over steep river gorges. Alt. 600 to 1100 m. Flowering details withheld.

DISTRIBUTION IN BORNEO: KALIMANTAN.

GENERAL DISTRIBUTION: Endemic to Borneo.

NOTES: *P. topperi* is here reduced to synonymy within *P. kolopakingii*. The type of *P. topperi*, which is well illustrated in the journals cited above, has broader leaves, a more boldly striped and erect dorsal sepal, and a lip that is somewhat retuse at the apex but otherwise agrees well with that of *P. kolopakingii*.

DERIVATION OF NAME: The generic name is derived from the Greek *Paphia*, of Paphos; epithet of Venus or Aphrodite (Paphos is a town in S.W. Cyprus near where the Goddess of Love was supposed to have been born) and *pedilon*, a sandal, alluding to the slipper-shaped lip, i.e. slipper of Venus. The specific epithet honours A. Kolopaking (Liem Khe Wie), proprietor of Simanis Orchids in East Java, who cultivated the type plant.

85. PAPHIOPEDILUM LOWII (Lindl.) Stein

Paphiopedilum lowii (*Lindl.*) *Stein*, Orchideenbuch: 476 (1892). Type: Borneo, *Low* s.n. (holotype K).

Cypripedium lowii Lindl. in Gard. Chron. 1847: 765 (1847).

Cordula lowii (Lindl.) Rolfe in Orchid Rev. 20: 2 (1912), as *C. lowiana*.

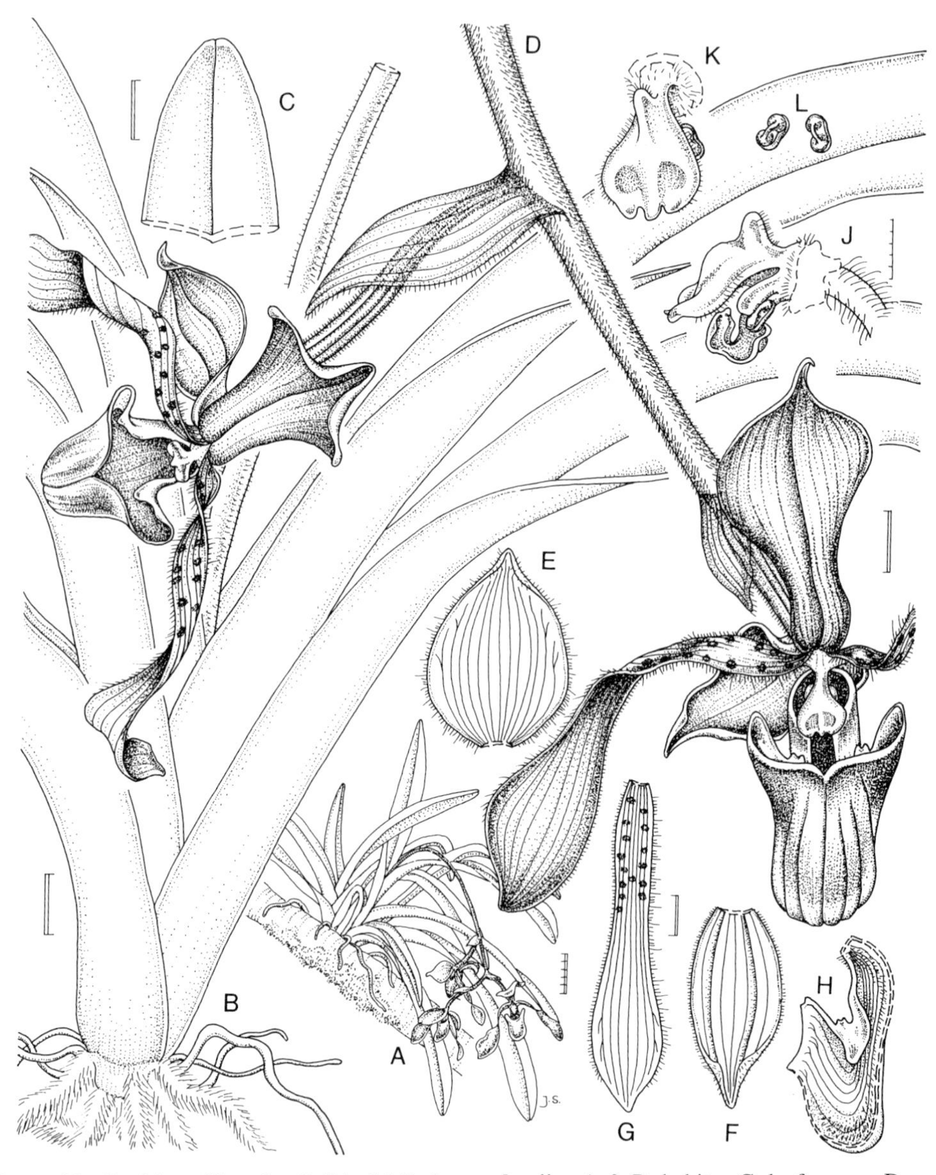

Figure 85. Paphiopedilum lowii (Lindl.) Stein var. **lowii**. - A & B: habit. - C: leaf apex. - D: part of inflorescence. - E: dorsal sepal. - F: synsepal.- G: petal. - H: lip, longitudinal section. - J: column, side view. - K: staminode, front view. - L: anthers. A–D drawn from *Chew & Corner* RSNB 7062 and *Lamb* 4, E–L from *Collenette* s.n. by Judi Stone. Scale: single bar divided into mm; double bar divided into cm.

var. **lowii**

Epiphyte, rarely ***lithophytic. Leaves*** 4–6, 22–40 × 2.8–6 cm, linear-ligulate, unequally roundly bilobed at apex, mid-green. ***Inflorescence*** erect-arcuate, 3- to 7-flowered; peduncle up to 50 cm long, green, mottled purple, shortly pubescent; floral bracts 2–4.5 × 2.2 cm, elliptic, obtuse, yellow, marked with purple, pubescent. ***Flowers*** 9–14 cm across; pedicel with ovary greenish; dorsal sepal pale green, mottled dark purple in the basal half; synsepal pale green; petals pale yellow with a purple apical third and maroon-spotted in the basal two-thirds; lip dull ochre-brown; staminode pale ochre to brownish green. ***Pedicel*** with ***ovary*** 4.5–7 cm long, pubescent, long rostrate. ***Dorsal sepal*** 3.3–5.5 × 2.5–3.2 cm, elliptic-ovate, obtuse, undulate and ciliate on margins, with recurved basal margins. ***Synsepal*** 2.2–4 cm, elliptic, obtuse, two-keeled on outer surface. ***Petals*** 5–9 × 1.5–2 cm, often once-twisted in middle, spathulate, subacute to obtuse, ciliate. ***Lip*** 3.5–4 × 2.7 cm. ***Staminode*** 10 × 7 mm, obovate, apically three-toothed with a long erect hook at the base. Plate 23A & B.

HABITAT AND ECOLOGY: Epiphytic in crutches of large branches of trees in riverine and lower montane forest; moss- or humus-filled hollows of rocks and boulders, especially limestone. Alt. 200 to 1700 m. Flowering details withheld.

DISTRIBUTION IN BORNEO: KALIMANTAN TIMUR, SABAH and SARAWAK.

GENERAL DISTRIBUTION: Peninsular Malaysia, Sumatra, Java, Borneo and Sulawesi.

NOTES: *P. lowii* is the most widespread of the multiflowered tropical slipper orchids. Despite its range, var. *lowii* is relatively uniform in the shape of its floral segments and coloration. Some populations in Borneo have leaves that are broader than usual, 4–5 cm wide, and rather deeper coloured flowers.

DERIVATION OF NAME: The specific epithet honours Sir Hugh Low who collected the type.

forma **aureum *P.J. Cribb* comb. & stat. nov.**

Paphiopedilum lowii (Lindl.) Stein var. *aureum* P.J. Cribb in Orchid Rev. 98 (1158): 109, fig. 74 (1990). Type: Borneo, Sarawak, Seventh Division, *c.* 1000 m., flowered in cultivation at Forest Research Institute Nursery March 1989, *Yii Puan Ching & Lai Shak Tek* in LST 1364 (holotype K).

This beautiful primrose-yellow flowered albino was first brought to the attention of Phillip Cribb by the late Miss Rena George, forestry officer in charge of the Forest Research Institute Nursery in Kuching. Albino variants of slipper orchids have been highly prized by generations of orchid growers, and it is to be hoped that this desirable plant can be successfully propagated and made widely available to the trade. Plate 23C.

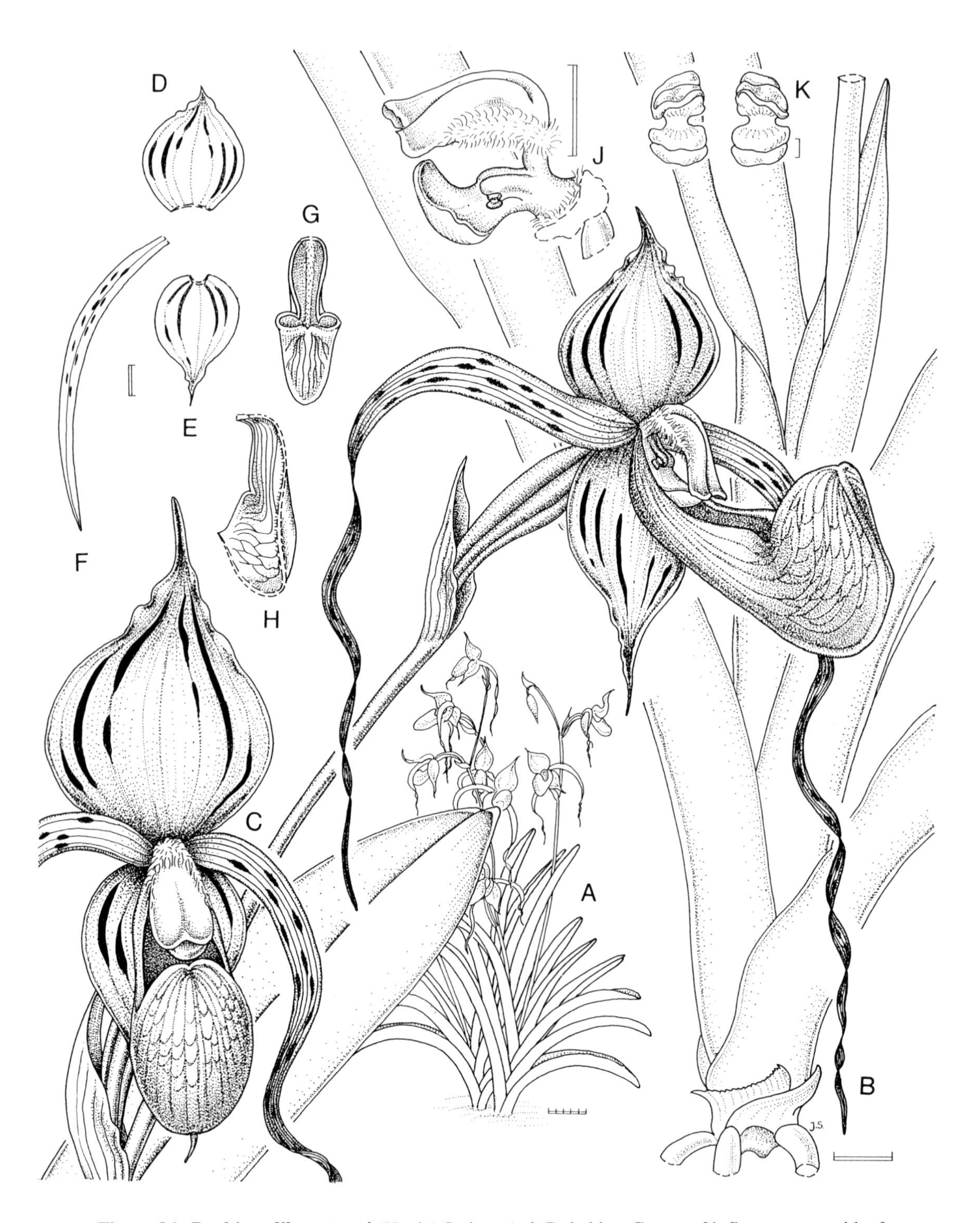

Figure 86. Paphiopedilum stonei (Hook.) Stein. - A & B: habit. - C: part of inflorescence and leaf. - D: dorsal sepal. - E: synsepal. - F: petal.- G: lip, front view. - H: lip, longitudinal section. - J: column and staminode, side view. - K: anthers. A & B drawn from *Lee* S. 40032 and C–K from *Cribb* 89/12 by Judi Stone. Scale: single bar divided into mm; double bar divided into cm.

86. PAPHIOPEDILUM STONEI (Hook.) Stein

Paphiopedilum stonei (*Hook.*) *Stein*, Orchideenbuch: 487 (1892). Type: Borneo, hort. *Low* (holotype K).

Cypripedium stonei Hook. in Bot. Mag. 88, t.5349 (1862).

C. stonei Hook. var. *platytaenium* Rchb.f. in Gard. Chron. 1867: 1118 (1867). Type: Borneo, cult. *Day* (holotype K).

Paphiopedilum stonei (Hook.) Stein var. *platytaenium* (Rchb.f.) Pfitzer in Engler, Pflanzenr. Orch. Pleon.: 64 (1903).

Cordula stonei (Hook.) Rolfe in Orchid Rev. 20 : 2 (1912).

C. stonei (Hook.) Rolfe var. *platytaenia* (Rchb.f.) Ames in Merr., Bibl. Enum. Born. Pl.: 137 (1921).

Lithophyte. Leaves *c.* 5, up to 70 × 4.5 cm, ligulate, rounded to obtuse, green. ***Inflorescence*** arcuate, up to 70 cm long, usually 2- to 4-flowered; peduncle 17–35 cm long, glabrous, purplish; floral bracts 3.5–5.5 × 1.6–2.2 cm, lanceolate, acute or acuminate. ***Flowers*** up to 12 cm across; sepals white usually lined with dark maroon; petals yellow, lined and spotted maroon, ± flushed with maroon in apical half; lip pale brown or pale yellow, flushed pink with darker veins; staminode yellow. ***Pedicel*** with ***ovary*** 4–7 cm long, glabrous. ***Dorsal sepal*** 4.5–5.7 × 3–4.4 cm, ovate, acuminate. ***Synsepal*** 3.7–5 × 2–3.4 cm, elliptic-ovate, acuminate. ***Petals*** 10–15 × 0.4–0.75 cm, (rarely more, up to 2 cm broad in var. *platytaenium*), arcuate-dependent, linear-tapering, straight or twisted in apical half. ***Lip*** 4.5–5.7 × 2.8 cm, pointing forwards, grooved on back. ***Staminode*** 14 × 11 mm, convex, subcircular, truncate or incised at apex, margins coarsely hairy. Plate 23D.

HABITAT AND ECOLOGY: Limestone cliffs, lightly shaded by the crowns of trees growing at the base of the cliffs. Alt. sea level to 500 m. Flowering details withheld.

DISTRIBUTION IN BORNEO: SARAWAK.

GENERAL DISTRIBUTION: Endemic to Borneo.

NOTES: *P. stonei* is one of the most highly prized of all species in the genus. It is unlikely to be confused with any other species but is, nevertheless, quite variable in nature and several extreme variations have received recognition. The most remarkable of these is var. *platytaenium* which has broad petals up to 1.5 or 2 cm wide. Plants which lack dark purple pigment in the sepals have been called var. *candidum* (Masters) Pfitzer.

DERIVATION OF NAME: Named in honour of a Mr. Stone, the gardener to John Day, a wealthy Victorian orchid grower who lived in Tottenham, London.

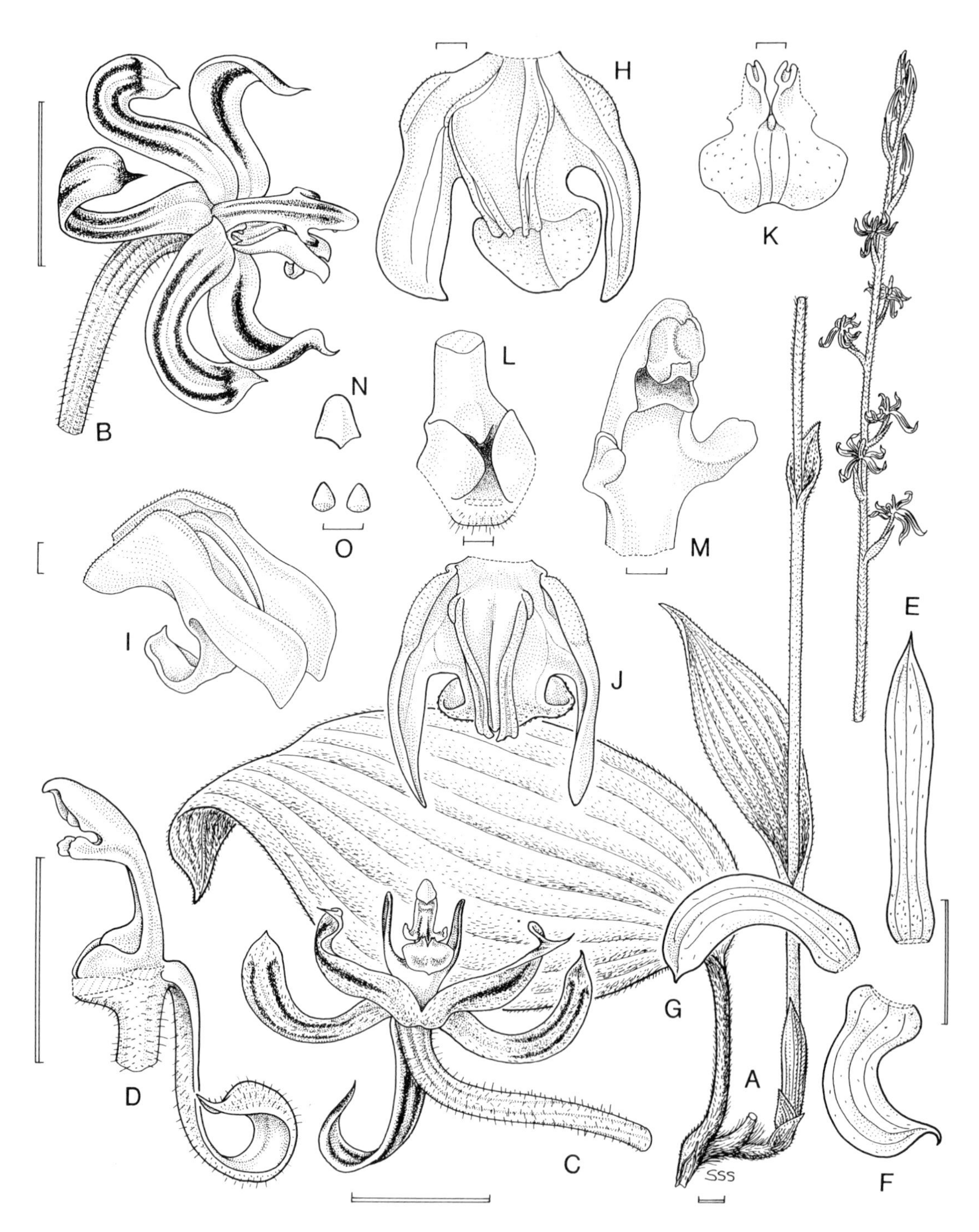

Figure 87. Pilophyllum villosum (Blume) Schltr. - A: habit. - B: flower, from above. - C: flower, front view. - D: upper portion of ovary, dorsal sepal and column. - E: dorsal sepal. - F: lateral sepal. - G: petal. - H: lip, flattened. - I: lip, side view. - J: lip, from above. - K: lip, mid-lobe, from above. - L: base of column. - M: column apex with anther-cap removed, oblique view. - N: anther-cap, front view. - O: pollinia. All drawn from *Lamb* AL 372/85 by Susanna Stuart-Smith. Scale: single bar = 1 mm; double bar = 1 cm.

87. PILOPHYLLUM VILLOSUM (Blume) Schltr.

Pilophyllum villosum (*Blume*) *Schltr.*, Die Orchideen: 131 (1914). Type: Java, Gede, *Blume* s.n. (holotype not located, lectotype *Blume*, Tab. & Plat., plate 7 (1825)).

Chrysoglossum villosum Blume, Bijdr.: 338 (1825).

Terrestrial 19–30 cm high; all parts, except inside of bracts and flowers, densely yellowish-brown hairy. ***Roots*** villous. ***Rhizome*** to 13 cm long, 5–6 mm in diameter, creeping, simple. ***Pseudobulbs*** 3.5–10 × 0.5 cm, borne 5–15 mm apart, 1-leaved, alternatingly 1–4 bearing a leaf and 1 bearing an inflorescence, leaf-bearing ones somewhat quadrangular, tapering at apex, inflorescence-bearing ones cylindrical. ***Leaves*** convolute; blade 16–24 × 7.5–12 cm, oblong to ovate-oblong, acuminate, main veins 5; petiole 2–20 mm long, semi-orbicular, sulcate. ***Inflorescence*** erect, laxly 15- to 25-flowered; peduncle 24–42 cm long, with several tubular sheaths 2–3 cm long; rachis to 9 cm long; floral bracts 1.6–2.5 × 0.45–0.8 cm, lanceolate, acute to acuminate, inner surface glabrous. ***Flowers*** non-resupinate, a few open simultaneously, widely opening; pedicel with ovary purple; sepals and petals orange-yellow stained purple; lip white flushed with pink; column white with orange stelidia. ***Pedicel*** with ***ovary*** 10–11.5 mm long. ***Dorsal sepal*** 14 × 3 mm, lanceolate, acute to acuminate, inner surface glabrous. ***Lateral sepals*** 8.5–11 × 2.5–3 mm, oblong-lanceolate, falcate, acuminate, inner surface glabrous. ***Petals*** 12–12.5 × 3 mm, obovate-lanceolate, slightly falcate, acuminate to cuspidate, glabrous. ***Lip*** *c.* 8.5 mm long, mobile, glabrous, rather fleshy, 3-lobed; hypochile: 5.5 × 5.5 mm, parallel to column, base semi-orbicular, lateral margins somewhat folded; front part on either side with an obliquely ligulate, falcate lobe *c.* 3 × 1.5 mm; disc with two erect, fleshy keels borne near base and terminating *c.* 1 mm from top of epichile; epichile: *c.* 3 × 3 mm, transversely elliptic, recurved, top emarginate, apex acute, with a small keel between distal part of lateral keels. ***Column*** 10 mm long, erect, glabrous; foot *c.* 2.5 mm long, flat, with a semi-orbicular fleshy lobe on either side at the front; with a central fleshy basal keel; stelidia 3 × 0.5 mm, ligulate, obtuse, fleshy, forward-projecting; anther-cap *c.* 1.5 × 1 mm, obovate, glabrous; pollinia two. Plate 23E.

HABITAT AND ECOLOGY: Lower montane ridge-top forest, often in open places. Alt. 700 to 1650 m. Flowering observed in June and December.

DISTRIBUTION IN BORNEO: SABAH: Crocker Range, Mt. Kinabalu. SARAWAK: Bario, Batu Buli.

GENERAL DISTRIBUTION: Peninsular Malaysia, Java, Borneo, Seram, Philippines, New Guinea and Solomon Islands.

DERIVATION OF NAME: The generic name is derived from the Greek *pilos*, felt, and *phyllon*, leaf, describing the felt-like hairy covering on the leaves, etc. The specific epithet is derived from the Latin *villosus*, i.e. shaggy with fairly long soft hairs.

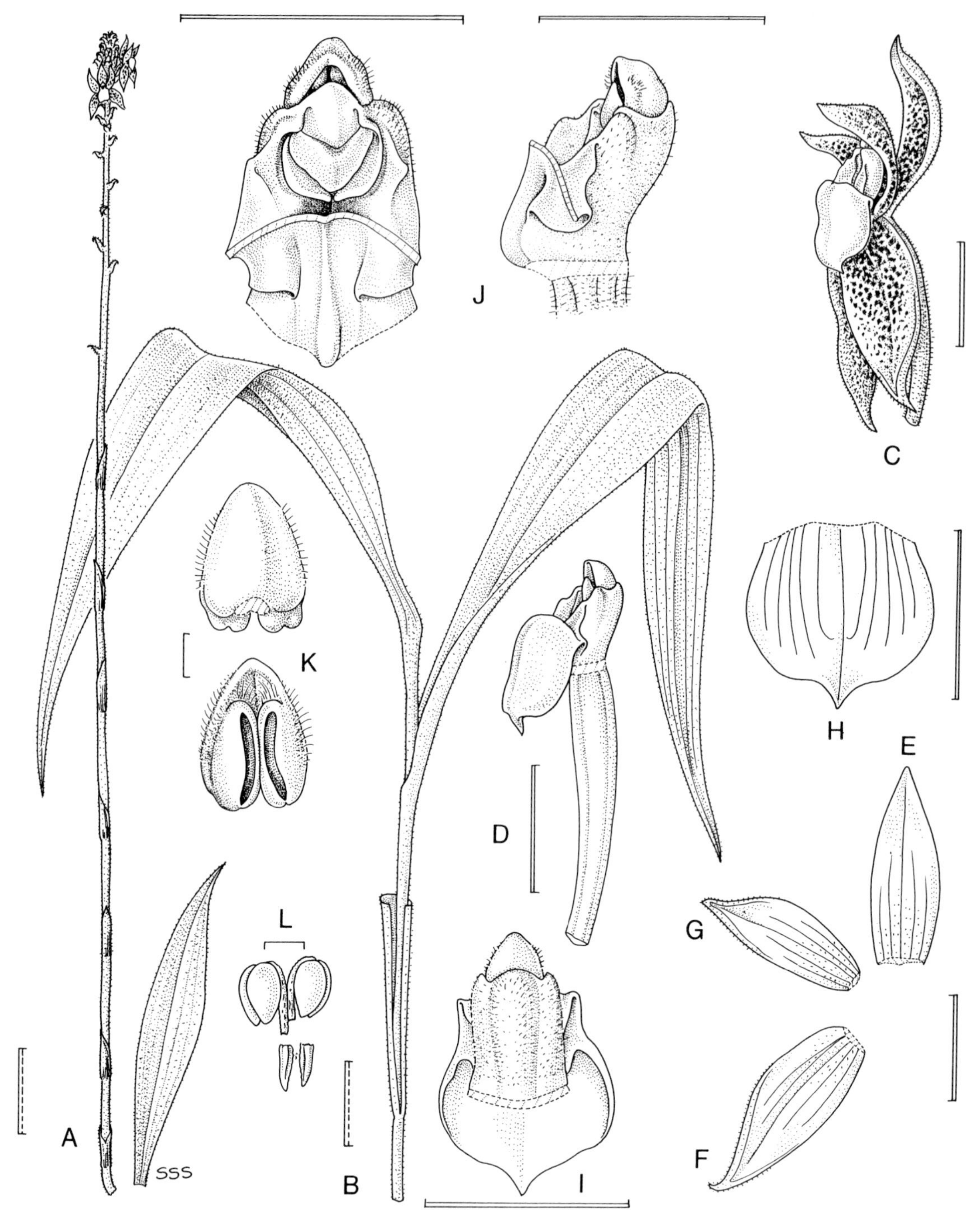

Figure 88. Plocoglottis gigantea (Hook.f.) J.J. Sm. - A & B: habit. - C: flower, side view. - D: pedicel with ovary, lip and column, side view. - E: dorsal sepal. - F: lateral sepal. - G: petal. - H: lip, flattened. - I: lip, column and anther-cap, back view. - J: column and anther-cap, front and side views. - K: anther-cap, back and inner views. - L: pollinarium. A & B drawn from *Robinson & Kloss* s.n., B from *J. & M.S. Clemens* 26519 and C–L from *Lohok* 10 by Susanna Stuart-Smith. Scale: single bar = 1 mm; double bar (A & B) = 6 cm; double bar = 1 cm.

88. PLOCOGLOTTIS GIGANTEA (Hook. f.) J.J. Sm.

Plocoglottis gigantea (*Hook. f.*) *J.J. Sm.* in Repert. Spec. Nov. Regni Veg. 32: 228(1933). Types: Peninsular Malaysia, Perak, Ulu Bubong, *King's collector* 10277 (syntype K); Perak, Assam Kumbong, *Wray* 2932 (syntype K).

Calanthe gigantea Hook.f., Fl. Brit. Ind. 5: 856 (1890).

Plocoglottis foetida Ridl. in J. Linn. Soc., Bot. 32: 319 (1896). Types: Singapore, Bukit Ṭimah, *Ridley* (syntype SING); Peninsular Malaysia, Joḥore, Tanjong Kopang, near Johore, *Ridley* (syntype SING); Malacca, Jus, *Ridley* (syntype SING); Selangor, Kwala Lumpur, *Ridley* (syntype SING).

Erect ***terrestrial*** to 100 cm tall. ***Rhizome*** fleshy. ***Roots*** fleshy, lanuginose. ***Stems*** leafy, to 1.5 cm thick, the basal 40 cm or so covered in many imbricate sheaths, the remaining distal portion leafy. ***Leaves*** 6–18, alternate, distichous; blade 22–25 × 2.5–5 cm, lanceolate or elliptic-lanceolate, acute to acuminate, plicate, with 3–5 prominent main veins, cuneate at base and narrowed to a 0.3–1 cm long sulcate petiole; sheath 4–6 cm long. ***Inflorescence*** up to 1 metre or more tall, from base of leafy stem, stout, densely 20- to 50 or more-flowered; rachis 20–40 cm long; peduncle, rachis and floral bracts densely shortly pubescent; floral bracts 1–1.5 × 0.6–0.8 cm, ovate, acute, subulate, distally recurved , semi-amplexicaul, rigid, persistent. ***Flowers*** *c.* 1.5 cm across; having a foetid odour; sepals and petals waxy, yellow or orange-yellow, speckled crimson; lip creamy white fading to pale yellow; column yellow stained reddish crimson. ***Pedicel*** with ***ovary*** 2 cm long, narrowly clavate to fusiform, pubescent. ***Sepals*** and ***petals*** shortly pubescent, especially on abaxial surface. ***Dorsal sepal*** 1.5 –1.8 × 0.6–0.7 cm, oblong-ovate or oblong-elliptic, acute, slightly reflexed and curved, rigid, fleshy. ***Lateral sepals*** 1.9–2 × 0.7–0.8 cm, slightly obliquely ovate-elliptic, subfalcate, dorsally somewhat carinate at the acute to caudate-acuminate apex, deflexed and parallel with pedicel with ovary, margins touching in front in basal portion, distal portion spreading a little, rigid, fleshy. ***Petals*** 1.6–1.7 × 0.5–0.6 cm , narrowly elliptic, acute, curving forward, thinner than sepals, but rigid. ***Lip*** 0.8–0.9 × 0.8–0.9 cm, adnate and decurrent with column below, ovate to quadrate, truncate, apiculate, strongly decurved, side lobes reduced to coriaceous auricles, glabrous to puberulus; spur absent. ***Column*** 6 × 3–4 mm, stout, pubescent; anther-cap 2.5–3 × 2 mm, ovate, conical, cucullate, with a small patch of hairs on each side near the middle and a few more hairs above the loculi; pollinia four. Plate 24A.

HABITAT AND ECOLOGY: Hill-dipterocarp forest. Alt. 900 m. Flowering observed from August to October.

DISTRIBUTION IN BORNEO: SABAH: Mt. Kinabalu.

GENERAL DISTRIBUTION: Thailand, Peninsular Malaysia, Sumatra, Mentawai Islands (Siberut Island) and Borneo.

NOTES: The lip has an elastic joint which, unlike many other species, does not have a trigger mechanism. It is at first deflexed so as to expose its whole upper surface, but is not held in position by a sepal. If the lip is moved forwards a little it springs up into contact with the column, but no pollinating insect has been observed to effect this movement. If the lip is moved downwards again it will stay in its original position until again moved forwards.

DERIVATION OF NAME: The generic name is derived from the Greek *ploke*, binding together and *glotta*, tongue, alluding to the lip which is partially adnate to the base of the column, to which it is joined on each side by a membranous fold. The specific epithet refers to the large proportions of the plant.

89. PLOCOGLOTTIS HIRTA Ridl.

Plocoglottis hirta *Ridl.* in J. Straits Branch Roy. Asiat. Soc. 50: 137 (1908). Type: Borneo, Sarawak, Bidi, *Brookes* s.n. (holotype SING).

Robust erect ***terrestrial*** to 2 m tall. ***Rhizome*** not seen. ***Roots*** lanuginose. ***Stems*** leafy, up to 1.5 cm thick. ***Leaves*** several along distal portion of stem, softly downy-pubescent on adaxial and abaxial surfaces; blade 40–65 × 7–12 cm, broadly lanceolate, cuneate at base, acuminate, plicate, with 5 prominent main veins, thin-textured; petiole 2–4 cm long, V-shaped; sheath 10–17 × 2–3.5 cm, inflated. ***Inflorescence*** 60 cm to over 1 m tall, lateral, axillary, stout, densely many-flowered; peduncle and rachis softly downy-pubescent; peduncle 0.8–1 cm wide; rachis expanding to 60 cm long; floral bracts 2–3.3 cm long, 0.9–1.4 cm wide when flattened, ovate-lanceolate, clasping and cupulate at base, distally falcate-recurved, caudate-acuminate, shortly pubescent, green, the unexpanded grouped into a tuft at the inflorescence apex. ***Flowers*** foetid; pedicel with ovary greenish-cream, flushed brown; sepals and petals dark yellow with scattered purple-red markings; lip white on the auriculate base, otherwise pale yellow with a purple-red patch either side near the base; column yellow, flushed brick-red to purple; anther-cap yellow. ***Pedicel*** with ***ovary*** 2–3.3 cm long, cylindrical, densely shortly pubescent. ***Sepals*** and ***petals*** rigid, fleshy, thickened at apex, shortly pubescent. ***Dorsal sepal*** 1.7–2.2 × 0.5–0.6 cm, lanceolate, acute, erect. ***Lateral sepals*** 2.1–2.5 × 0.5–0.6 cm, obliquely lanceolate, acute, parallel at base, then bent and spreading. ***Petals*** 1.5–1.8 × 0.3–0.35 cm, obliquely narrowly lanceolate, acute, spreading. ***Lip*** *c.* 0.5 cm long, 3.5–7 mm wide at base, 9–15 mm wide across apical lobes, auriculate at base, subquadrate, distally with two decurved-falcate, obtuse lobes, each 4–5 × 2–2.5 mm, and a central decurved subulate tooth-like mucro; disc with two lower rounded ridges, sulcate between, whole surface glabrous to minutely papillose. ***Column*** 5–6 × 3 mm, shortly pubescent; anther-cap 3 × 1.6 mm, oblong, cucullate, with scattered bristly hairs. Plate 24B.

HABITAT AND ECOLOGY: Lowland forest on limestone; limestone boulder scree. Alt. sea level to 600 m. Flowering observed from January to April, and October.

DISTRIBUTION IN BORNEO: SARAWAK: Bau area; Bidi; Mt. Jambusan; Mt. Mulu National Park.

GENERAL DISTRIBUTION: Endemic to Borneo.

DERIVATION OF NAME: The specific epithet is derived from the Latin *hirtus*, hairy, a distinctive feature of this species.

Figure 89. Plocoglottis hirta Ridl. - A: habit. - B: close-up of portion of abaxial surface of leaf. - C: flower, oblique view. - D: upper portion of ovary, lip and column. - E: dorsal sepal. - F: lateral sepal. - G: petal. - H: lip and column, back view. - I: column with anther-cap removed, front view. - J: anther-cap, side view. - K: anther-cap, oblique view. - L: pollinia. A (habit) drawn from *Collenette* 2354 and *Nielsen* 967, B from *Collenette* 2354, and C–L from *Lamb* AL SAR 268/86 by Susanna Stuart-Smith. Scale: single bar = 1 mm; double bar = 1 cm, except habit A = 3 cm.

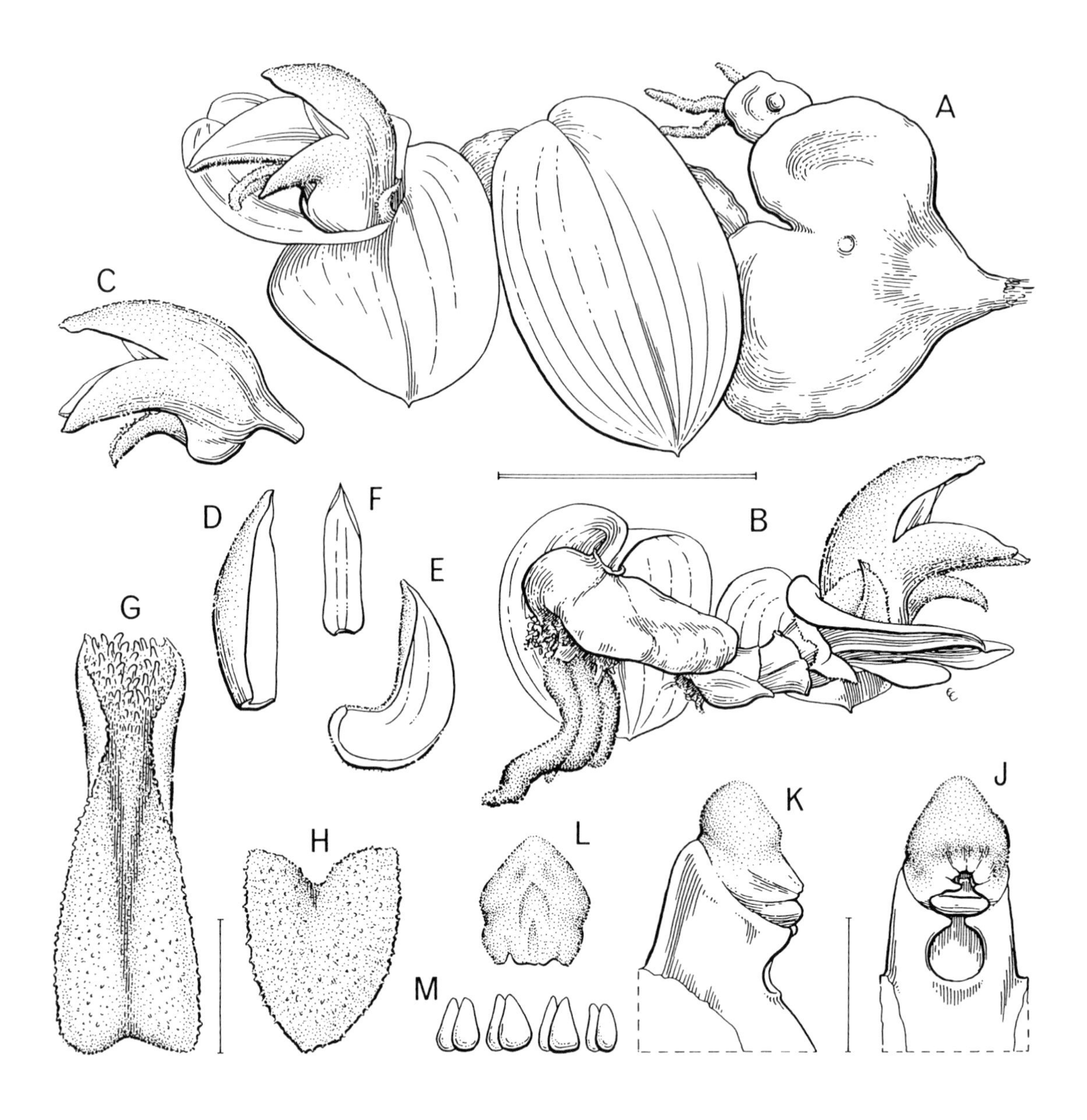

Figure 90. Porpax borneensis J.J. Wood & A. Lamb. - A & B: habit. - C: flower, side view. - D: dorsal sepal. - E: lateral sepal. - F: petal.- G: lip, front view. - H: lip apex - J: column and anther-cap, front view. - K: column and anther-cap, side view. - L: anther-cap, back view. - M: pollinia. All drawn from *Lamb* AL 1164/89 (holotype) by Eleanor Catherine. Scale: single bar = 1 mm; double bar = 1 cm.

90. PORPAX BORNEENSIS J.J. Wood & A. Lamb

Porpax borneensis *J.J. Wood & A. Lamb* in Wood, Beaman & Beaman, The Plants of Mt. Kinabalu 2, Orchids: 300, fig. 49, plate 74A (1993). Type: Borneo, Sabah, Sandakan District, Mt. Tawai (Tavai), south of Telupid, *Lamb* AL 1164/89 (holotype K).

Dwarf ***epiphyte,*** or ***lithophytic. Rhizome*** creeping. ***Pseudobulbs*** 1- to 2-leaved, 0.8 × 6–1.4 × 1 cm, 3–4 mm thick, flattened, ovoid, often bilobed, when young covered by a sheath which disintegrates with age leaving the pseudobulb naked. ***Leaves*** 0.9 × 0.8–1.5 × 1(–4.5 × 1) cm, oblong or oblong–ovate, apex minutely retuse or asymmetric, glabrous. ***Inflorescence*** proteranthous, 1-flowered, borne from base of pseudobulbs, consisting of a very short scape 4 mm long, bearing two leaves above and three imbricate, ovate, acute, carinate sheaths 2–3 mm long below; floral bract 3 mm long, 3 mm wide at base, ovate, acute, carinate, apex reflexed and minutely serrate, translucent creamy-white. ***Flower*** resupinate; dark cardinal-red, the sepals with pale yellow tips. ***Sepals*** 9–10 × 3 mm, connate at base, narrowly ovate-elliptic, acute, concave, minutely papillose on exterior. ***Mentum*** obtuse. ***Petals*** 1.5 × 8 mm, free, narrowly elliptic, acute, somewhat concave. ***Lip*** 4 mm long, 1.8–2 mm wide when flattened, entire, oblong-ligulate, obtuse, conduplicate, curved, margin irregularly serrulate, surface minutely papillose, especially along centre of undersurface, disc with several papillae of varying sizes toward base. ***Column*** minute; foot *c.* 2 mm long; anther-cap ovate, cucullate, minutely papillose; pollinia eight. Plate 24C.

HABITAT AND ECOLOGY: Low primary hill-forest on ultramafic substrate, often adpressed to rocks. Alt. 600 to 1000 m. Flowering observed in February, March, August and November.

DISTRIBUTION IN BORNEO: SABAH: Mt. Kinabalu; Mt. Tawai.

GENERAL DISTRIBUTION: Endemic to Borneo.

NOTES: *P. borneensis* appears to be most closely related to *P. ustulata* (C.S.P. Parish & Rchb.f.) Rolfe from Myanmar (Burma) and Thailand. It differs in having minutely papillose, never hirsute, sepals and an entire lip with an irregularly serrulate margin and several basal papillae. Seidenfaden (1977) recognised ten species, all confined to the Asian mainland, east as far as Indochina and south to Peninsular Malaysia. Six of these are represented in Thailand. *P. borneensis* brings the total number of species to eleven and also extends the range south-eastward. These plants are rarely collected on account of their tiny size; it seems likely that they may also be present in Sumatra and Java.

DERIVATION OF NAME: The generic name is Greek for the handle of a shield, probably alluding to the pair of small oblong leaves recurved at their point which arise from the pseudobulb, or, according to Kraenzlin, perhaps referring, as "handle of a shield", to the scutiform pseudobulbs. The specific epithet refers to the island of Borneo.

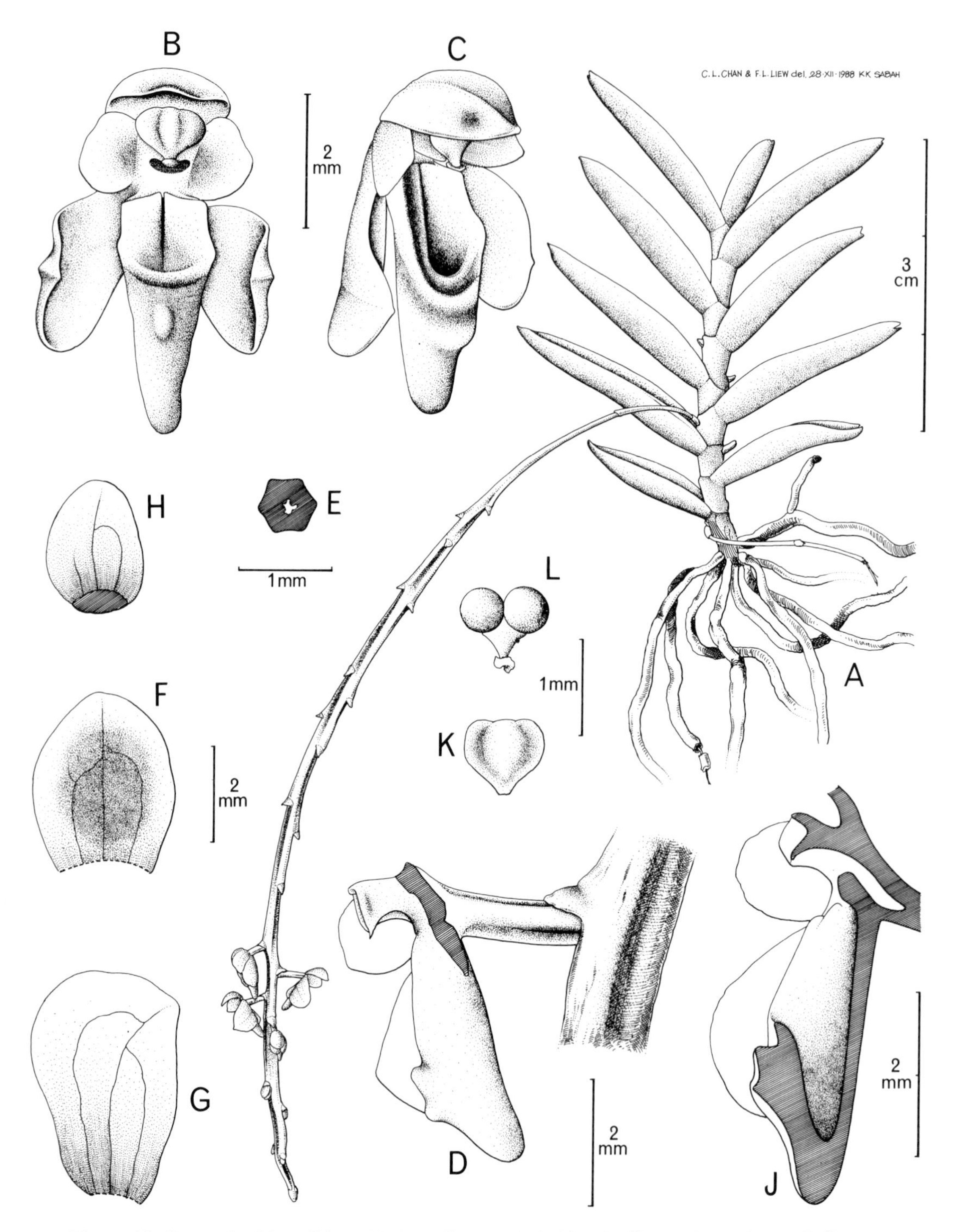

Figure 91. Porrorhachis galbina (J.J. Sm.) Garay. - A: habit. - B: flower, front view. - C: flower, oblique view. - D: ovary, transverse section. - E: rachis and flower with dorsal sepal, lateral sepal, petal and anther-cap removed, side view. - F: lip and column, with lateral sepal and petal attached, longitudinal section.- G: dorsal sepal. - H: lateral sepal. - I: petal. - J: anther-cap, back view. - K: pollinarium. All drawn from *Lamb* AL 1133/89 by C.L. Chan and Lucy F.L. Liew.

91. PORROHRACHIS GALBINA (J.J. Sm.) Garay

Porrorhachis galbina (*J.J. Sm.*) *Garay* in Bot. Mus. Leafl. Harv. Univ. 23(4): 191 (1972). Type: Java, Priangan, Pasir Angin (Tjadas Malang), Tjibeber, *Winckel* 339 (holotype L).

Saccolabium galbinum J.J. Sm. in Bull. Jard. Bot. Buitenzorg, ser. 2, 26: 97 (1918).

Erect to pendulous ***epiphyte. Stems*** 2.5–6 cm long, internodes 2–3.5 mm long, simple, straight or curved, leafy, rooting at base. ***Leaves*** spreading, *c.* 6–25; blade 1.6–2.6 × 0.3–0.7 cm, narrowly elliptic, apex unequally acutely or subacutely bidentate, with a conical, tooth-like mucro in the sinus, semi-amplexicaul, thick and fleshy, but not rigid, longitudinally sulcate to canaliculate, shiny dark green; sheaths 2–3.5 × *c.* 2.5 mm, tubular, transversely oval in section. ***Inflorescence*** emerging from leaf sheath opposite blade, spreading to pendulous, laxly many-flowered, 2 or 3(–4) flowers open at a time in succession; flowers borne 1–2 mm apart; peduncle 1.5–2.5 cm long, filiform, becoming expanded below rachis, with 2 tubular sterile bracts to *c.* 1 mm long at base and roughly midway along; rachis 2–15 cm long, thickened and fleshy, laterally sulcate; floral bracts 0.5–0.8 mm long, ovate-triangular, acute or subacute, concave, adpressed to pedicel with ovary, green. ***Flowers*** ephemeral; spreading, quaquaversal, not opening widely, laterally compressed; unscented; pedicel with ovary pale green; remainder of flower greenish yellow to shiny dark yellow with a whitish spur, fading to orange or ochre. ***Pedicel*** with ***ovary*** 2.5–3 mm long, straight, not twisted, apex oblique. ***Dorsal sepal*** 2.5 × 1.7 mm, ovate or subovate-quadrangular, obtuse, concave, curving forward over column. ***Lateral sepals*** 2.6 × 2 mm, obliquely obovate-quadrangular, obtuse concave, porrect, parallel, adpressed to lip. ***Petals*** 1.6 × 1.2–1.3 mm, obliquely ovate, obtuse, concave, porrect. ***Lip*** immobile, straight, narrow, excavated, obscurely 3-lobed, laterally compressed, convex below, linear-oblong in outline when viewed from above, 3 × 1 mm, glabrous; side lobes very short, broad at base, gently obtusely angled, fleshy, connected with base of mid-lobe by a round emarginate transverse crest; mid-lobe reduced to an obtuse transverse calliform structure; spur forming the top of the lip, shortly conical, obtuse. ***Column*** 1 mm long, semi-terete, obtuse, straight to gently inclined; stigmatic cavity large, suborbicular; rostellum short, bidentate; anther-cap reniform, apex shortly obtuse; pollinia two, entire. ***Fruit*** narrowly fusiform. Plate 24D & E.

HABITAT AND ECOLOGY: Lower montane forest; mossy forest; montane scrub on sandstone ridges; epiphytic on branches of shrubs and small trees. Alt. 1100 to 1500 m. Flowering observed in January, February, May, June, October and November.

DISTRIBUTION IN BORNEO: SABAH: Crocker Range, Kimanis road, Sinsuron road; Mt. Alab; Mt. Kinabalu.

GENERAL DISTRIBUTION: Java and Borneo.

DERIVATION OF NAME: The generic name is derived from the Greek *porrho*, forward, and *rhachis*, rachis, referring to the position of the inflorescences which are inserted at a right angle to the main stem. The specific epithet is derived from the Latin *galbanus*, the colour of gum galbanum, which is greenish yellow, and refers to the flower colour.

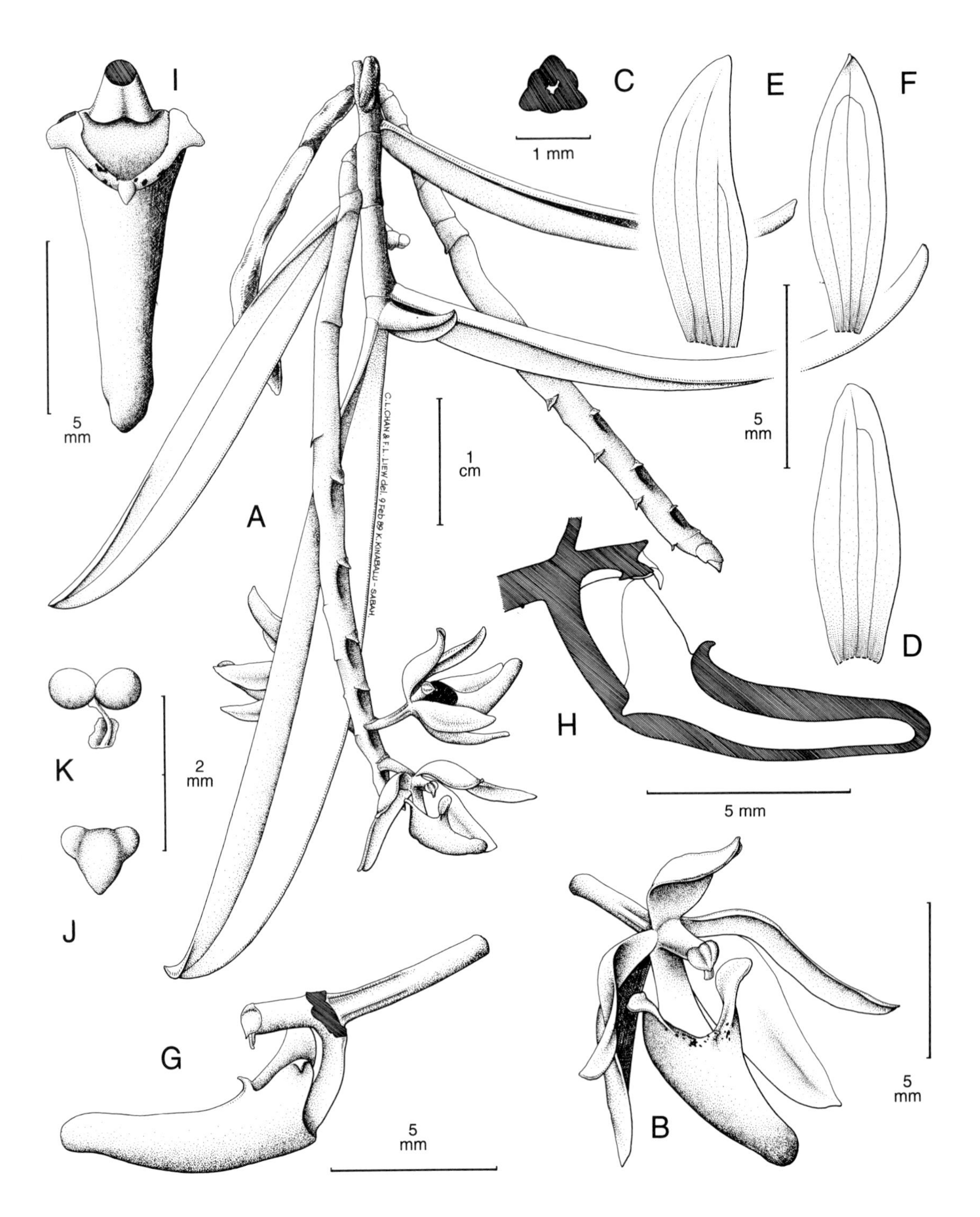

Figure 92. Pteroceras biserratum (Ridl.) Holttum. - A: habit. - B: flower, oblique view. - C: ovary, transverse section. - D: dorsal sepal. - E: lateral sepal. - F: petal.- G: pedicel with ovary, lip and column, side view. - H: lip and column, longitudinal section. - I: lip, front view. - J: anther-cap, back view. - K: pollinarium. All drawn from a living plant cultivated at Tenom Orchid Centre (TOC) by C.L. Chan and Lucy F.L. Liew.

92. PTEROCERAS BISERRATUM (Ridl.) Holttum

Pteroceras biserratum (*Ridl.*) *Holttum* in Kew Bull. 14: 269 (1960). Type: Peninsular Malaysia, Perak, Ipoh, *Ridley* 10156 (holotype SING).

Sarcochilus biserratus Ridl. in J. Bot. 38: 73 (1900).

S. pachyrhachis Schltr. in Bull. Herb. Boissier, ser. 2, 6: 468 (1906). Type: Kalimantan Barat, Koetai, Sungai Penang, Samarinda, *Schlechter* 13348 (holotype B, destroyed).

S. bipennis J.J. Sm. in Bull. Jard. Bot. Buitenzorg, ser. 3, 2: 105 (1920). Type: Sumatra, Sumatera Barat, Kajutanam, *Groeneveldt* cult. 1377 (holotype not located).

Epiphyte. Stem 2–16.5 cm long, prostrate, curved, internodes 4–8 mm long. ***Leaves*** 3–5 extant on a flowering shoot, 2.7–17.2 × 0.3–1.3 cm, linear–lanceolate to lanceolate, sometimes slightly falcate, acute to minutely bifid. ***Inflorescences*** up to 18 per shoot, 2.2–18 cm long, suberect to pendent; peduncle 0.6–3 cm long, with 1–3 tiny basal sheaths; rachis 1.4–15 cm long, laterally compressed, at each flower node widened into 2 parallel, entire wings or keels; floral bracts 1–1.7 × 2.4–3.2 mm, broadly triangular, obtuse, concave, scale-like. ***Flowers*** 9–53 per rachis, 1–7 open at a time; resupinate; sweetly-scented; sepals and petals white to transparent yellowish green; lip white with a violet-speckled mid-lobe; column white; anther-cap yellowish to orange, usually with a greenish yellow bar. ***Pedicel*** with ***ovary*** 3.4–5.6(–7) mm long. ***Dorsal sepal*** 5.8–8 × 2.4–3 mm, broadly lanceolate, obtuse to acute, slightly concave. ***Lateral sepals*** 5–7.5 × 1.7–3 mm, adnate to column-foot for up to a quarter of its total length, obliquely broadly lanceolate, acute to obtuse, slightly concave. ***Petals*** 5.2–7 × 1.4–2 mm, linear-lanceolate, acute to obtuse, sometimes slightly oblique. ***Lip*** 4.8–7 mm long, sessile; side lobes 1.5–2.1 × 0.6–1.2 mm, erect, apically slightly outcurved, narrowly triangular, subacute; mid-lobe reduced to a tiny recurved tooth just above spur entrance; spur 1.2–2.2 mm in outer diameter, obtuse, slightly upcurved. ***Column*** 1.9–2.4 mm long; foot 2.5–3 mm long; anther-cap transversely elliptic; pollinia two, slightly cleft. Plate 25A.

HABITAT AND ECOLOGY: Lowland dipterocarp forest. Alt. sea level to 500 m. Flowering observed in March, July and August.

DISTRIBUTION IN BORNEO: KALIMANTAN BARAT: Samarinda area. SABAH: Lahad Datu area.

GENERAL DISTRIBUTION: Peninsular Malaysia, Sumatra and Borneo.

DERIVATION OF NAME: The generic name is derived from the Greek *pteron*, feather or quill and *keras*, horn, in reference to the two narrow wing-like side lobes on the lip. The specific epithet is derived from the Latin *bi-*, two, and *serratus*, saw-edged, referring to the rachis structure.

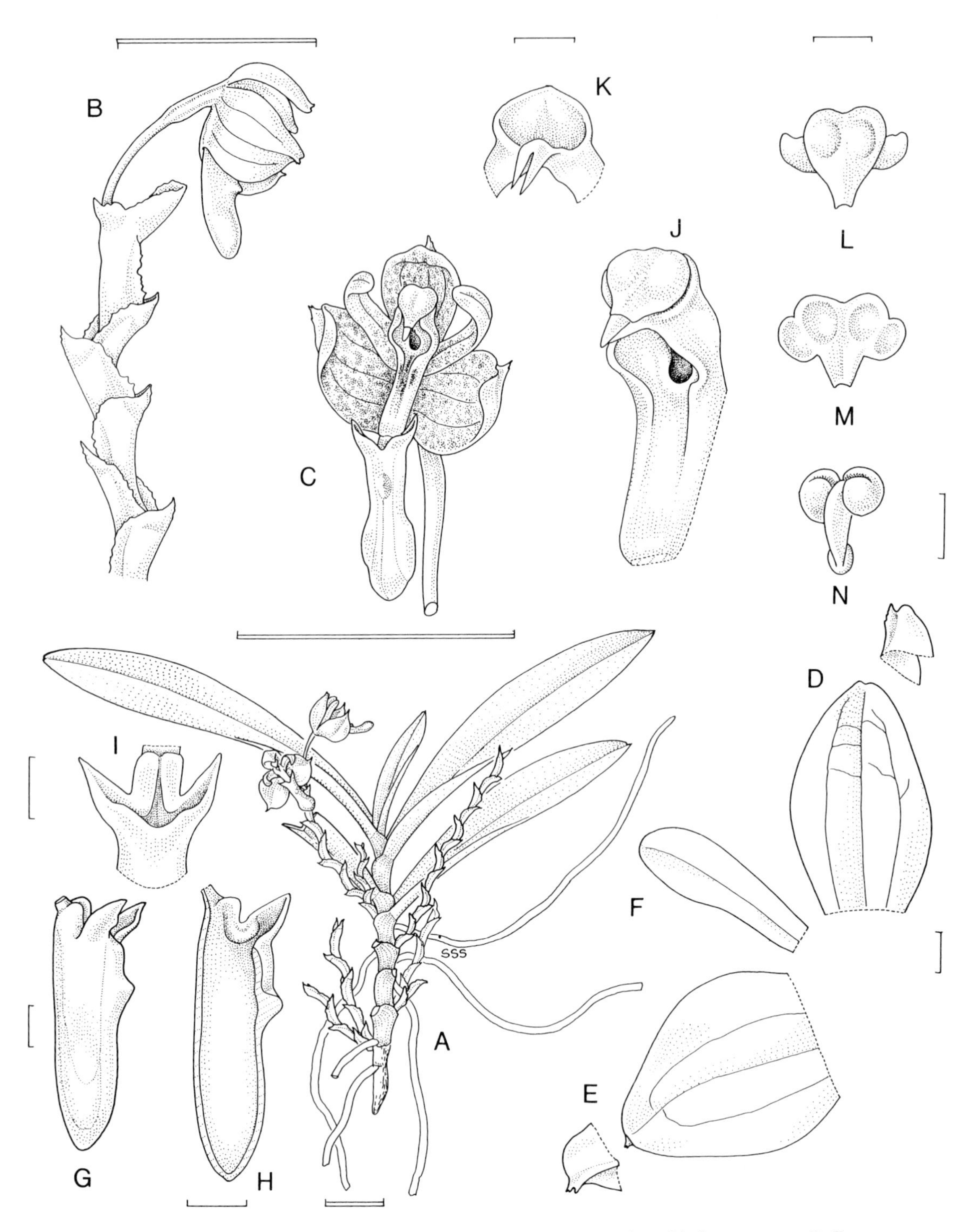

Figure 93. Pteroceras erosulum H.Ae. Peders. - A: habit. - B: portion of inflorescence. - C: flower, oblique view. - D: dorsal sepal, with close-up of keeled apex. - E: lateral sepal, with close-up of keeled apex. - F: petal. - G: lip, side view. - H: lip, longitudinal section. - I: upper portion of lip, front view. - J: column with anther-cap, oblique view. - K: column apex, anther-cap removed. - L: anther-cap, outer surface, spread out. - M: anther-cap, inner surface, spread out. - N: pollinarium. All drawn from *Wood* 676 (holotype) by Susanna Stuart-Smith. Scale: single bar = 1 mm; double bar = 1 cm.

93. PTEROCERAS EROSULUM H. Ae. Peders.

Pteroceras erosulum *H. Ae. Peders.* in Opera Bot. 117: 45, fig. 17 (1993). Type: Borneo, Sabah, Sipitang District, 8km NW of Long Pa Sia, *Wood* 676 (holotype K).

Epiphyte. Stem 4–5 cm long, straight, internodes 3–5 mm long. ***Leaves*** 6–7 extant on flowering shoot, 4.8–6.5 × 0.9–1.1 cm, obliquely lanceolate, minutely bilobed, lobes rounded, finely serrate-erose. ***Inflorescences*** 3–5 per shoot, 1–5.2 cm long, horizontal to erect; peduncle 5–6 mm long, with 2 tiny basal sheaths; rachis 0.6–4.6 cm long, laterally compressed, somewhat fractiflex, at each flower node widened into 2 parallel, distinctly erose wings or keels; floral bracts 1.7–2.5 × 3.3–4.2 mm, broadly triangular, acuminate, scale-like. ***Flowers*** 3–14 per rachis, 1–3 open at a time; resupinate; sepals and petals pale yellow speckled with red; lip very pale yellow, the spur white. ***Pedicel*** with ***ovary*** 8–9 mm long. ***Dorsal sepal*** 5 × 3 mm, elliptic, rounded to minutely emarginate, with a short, acute, finely serrate apical keel, slightly concave. ***Lateral sepals*** 5 × 4 mm, adnate to column-foot for its whole length, obliquely broadly elliptic, rounded, with a short, acute, finely serrate apical keel. ***Petals*** 4.5 × 1.3 mm, linear to narrowly subspathulate, rounded. ***Lip*** *c.* 5.5 mm long, subsessile; side lobes *c.* 1 × 0.6–0.7 mm, narrowly triangular, acute, erect, pointing upwards/backwards; mid-lobe *c.* 1 mm long, *c.* 0.4 mm wide, *c.* 0.4 mm high, cone-shaped, somewhat laterally compressed; spur 1.2–1.9 mm in outer diameter, cylindric, obtuse, somewhat widened distally, with a callus on the adaxial wall. ***Column*** 3 mm long; foot *c.* 2.8 mm long; anther-cap transversely elliptic; pollinia two, deeply cleft. Plate 25B.

HABITAT AND ECOLOGY: Lower montane forest. Alt. 1260 m. Flowering observed in October.

DISTRIBUTION IN BORNEO: SABAH: Sipitang District. SARAWAK: see photograph by A. Vogel in Orchideeën 5: 78 (1996).

GENERAL DISTRIBUTION: Endemic to Borneo.

DERIVATION OF NAME: The specific epithet is derived from the Latin *erosus*, having an irregularly toothed or apparently gnawed margin, in reference to the keels on the rachis.

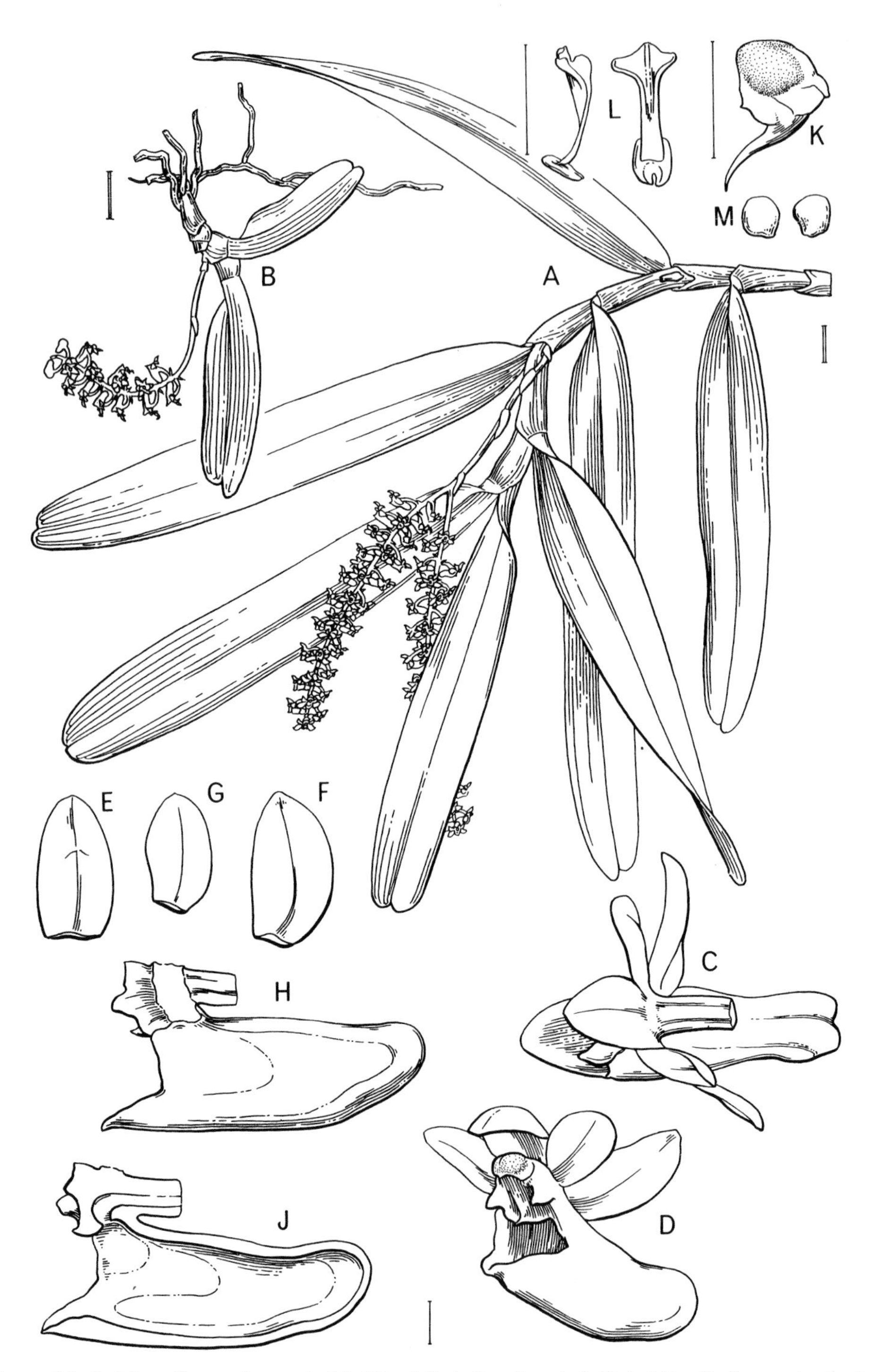

Figure 94. Robiquetia crockerensis J.J. Wood & A. Lamb. - A & B: habit. - C: flower, back view. - D: flower, oblique view. - E: dorsal sepal. - F: lateral sepal.- G: petal. - H: pedicel with ovary, lip and column, side view. - J: pedicel with ovary, lip and column, longitudinal section. - K: anther-cap, side view. - L: stipes and viscidium, front and side views. - M: pollinia. A & B drawn from *Carr* 3107, SFN 26473 (holotype) and C–M from *Lamb* AL 31/82 by Eleanor Catherine. Scale: single bar = 1 mm; double bar = 1 cm.

94. ROBIQUETIA CROCKERENSIS J.J. Wood & A. Lamb

Robiquetia crockerensis *J.J. Wood & A. Lamb* in Wood, Beaman & Beaman, The Plants of Mt. Kinabalu 2, Orchids: 306, fig. 51, plate 76C (1993). Type: Borneo, Sabah, Mt. Kinabalu, Mahandei River, *Carr* 3107, SFN 26473 (holotype K, isotypes A, C, L, LAE, SING).

Epiphyte. Stems (2–)6–20 cm long, rooting at the base, spreading to pendulous. ***Leaves*** (4–)10–18 × 1.2–2.3 cm, ligulate, margins undulate, apex obtusely unequally bilobed, purple on the reverse, sheaths 1–2 cm long. ***Inflorescence*** simple and racemose, or with one branch, densely many-flowered; peduncle 1.5–4.5 cm long; sterile bracts 3–6 mm long, ovate, obtuse, amplexicaul; rachis 3–10 cm long, branch (when present) 2–9 cm long; floral bracts 3–4 mm long, triangular, acuminate. ***Flowers*** unscented; pedicel with ovary purple-green; sepals and petals translucent greenish white or yellow, spotted with purple; lip and spur white with some purple spots, slightly ramentaceous at base. ***Pedicel*** with ***ovary*** 5–6 mm long, slender, straight, slightly ramentaceous. ***Dorsal sepal*** 3 × 2 mm, ovate, obtuse, strongly concave, curving over column. ***Lateral sepals*** 4 × 2 mm, slightly obliquely oblong-elliptic, obtuse, spreading. ***Petals*** 2–2.5 × 1–1.5 mm, oblong, obtuse, spreading. ***Lip*** fleshy; side lobes 1.2–1.5 mm wide, irregularly triangular, acute, erect, clasping base of column; mid-lobe 1.5–1.6 mm long, 1.4–1.5 mm wide at base, oblong-ovate, obtuse, fleshier than side lobes, flat; spur 4–5 × 2–2.1 mm, saccate, apex obtuse, slightly retuse, parallel to, and with the apex sometimes touching the pedicel with ovary, interior unadorned. ***Column*** 1.1–1.5 mm long; anther-cap extinctoriform, tail flat; pollinia two, cleft, pubescent; stipes spathulate; viscidium small. Plate 25C.

HABITAT AND ECOLOGY: Hill forest; lower montane forest, sometimes on ultramafic substrate. Alt. 800 to 1900 m. Flowering observed in February, March, September and December.

DISTRIBUTION IN BORNEO: SABAH: Crocker Range, Mt. Alab; Mt. Kinabalu.

GENERAL DISTRIBUTION: Endemic to Borneo.

NOTES: *R. crockerensis* is related to *R. pachyphylla* (Rchb. f.) Garay from Myanmar (Burma) and Thailand, but is at once distinguished by its thinner leaves, which are never thick and fleshy, and translucent greenish white or yellow flowers which are spotted with purple. The lip has more prominent side lobes and a shorter spur lying parallel to the pedicel with ovary, and the stipes is spathulate.

DERIVATION OF NAME: The generic name honours Pierre Robiquet, a French chemist who, among other things, discovered caffeine and morphine. The specific epithet refers to the Crocker Range, which extends down the entire length of western Sabah and of which Mt. Kinabalu is a part.

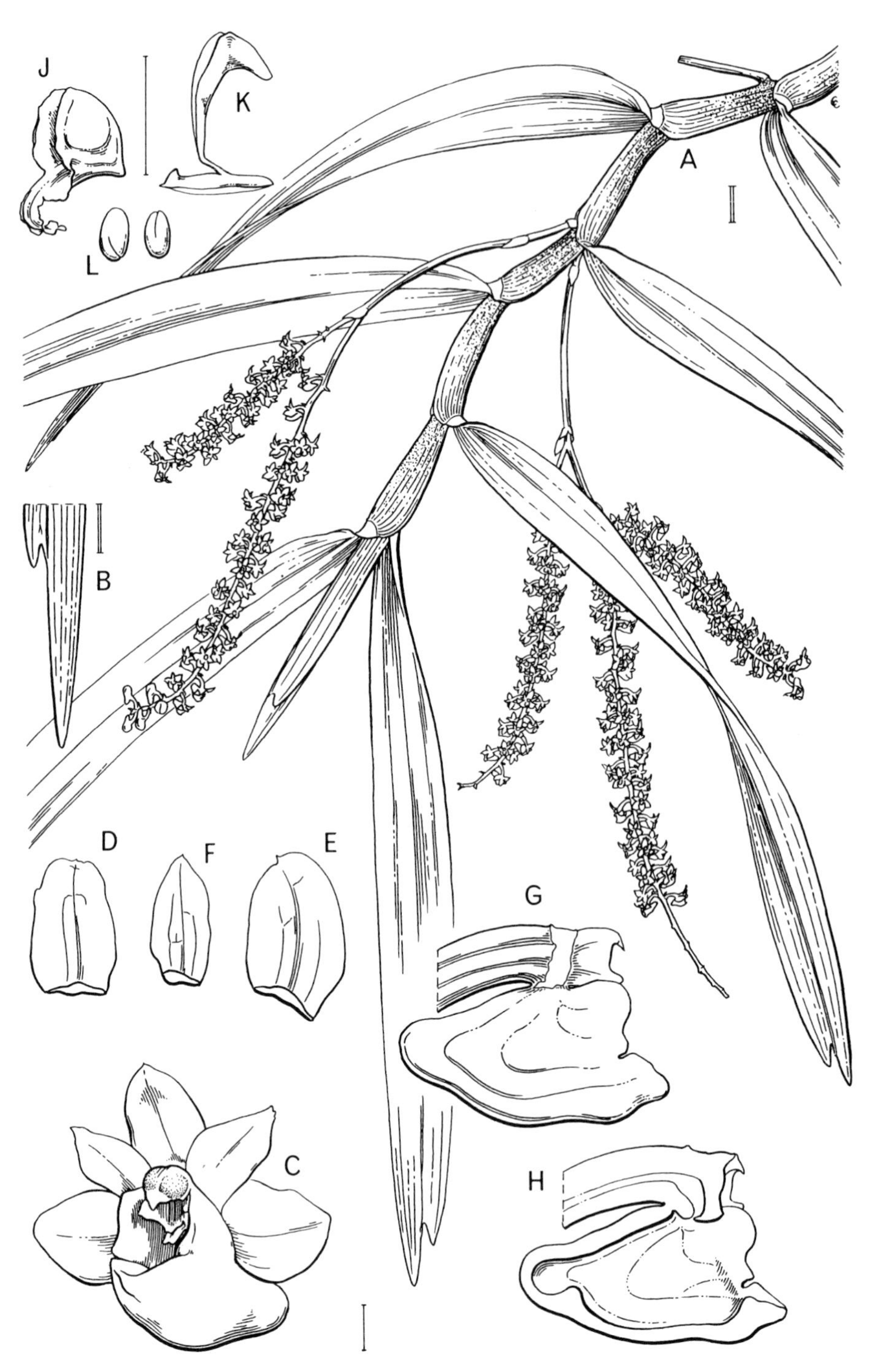

Figure 95. Robiquetia pinosukensis J.J. Wood & A. Lamb. - A: habit. - B: leaf apex. - C: flower, oblique view. - D: dorsal sepal. - E: lateral sepal. - F: petal.- G: pedicel with ovary, lip and column, side view. - H: pedicel with ovary, lip and column, longitudinal section. - J: anther-cap, side view. - K: stipes and viscidium, side view. - L: pollinia. A (habit) drawn from *Chow & Leopold* SAN 74504 (holotype), B (leaf apex) from *Beaman* 7227 and C–L from *Brentnall* 156 by Eleanor Catherine. Scale: single bar = 1 mm; double bar = 1 cm.

95. ROBIQUETIA PINOSUKENSIS J.J. Wood & A. Lamb

Robiquetia pinosukensis *J.J. Wood & A. Lamb* in Wood, Beaman & Beaman, The Plants of Mt. Kinabalu 2, Orchids: 308, fig. 52 (1993). Type: Borneo, Sabah, Mt. Kinabalu, Mesilau Trail, *Chow & Leopold* SAN 74504 (holotype K, isotypes A, L, SAN, SAR, SING).

Epiphyte. Stems 25–40 cm long, producing long verruculose roots toward the base, spreading to pendulous. ***Leaves*** 13–28 × (1–)1.5–2.1 cm, ligulate, coriaceous, apex very unequally bilobed, one lobule up to 4 cm long, the other only 0.2–0.4 cm long and tooth-like; sheaths 2–3.5 cm long, usually stained or speckled with purple. ***Inflorescence*** usually paniculate, with up to 3 branches, sometimes simple and racemose, densely many-flowered; peduncle 5–11.5 cm long; sterile bracts 2–3, remote, 0.4–1 cm long, tubular, obtuse, amplexicaul; rachis 8–15 cm long, branches 4.5–9 cm long; floral bracts, 2–3 × 1.5 mm, ovate, aristate. ***Flowers*** unscented or smelling faintly of cucumber; variously described as having orange-yellow sepals and petals spotted with red, the lip pale yellow, speckled with pale purple apically, the spur apex greenish yellow, or greenish yellow spotted purple, or cream, spotted with red, or orange-white, or lemon, spotted with purple. ***Pedicel*** with ***ovary*** 3 mm long, slender, gently curved. ***Sepals*** and ***petals*** spreading in mature flowers. ***Dorsal sepal*** 2–3 × 2 mm, ovate or oblong, obtuse, concave. ***Lateral sepals*** 4 × 2 mm, oblong, slightly oblique, apex obtuse and minutely erose. ***Petals*** 2–3 × 1 mm, oblong to elliptic, apex obtuse and minutely erose. ***Lip*** fleshy, lip wall swollen and callus-like at junction between side lobes and mid-lobe; side lobes very low, shallowly unequally bilobed, 1.5–2 mm wide, clasping base of column; mid-lobe 1–1.2 mm long, 1–1.2 mm wide at base when unflattened, triangular–ovate, subacute, concave and V-shaped in cross section; spur 3–3.5 mm long, 3 mm wide at base, saccate, broadly conical, ventral surface slightly constricted below middle; apex obtuse, not retuse, interior unadorned, borne at an acute angle to the pedicel with ovary. ***Column*** 0.5 mm long; anther-cap extinctoriform, pointed at the top, elongated into a little tail which is upturned at the apex; pollinia glabrous; stipes uncinate; viscidium large.

HABITAT AND ECOLOGY: Lower montane forest. Alt. 1500 to 2000 m. Flowering observed in January, April, September, October and December.

DISTRIBUTION IN BORNEO: SABAH: Mt. Kinabalu.

GENERAL DISTRIBUTION: Endemic to Borneo.

NOTES: *R. pinosukensis* is distinguished from all other species by its very unequally bilobed leaf tips with one lobule up to 4 cm long, the other only 0.2–0.4 cm long and tooth-like, and purple–stained or speckled leaf sheaths. The lip has a broadly conical spur to 3.5 mm long and the extinctoriform, i.e. candle snuffer-shaped, anther-cap has an elongated 'tail', which is upturned at the apex.

DERIVATION OF NAME: The specific epithet refers to the Pinosuk Plateau area of Mt. Kinabalu, from where several specimens have been collected.

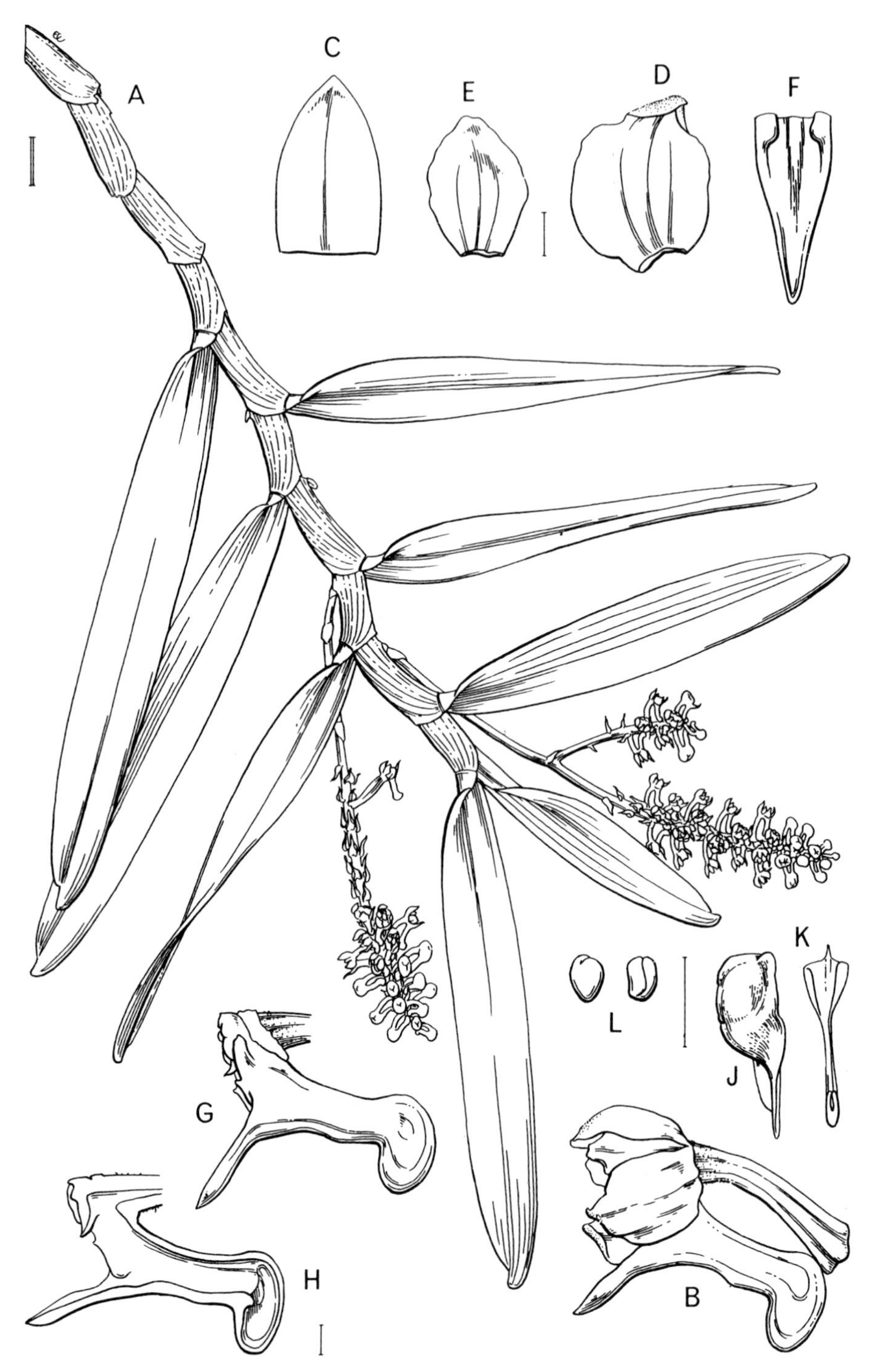

Figure 96. Robiquetia transversisaccata (Ames & C. Schweinf.) J.J. Wood. - A: habit. - B: flower, side view. - C: dorsal sepal. - D: lateral sepal. - E: petal. - F: lip, spur removed, front view.- G: pedicel with ovary, lip and column, side view. - H: pedicel with ovary, lip and column, longitudinal section. - J: anther-cap, side view. - K: stipes and viscidium, front view. - L: pollinia. A (habit) drawn from *J. & M.S. Clemens* 50340 and B–L from *Lamb* AL 196/84 by Eleanor Catherine. Scale: single bar = 1 mm; double bar = 1 cm.

96. ROBIQUETIA TRANSVERSISACCATA (Ames & C. Schweinf.) J.J. Wood

Robiquetia transversisaccata (*Ames & C. Schweinf.*) *J.J. Wood* in Wood, Beaman & Beaman, The Plants of Mt. Kinabalu 2, Orchids: 310, fig. 53 (1993). Type: Borneo, Sabah, Mt. Kinabalu, Kiau, *J. Clemens* 166 (holotype AMES).

Malleola transversisaccata Ames & C. Schweinf., Orchidaceae. 6: 228 (1920).

Epiphyte. Roots 1–2 mm in diameter, rigid, flexuous, sulcate, sparsely branching. ***Stem*** (6–)15–26(–30) cm long, merging into a root-like portion below, entirely enclosed by leaf sheaths, arcuate, fractiflex. ***Leaves*** 6–10, spreading; sheaths 1.2–1.5 cm long, lowermost becoming fibrous, the upper loose, infundibuliform, laterally flattened, coriaceous, striate-veined; blade 8.5–10.6 × 1.25–2.35 cm, oblong-ligulate, elliptic-oblong or linear, apex obtusely unequally bilobed with vertically semi-orbicular, minutely erose lobules, base complicate, margins sometimes undulate. ***Inflorescence*** densely many-flowered; peduncle 2.6–6 cm long, emerging through the middle or upper sheaths opposite the leaf bases, spreading to arcuate or decurved, with *c.* 3 scarious sterile bracts *c.* 4 mm long; rachis 4–7 cm long; floral bracts 2–3.5 mm long, triangular, cucullate, acuminate, reflexed. ***Flowers*** *c.* 7–8 mm across; pedicel with ovary greenish white; sepals and petals dark yellow spotted with red; lip yellow stained with orange-red, the spur white, pale greenish distally; column white. ***Pedicel*** with ***ovary*** 1–1.5 cm long, very slender. ***Dorsal sepal*** 4 × 2.8 mm, ovate, broadly rounded, sometimes obliquely bilobed, strongly concave, minutely denticulate. ***Lateral sepals*** 4.5 × 3 mm, obliquely obovate-oblong, obtuse, slightly cucullate or apically dorsally cornute. ***Petals*** 3.5 × 2 mm, oblong-spathulate, broadly rounded, minutely denticulate above middle. ***Lip*** 3-lobed, prominently spurred; side lobes 2–2.5 mm across apex, transversely oblong, obscure, truncate, erect; mid-lobe 1.5 mm long, triangular, obtuse, upcurved, fleshy and thickened within below apex; spur 5–11 mm long, cylindric, apex obtuse, incurved, provided on anterior wall below base of mid-lobe with a small callus and on posterior wall about 4 mm from apex with an upcurved ligulate, obtuse callus that is longitudinally sulcate beneath. ***Column*** 3 mm long; foot short; anther-cap triangular, cucullate, prolonged in front into a triangular beak. Plate 25D.

HABITAT AND ECOLOGY: Hill and lower montane forest, sometimes on ultramafic substrate. Alt. 900 to 1500 m. Flowering observed in April and November.

DISTRIBUTION IN BORNEO: SABAH: Crocker Range, Sinsuron road; Mt. Kinabalu.

GENERAL DISTRIBUTION: Endemic to Borneo.

DERIVATION OF NAME: The specific epithet is derived from the Latin *transversus*, lying crosswise, and *saccatus*, pouched or bag-shaped, in reference to the distal part of the spur.

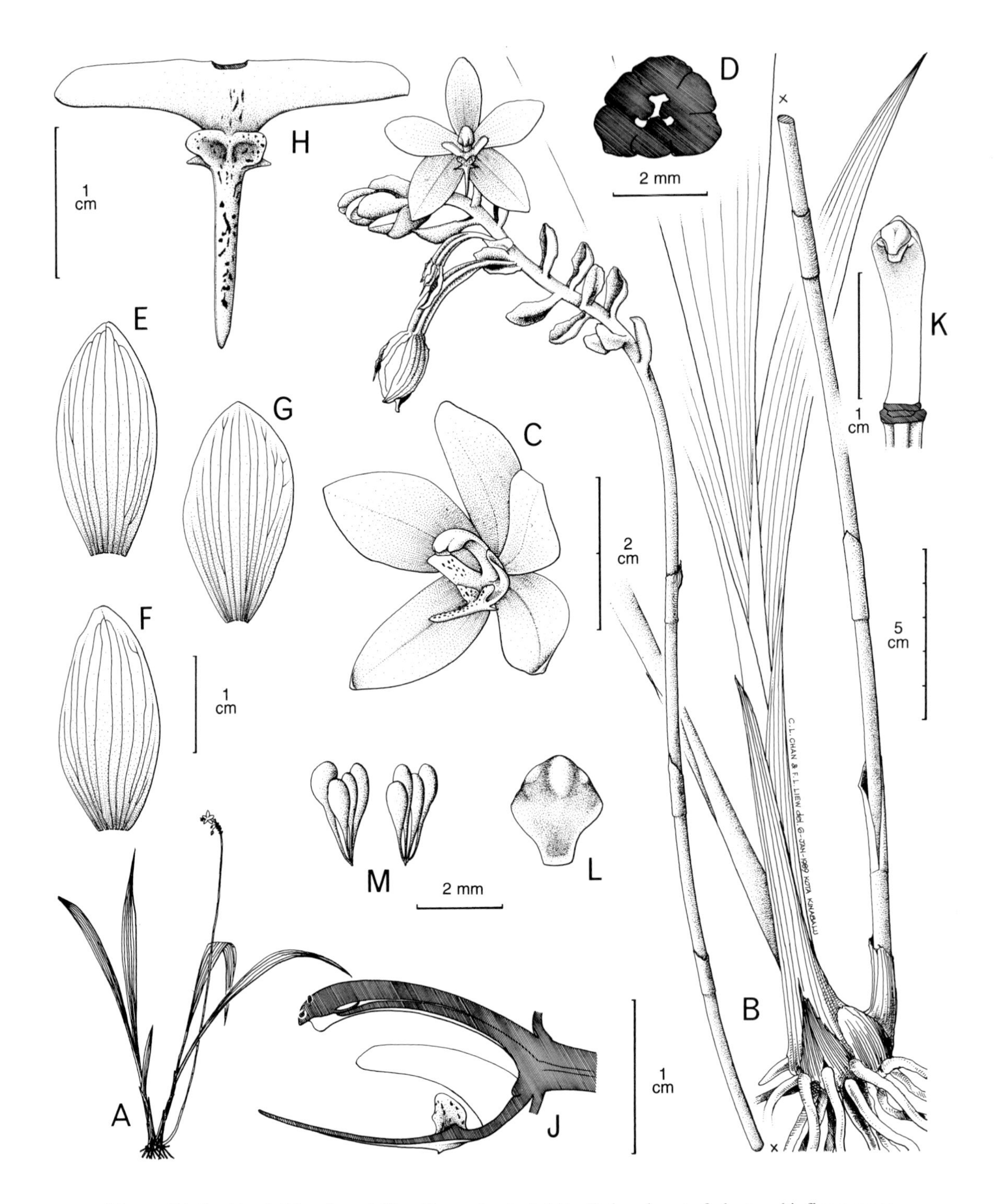

Figure 97. Spathoglottis microchilina Kraenzl. - A: habit. - B: basal part of plant and inflorescence. - C: flower, oblique view. - D: ovary, transverse section. - E: dorsel sepal. - F: lateral sepal. - G: petal. - H: lip, flattened, front view - J: pedicel with ovary, column and lip, longitudinal section. - K: column, front view. - L: anther-cap, back view. - M: pollinia. All drawn from *Chan* 127 by C.L. Chan and Lucy F.L. Liew.

97. SPATHOGLOTTIS MICROCHILINA Kraenzl.

Spathoglottis microchilina *Kraenzl.* in Bot. Jahrb. Syst. 17: 484 (1893). Type: Sumatra, mountains near Padang, *Micholitz* s.n. (holotype B).

Terrestrial. Pseudobulbs 1–2 × 1–1.4 cm, ovoid, enclosed in sheaths. ***Cataphylls*** 3–6, 1.5–18 cm long, acute to acuminate, imbricate. ***Leaves*** 28–50 × 2–5.5 cm, linear-lanceolate to lanceolate, acuminate, plicate, long-petiolate, thin-textured, many veined. ***Inflorescence*** emerging from base of pseudobulb, laxly few- to *c.* 45-flowered; peduncle 45-70 cm long, erect, brownish-green; non-floriferous bracts 1.2–3.5 cm long, ovate, obtuse to acute, sheathing, upper remote, the lower two a short distance apart; rachis (5–)8–21cm long; floral bracts 0.7–1.5 cm long, oblong-elliptic, obtuse to acute, strongly concave-cymbiform, olive-green. ***Flowers*** opening 1 or 2 in succession, borne 0.2–1.5 cm apart, 3–4 cm in diameter, unscented, sepals and petals bright yellow, lateral sepals sometimes slightly spotted with red in the centre at the base, lip darker yellow, stained, flecked and speckled with purplish red. ***Pedicel*** with ***ovary*** 3 cm long, narrowly clavate. ***Sepals*** shortly hirsute at base, spreading. ***Dorsal sepal*** 2.5 × 1.1 cm, oblong-elliptic, obtuse. ***Lateral sepals*** 2.4 × 1.1–1.2 cm, slightly obliquely elliptic, obtuse. ***Petals*** 2.3–2.4 × 1.2 cm, elliptic, obtuse. ***Lip*** 1.6–1.8 cm long; side lobes 1.1–1.2 cm long, 3–4 mm wide at base, *c.* 2.1 mm wide distally, narrowly oblong, obtuse, erect-incurved; mid-lobe 1.3–1.4 cm × 1.6–1.8 mm, linear-lanceolate, acute, margins somewhat revolute, with sparsely pilose triangular-acute basal auricles 1.6 × 1.1 mm; disc at base of mid-lobe bearing a bilobed fleshy callus, each half obovate, obtuse, divergent and *c.* 3 × 2 mm, with a few tufts of weak hairs below. ***Column*** 1.5–1.6 cm long, slender, curved; anther-cap 2.5–2.6 × 2.5 mm, oblong-ovate, truncate; pollinia eight, in two groups of four. Plate 26A.

HABITAT AND ECOLOGY: Hill forest; lower montane forest; on roadside banks and ditches; margins of secondary vegetation; among mossy rocks beside rivers. Alt.900–1700 m. Flowering recorded throughout the year.

DISTRIBUTION IN BORNEO: BRUNEI: Belait District. KALIMANTAN: Mt. Njapa, Kelai River. SABAH: Crocker Range; Mt. Kinabalu; Mongaya River; Tawai Forest Reserve, Telupid. SARAWAK: Mt. Aping; Lawas River.

GENERAL DISTRIBUTION: Peninsular Malaysia, Sumatra and Borneo.

NOTES: This species is common along ditch banks in the Kinabalu Park where it can be seen around Park Headquarters pushing up through the dense tangle of tall ferns such as *Dicranopteris* and *Diplopterygium.* The flowers open widely initially on certain days, but close after pollination and remain closed but fresh for a long time which has led to reports of cleistogamy. Its relationship with *S. aurea* Lindl., distributed in Peninsular Malaysia, Sumatra, Java and Borneo, requires further study. It may be that they represent one variable species, although *S. aurea* generally has a lip with a broader, obtuse mid-lobe up to 4 mm wide.

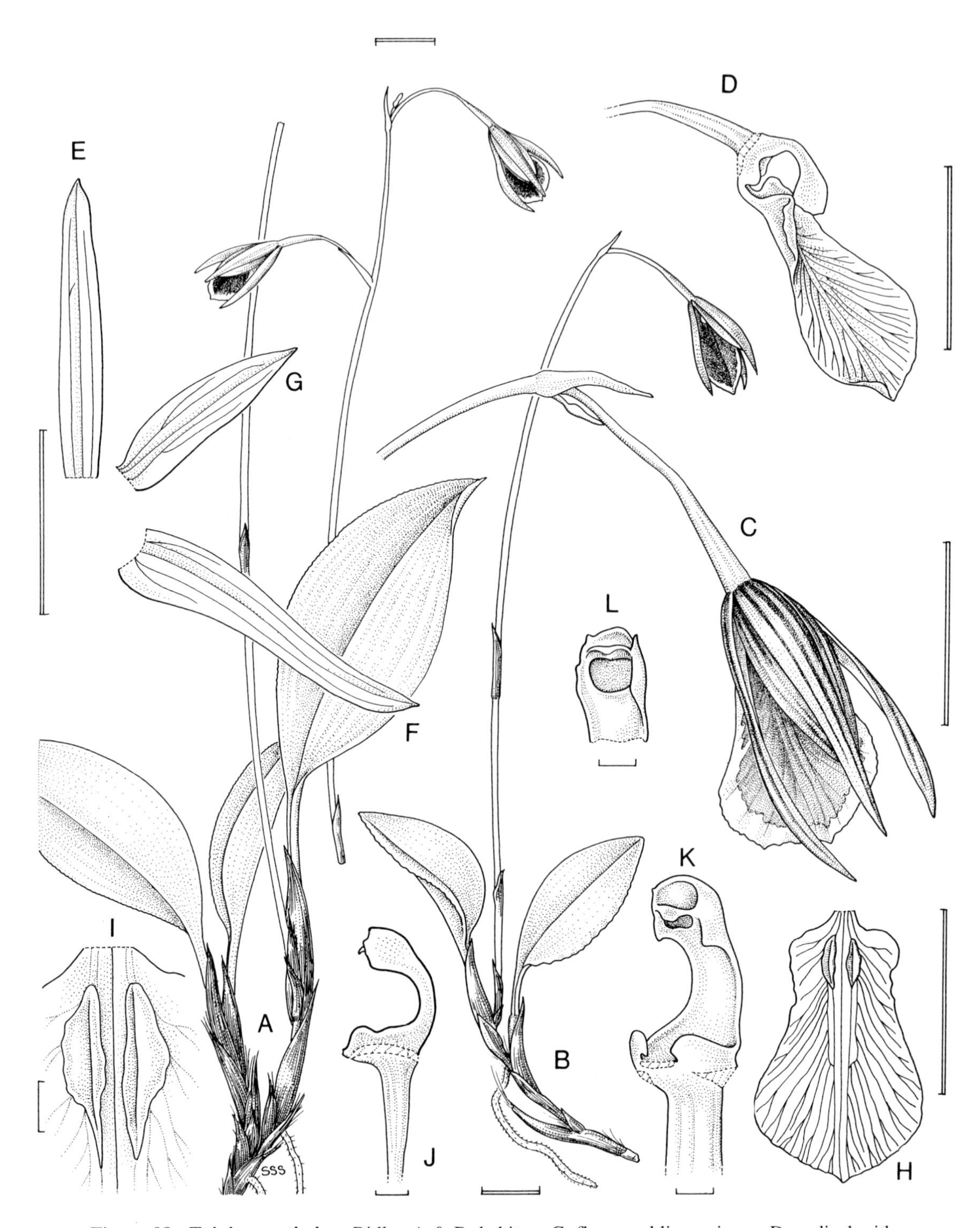

Figure 98. Tainia vegetissima Ridl. - A & B: habits. - C: flower, oblique view. - D: pedicel with ovary, lip and column, side view. - E: dorsal sepal. - F: lateral sepal. - G: petal. - H: lip, flattened. - I: base of lip showing keels. - J: ovary and column, side view. - K: upper portion of ovary and column, oblique view. - L: column apex, front view. A (habit) drawn from *Ridley* s.n. and B–L from *Lamb* AL 1428/92 by Susanna Stuart-Smith. Scale: single bar = 1 mm; double bar = 1 cm.

DERIVATION OF NAME: The generic name is derived from the Greek *spatha,* broad and *glotta,* tongue, alluding to the broad mid-lobe of the lip found in some species such as *S. gracilis* Rolfe ex Hook. f. and *S. plicata* Blume. The Greek specific epithet is derived from *micro,* small or little, and *chilus,* lipped.

98. TAINIA VEGETISSIMA Ridl.

Tainia vegetissima *Ridl.* in J. Linn. Soc., Bot.: 38: 328 (1908). Type: Peninsular Malaysia, Mt. Tahan, *Robinson* 5314 (holotype BM).

Terrestrial. Pseudobulb 1.1–1.7 × 0.1–0.2 cm, cylindrical to slightly ovoid. ***Leaf blade*** 2.4–5.8 × 1.1–2.2 cm, ovate, acute, margin somewhat thickened, one margin sometimes strongly crenulate, the other margin not or only slightly crenulate, rather fleshy, 11- to 15-veined, shiny purplish brown; petiole 0.5–2 cm long, crimson. ***Inflorescence*** terminal, 19–27 cm long, 1- to 4-flowered, arising *c.* 4 mm above base of pseudobulb; peduncle 13–24.8 cm long, with 4 scales up to 1.8 cm long; rachis 3.2–5.6 cm long; floral bracts 4–10 mm long, triangular, spreading. ***Flowers*** mostly opening simultaneously; pale yellow closely lined crimson, the lip crimson, edged with yellow. ***Pedicel*** with ***ovary*** 0.8–1.6 cm long. ***Dorsal sepal*** 11–14 × 1.5–2 mm, elliptic, acute. ***Lateral sepals*** 11.5–13 × 2–2.5 mm, obliquely elliptic, slightly falcate, acute. ***Petals*** 8–12.5 × 2–2.5 mm, obliquely elliptic, straight to slightly falcate, acute. ***Lip*** entire, pandurate in outline, 10.5–15 × 7.5–9 mm, obtuse; disc with two keels 1.5–2.5 mm long, raised and plate-like from near base, highest halfway. ***Column*** 3.5–6 mm long; stelidia absent; apex truncate, margin entire; foot 1 mm long; anther-cap 1 × 1.5 mm; pollinia eight. Plate 26B.

HABITAT AND ECOLOGY: Lower montane mossy forest. Alt. 1200 to 1300 m. Flowering observed in September.

DISTRIBUTION IN BORNEO: SARAWAK: Mt. Dulit.

GENERAL DISTRIBUTION: Peninsular Malaysia and Borneo.

DERIVATION OF NAME: The generic name is Greek and means a fillet, possibly referring to the long narrow leaves with their usually elongated petioles, or perhaps to the elevated keels borne longitudinally on the lip. The specific epithet is derived from the Latin *vegitus*, fresh or vigorous, in reference to the shiny leaves.

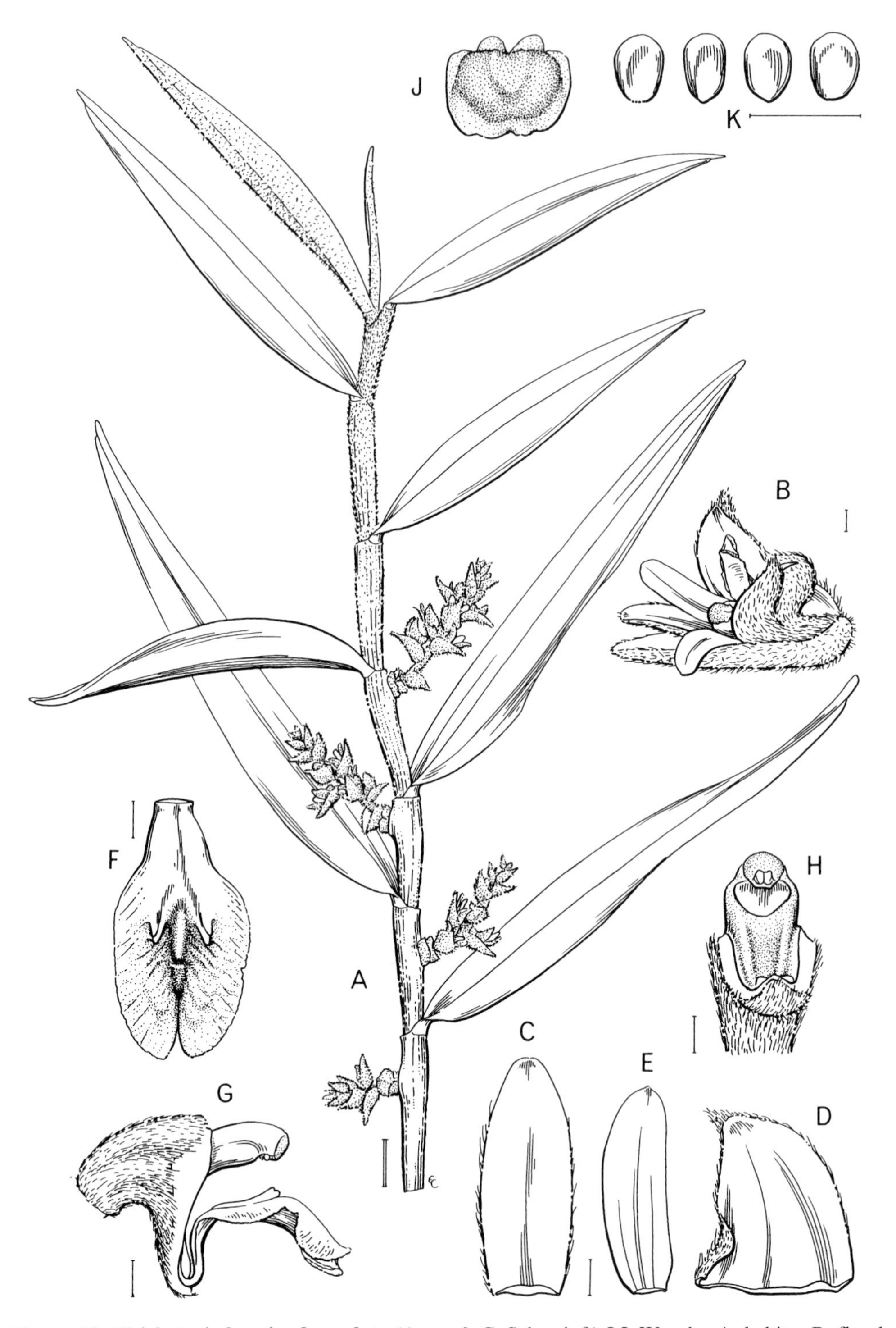

Figure 99. Trichotosia brevipedunculata (Ames & C. Schweinf.) J.J. Wood. - A: habit. - B: floral bract and flower, side view. - C: dorsal sepal. - D: lateral sepal. - E: petal. - F: lip, flattened.- G: pedicel with ovary, lip and column, side view. - H: column with anther-cap, front view. - J: anther-cap, back view. - K: four of eight pollinia. A (habit) drawn from *J. & M.S. Clemens* 26837 and B–K from *Wood* 619 by Eleanor Catherine. Scale: single bar = 1 mm; double bar = 1 cm.

99. TRICHOTOSIA BREVIPEDUNCULATA (Ames & C. Schweinf.) J.J. Wood

Trichotosia brevipedunculata (*Ames & C. Schweinf.*) *J.J. Wood* in Wood, Beaman & Beaman, Plants of Mt. Kinabalu 2, Orchids: 333 (1993). Type: Borneo, Sabah, Mt. Kinabalu, Marai Parai, *J. Clemens* 255 (holotype AMES).

Eria brevipedunculata Ames & C. Schweinf., Orchidaceae 6: 118 (1920).

Epiphyte or ***terrestrial. Roots*** much branched, finely brown-tomentose. ***Stems*** to 1 m long, more or less approximate, entirely covered by leaf sheaths, terete, 5–6 mm in diameter, erect to gently curved. ***Leaf sheath*** 2.5–4.5 cm long, tubular, thick, striate, densely adpressed cinnamon-pilose, the hairs mostly disappearing on the older sheaths, but persistent at the slightly swollen nodes. ***Leaf blade*** 4.5–15.5 × 1–2(–2.5) cm, narrowly elliptic, narrowed to a very unequally and obscurely bilobed obtuse apex, sometimes oblique, cuneate at base, very thick, coriaceous, rigid, densely cinnamon-pilose when young, ascending, many-veined; petiole 2–5 mm long. ***Inflorescences*** 1–2, borne opposite the bases of the upper leaves, subdensely several-flowered; peduncle and rachis densely spreading brown-pilose; peduncle 5–8 mm long, emerging from 2 amplexicaul basal bracts 5–8 mm long; rachis 2–4 cm long; floral bracts 8–12 mm long, 4–9 mm wide at base when flattened, ovate-lanceolate, acuminate, deeply concave, distichous, spreading to slightly reflexed, green, spreading brown or reddish-brown pilose on the exterior. ***Flowers*** small; densely reddish brown pilose; unscented; sepals and petals greenish or greenish yellow; lip lemon-coloured, sometimes with a reddish median line; column pale green; anther-cap cream. ***Pedicel*** with ***ovary*** 3 mm long. ***Dorsal sepal*** 5–6.5 × 2–3 mm, oblong-elliptic, concave, apex obtuse and thickened. ***Lateral sepals*** 4.5–6 mm long, 3–5 mm wide at base, ovate-triangular, dorsally carinate and much thickened at the obtuse apex, oblique, falcate. ***Mentum*** obtuse, saccate. ***Petals*** 4.7–8 × 1–1.9 mm, oblong-linear, sometimes falcate, obtuse, thin-textured, sparsely ciliate. ***Lip*** 5.5–6 mm long, 3.5–4 mm wide across side lobes, obovate-spathulate in outline, claw rather stout with involute margins, gradually dilated into the sharply reflexed blade; side lobes obscure, semi-orbicular; mid-lobe reniform, deeply bilobed into two suborbicular lobules *c.* 2 mm across, with a minute blunt apicule in the sinus; disc with a pair of short tooth-like keels rising from the centre, with a farinose central callus nearer the apex. ***Column*** 1–2 mm long; foot 3–3.5 mm long, stout; pollinia eight. Plate 26C.

HABITAT AND ECOLOGY: Lower montane forest; oak-laurel forest, with *Agathis*; sometimes on ultramafic substrate. Alt. 1200 to 2100 m. Flowering observed in January, February, June, July and September to December.

DISTRIBUTION IN BORNEO: SABAH: Crocker Range, Sinsuron road; Mt. Kinabalu; Mt Trus Madi.

GENERAL DISTRIBUTION: Endemic to Borneo.

DERIVATION OF NAME: The generic name is derived from the Greek *trichotos*, hairy, referring to the hairy leaves, inflorescences and flowers of the majority of species. The specific epithet is derived from the Latin *brevis*, short, and *pedunculatus*, provided with a peduncle, referring to the shortly pedunculate inflorescence.

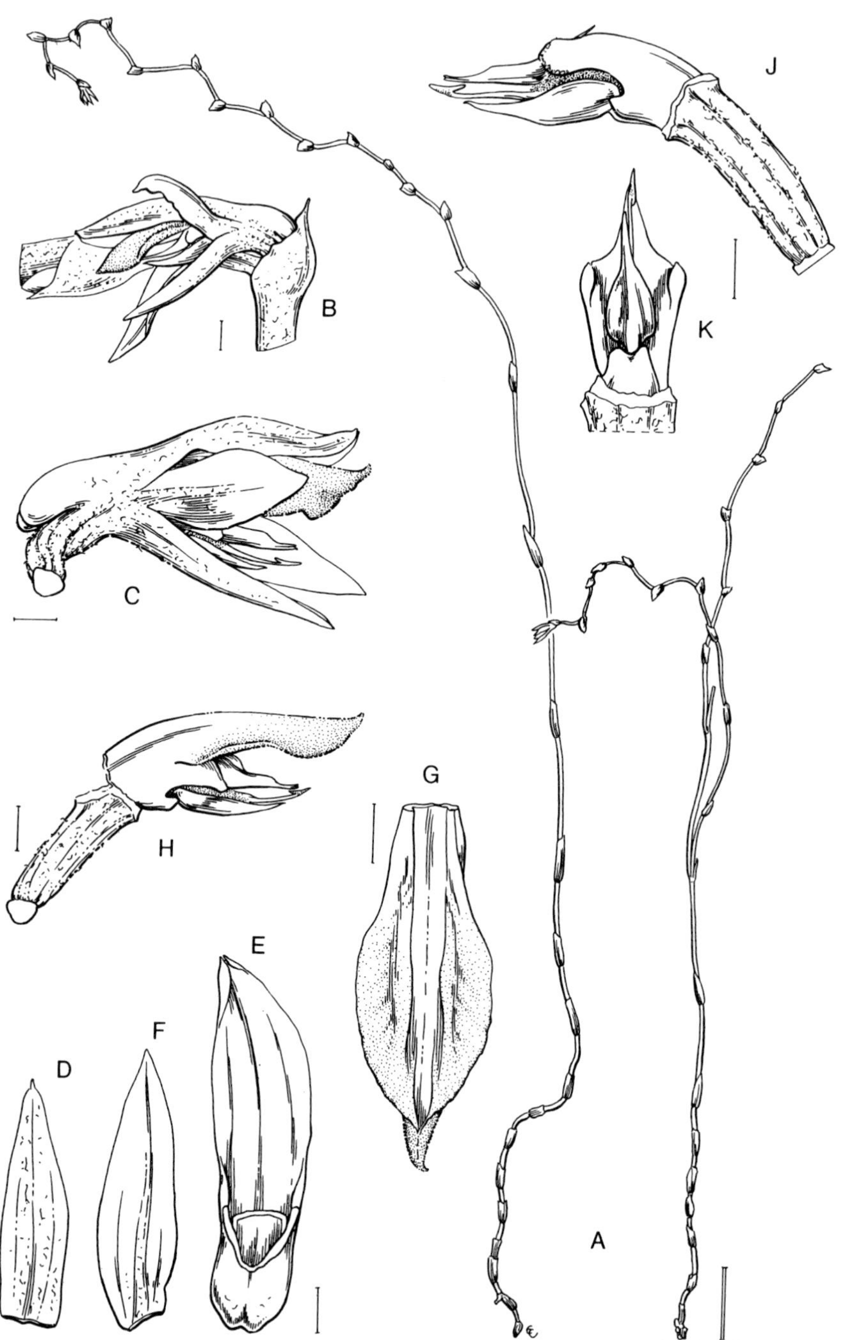

Figure 100. Tropidia connata J.J. Wood & A. Lamb. - A: habit. - B: part of inflorescence. - C: flower, side view. - D: dorsal sepal. - E: synsepal and spur, lip detached, front view. - F: petal.- G: lip, front view. - H: pedicel with ovary, column and lip, side view. - J: pedicel with ovary and column, side view. - K: column and anther-cap, back view. All drawn from *Lamb, Surat & Lim* in *Lamb* AL 1512/92 (holotype) by Eleanor Catherine. Scale: single bar = 1 mm; double bar = 1 cm.

100. TROPIDIA CONNATA J.J. Wood & A. Lamb

Tropidia connata *J.J. Wood & A. Lamb* in Wood & Cribb, Checklist Orchids Borneo: 47, fig. 4 (1994). Type: Borneo, Sabah, Sipitang District, Mt. Lumaku, *Lamb, Surat & Lim* in *Lamb* AL 1512/92 (holotype K).

Erect, clump-forming ***saprophyte. Stems*** 10 to 15 per clump, 15–20 cm high, wiry, simple or branched, internodes 0.8–2.2 cm long, 1 mm wide, off-white to cream-coloured, glabrous above, sparsely ramentaceous on lower internodes, bearing 5–10 ovate, acute to acuminate, adpressed, sparsely ramentaceous, brown sterile sheaths each 3–5 mm long. ***Inflorescence*** laxly 10- to 15-flowered, 1 or 2 flowers open at a time; rachis 3–9 cm long, fractiflex, sparsely ramentaceous; floral bracts 3 mm long, ovate, acute, off-white tipped brown, sparsely ramentaceous. ***Flowers*** non-resupinate; 1 cm across; white, apex of the lip yellow. ***Pedicel*** with ***ovary*** 2 mm long, sparsely ramentaceous. ***Sepals*** and ***petals*** spreading, the sepals sparsely ramentaceous on reverse. ***Dorsal sepal*** 5 × 1.5 mm, narrowly elliptic, acute. ***Lateral sepals*** connate into a 4–5 × 2.5 mm ovate-elliptic, cymbiform synsepal, its apex minutely bifid. ***Petals*** 6 × 1.5 mm, narrowly elliptic, acute. ***Lip*** 5 mm long, 2 mm wide when flattened, entire, elliptic, acute, concave, rather fleshy, margin minutely erose, curving back to a horizontal position, spur saccate, obtuse, enclosed by synsepal, *c.* 1.1 mm long. ***Column*** 1 mm long; rostellum *c.* 1.5 mm long; anther-cap rostrate; pollinia 2. Plate 26D.

HABITAT AND ECOLOGY: Mixed hill-dipterocarp forest with isolated patches of podsol forest on sandstone. Alt. 400 to 600 m. Flowering observed in June.

DISTRIBUTION IN BORNEO: SABAH: Mt. Lumaku.

GENERAL DISTRIBUTION: Endemic to Borneo.

NOTES: This curious saprophyte is easily distinguished from *T. saprophytica* J.J. Sm. (see Vol. 1, figure 95), the only other saprophytic species, by the fractiflex rachis, connate lateral sepals forming a synsepal, and shortly spurred lip. The flowers recall those of the non-saprophytic *T. angulosa* (Lindl.) Blume, but are much smaller and have a shorter spur.

DERIVATION OF NAME: The generic name is derived from the Greek *tropideion*, a keel, in reference to the boat-shaped lip common in this genus. The specific epithet is derived from the Latin *connatus*, fused so as not to be separated without injury, and referring to the fused lateral sepals.

REFERENCES

Introduction

Brieger, F.G. (1981). Subtribus Dendrobiinae. In Brieger, F.G., Maatsch, R. & Senghas, K. (eds.), a revision of Schlechter, R., *Die Orchideen*, ed. 3. Aufl. 1: 636–752. Paul Parey, Berlin.

Kraenzlin, F. (1910). Orchidaceae–Monandrae–Dendrobiinae 1. In Engler, A., *Das Pflanzenreich*, Heft. 45.

Pedersen, H. Ae., Wood, J.J., and Comber, J.B. (1996). A revised subdivision and bibliographical survey of *Dendrochilum* (Orchidaceae). *Opera Bot.* XXX: 1–84.

Schlechter, R. (1912 [1911–1914]). Die Orchidaceen von Deutsch-Neu-Guinea. *Repert. Spec. Nov. Regni Veg., Beih.* 1: 1–1079. The Orchidaceae of German New Guinea (English translation of above) (1982). The Australian Orchid Foundation, Melbourne.

van Steenis, C.G.G.J. (1979). Plant-geography of east Malesia. *J. Linn. Soc., Bot.* 79: 97–178.

Wood, J.J. (1990). The diversity of *Dendrobium* in Borneo. *Malayan Orch. Rev.* 24: 33–37.

Descriptions and Figures

Beccari, O. (1902). *Nelle Foreste di Borneo—viaggi e ricerche di un naturalista.* Salvadore Landi, Firenze.

Christenson, E.A. (1992). An enigmatic blue. *Amer. Orchid Soc. Bull.* 61: 242–247.

Cribb, P.J. (1987) *The genus Paphiopedilum.* Collingridge, London.

Du Puy, D. and Cribb, P.J. (1988). *The genus Cymbidium.* Christopher Helm, London.

Kitayama, K. (1991). *Vegetation of Mount Kinabalu Park, Sabah, Malaysia. Map of physiognomically classified vegetation.* East-West Center, Honolulu, Hawaii.

Seidenfaden, G. (1977). Contributions to the Orchid Flora of Thailand VIII. *Bot. Tidsskr.* 72, 1: 1–14.

Turner, H. (1992). A revision of the Orchid Genera *Ania, Hancockia, Mischobulbum, Tainia. Orchid Monographs* 6. Rijksherbarium and Hortus Botanicus, Leiden.

Wood, J.J., Beaman, R.S. and Beaman, J.H. (1993). *The Plants of Mount Kinabalu 2, Orchids.* Royal Botanic Gardens, Kew.

Wood, J.J. and Cribb, P.J. (1994) *A Checklist of the Orchids of Borneo.* Royal Botanic Gardens, Kew.

IDENTIFICATION LIST

This list is based on a selection of specimens. Each species is arranged according to the descriptive part of the volume. Collections are cited, under each locality, in alphabetical order using the collector's name. Precise locality details for some species are withheld for reasons of conservation.

1. **Acanthephippium eburneum**: SARAWAK: Known only from type collections.

2. **Acriopsis gracilis**: SABAH: Tenom District: 300 m: *Lamb* AL 34/82 (K). Nabawan, 450 m: *Lamb* AL 324/85 (K) & 1110/89 (K). Kinabatangan District: Maliau Basin, 1000 m: *Lamb* AL 915/88 (K).

3. **Agrostophyllum laterale:** SARAWAK: Bario, Sungai Menalio, Pa Dappur, 950 m: *Lee* S. 51065 (AAU, K, L, MEL, SAR, SING).

4. **Appendicula fractiflexa**: SABAH: Kinabatangan District: Maliau Basin, eastern rim area, 1350–1400 m: *Lamb* AL 1409/92 (K). Mt. Kinabalu, Penibukan, 1200–1500 m: *J. & M.S. Clemens* 31567 (BM).

5. **Appendicula longirostrata**: SABAH: Keningau District: Crocker Range, Kimanis road: *Abas* SAN 85938 (K, L, SAN, SAR). Penampang District, Kota Kinabalu to Tambunan road, 1350–1500 m: *Beaman* 10496a (K). Crocker Range: between Mt. Alab & Mt. Emas, 1560 m: *Wood* 792 (K). Mt. Kinabalu: Tenompok, 1500 m: *Carr* SFN 27344 (SING); Tahubang River, 1200 m: *J. & M.S. Clemens* 30704 (BM); Kinateki River Head, 2100 m: *J. & M.S. Clemens* 31788 (BM); Tinekuk Falls, 1800 m: *J. & M.S. Clemens* 40906 (BM, K); Mesilau River, 2700 m: *J. & M.S. Clemens* 51434 (BM, K) & 51599 (BM, K); Pig Hill, 1700–2000 m: *Sutton* 11 (K). SARAWAK: Mt. Penrissen, 1250 m: *Othman* S. 61338 (K, SAR).

6. **Appendicula uncata subsp. sarawakensis**: BRUNEI: Belait District, Melilas, 30 m: *Atkins* 571 (K). SABAH: Lahad Datu District, Mt. Nicola, 700 m: *Lamb* AL 1457/92 (K). SARAWAK: Bako National Park, Telok Asam, 100 m: *Purseglove* P. 4900 (K). Mt. Matang, 450 m: *J. & M.S. Clemens* 22407 (K). Matang road: *Ridley* s.n. (K). Baram: *Hose* 149 (K). Mt. Mulu National Park, Bukit Berar, 275 m: *Nielsen* 772 (AAU). Simanggang District, Tenso, Sungai Senang Besar: *Anderson* 9808 (K, SAR).

7. **Arachnis grandisepala**: SABAH: Known only from the type.

8. **Arachnis longisepala**: SABAH: Known only from the type.

9. **Ascochilopsis lobata**: SABAH: Known only from the type.

10. **Brachypeza zamboangensis**: SABAH: Batu Urun, 500 m: *Lamb* AL 374/85 (K).

11. **Bulbophyllum beccarii**: SABAH: Nabawan, 450 m: *Lamb* AL 1212/90 (K). Upper Kinabatangan, near Lanas, 600 m: *Kidman Cox* 170 (K). SARAWAK: Betong, Saribas Forest Reserve, near sea level: *Anderson* 13258 (K, SAR).

12. **Bulbophyllum kemulense**: KALIMANTAN TIMUR: Mt. Kemal, 1800 m: *Endert* 4278 (BO). SABAH: Mt. Alab, 1600 m: *Lamb* AL 1324/91 (K).

13. **Bulbophyllum polygaliflorum**: SABAH: Maliau Basin: sight record by *Lamb*. SARAWAK: Mt. Mulu National Park, 1700 m: *Hansen* 499 (C, K).

14. **Calanthe undulata**: KALIMANTAN TIMUR: Apokayan area, sight record by *Cribb*. SABAH: Nabawan, 450 m: *Lamb* AL 632/86 (K).

15. **Cleisocentron merrillianum**: SABAH: Mt. Kinabalu: Marai Parai Spur, 1600 m: *Bailes & Cribb* 819 (K); Bundu Tuhan, 1100 m: *Carr* SFN 27893 (K); Tahubang River, 1400 m: *J. & M.S. Clemens* 40344 (BM); Penibukan, 1200 m: *J. & M.S. Clemens* 40634 (BM); Little Mamut River, 1400 m: *Collenette* 1020 (K); Pinosuk Plateau, 1500 m: *Lamb* SAN 89677 (K, SAN).

16. **Corybas muluensis**: SABAH: Crocker Range, Kimanis road, 1350 m: *Wood* 823 (K). Sarawak: Mt. Mulu, 2000 m: *Lewis* 352 (K).

17. **Cymbidium borneense**: SABAH: Crocker Range, Mt. Alab, 1050–1200 m: *Lamb* C 18 (K). Mt. Kinabalu: Penataran Ridge, 1200 m: *Lamb* SAN 93357 (K, SAN). SARAWAK: Mt. Mulu National Park, 240 m: *Lewis* 314 (K).

18. **Dendrobium alabense**: SABAH: Mt. Alab, 1800 m: *Wood* 777 (K). Mt. Kinabalu: Pig Hill, 2000–2300 m: *Beaman* 9890 (K); Marai Parai, 1600 m: *Carr* 3123, SFN 26557 (SING); Kilembun River, 1500–1800 m: *J. & M.S. Clemens* 32608 (BM); Gurulau Spur, 1800–2100 m: *J. & M.S. Clemens* 50819 (BM); Tenompok, 1500 m: *J. & M.S. Clemens* s.n. (BM).

19. **Dendrobium beamanianum**: SABAH: Known only from the type.

20. **Dendrobium cymboglossum**: SABAH: Tawao: *Elmer* 21889 (K); Sepilok Forest Reserve: *Lamb* SAN 93402 (K, SAN).

21. **Dendrobium hamaticalcar**: SABAH: Mt. Kinabalu, Kaung, 400 m: *Carr* 3019 (K, SING). Mt. Trus Madi: *Bacon* 110 (E).

22. **Dendrobium kiauense**: SABAH: Mt. Kinabalu: Bundu Tuhan, 1400 m: *Carr* SFN 27235 (SING); Gurulau Spur: *J. Clemens* 304 (AMES, BM, K); Tenompok, 1500 m: *J. & M.S. Clemens* 28396 (BM, G, K); Hempuen Hill, 800–1000 m: *Cribb* 89/31 (K). Sipitang District, Ulu Long Pa Sia, 1290 m: *Wood* 693 (K).

23. **Dendrobium lambii**: SABAH: Mt. Alab, 1800 m: *Lamb* LMC 2251, SAN 92326 (K, SAN). Maliau Basin, 1000 m: *Lamb* AL 1427/92 (K).

24. **Dendrobium lancilobum**: SABAH: Mt. Alab, 1650 m: *Cribb* 89/46 (K). Sipitang District, Ulu Long Pa Sia, beside Maga River, 1260 m: *Wood* 684 (K).

25. **Dendrobium limii**: SABAH: Known only from the type.

26. **Dendrobium lowii**: SABAH: Tawau: *Lamb* s.n. (K).

27. **Dendrobium nabawanense**: SABAH: Nabawan, 540 m: *Collenette* 2292 (K). Lahad Datu District, Mt. Nicola, 700 m: *Lamb* AL 1452/92 (K).

28. **Dendrobium oblongum**: KALIMANTAN TIMUR: Mt. Beratus, 900 m: *Meijer* 711 (BO, K). SABAH: Mt. Kinabalu: Mahandei River, 1100 m: *Carr* 3037, SFN 26372 (BM, SING); Tenompok, 1500 m: *J. & M.S. Clemens* 28171 (BM, G, K); Marai Parai, 1500 m: *J. & M.S. Clemens* 32254 (BM); Gurulau Spur, 1700 m: *J. & M.S. Clemens* 50700 (BM). Crocker Range, Kota Kinabalu to Tambunan road, 1350–1500 m: *Beaman* 10489 (K). Mt. Trus Madi, 1560 m: *Wood* 883 (K). SARAWAK: Mt. Penrissen, 1140 m: *Museum collector* s.n. (SAR, K).

29. **Dendrobium panduriferum**: SABAH: Mt. Kinabalu, Lohan River, 600 m: *Lamb* AL 1109/89 (K).

30. **Dendrobium patentilobum**: SABAH: Mt. Kinabalu: Hempuen Hill, 800–1200 m: *Beaman* 7715 (K); Mamut Copper Mine, 1400–1500 m: *Beaman* 10321 (K); Penibukan: *Carr* 3044, SFN 26367 (SING); Numeruk Ridge, 1400 m: *J. & M.S. Clemens* 40060 (BM). Maliau Basin, 1100 m: *Lamb* AL 1424/92 (K). Mt. Nicola, 600 m: *Lamb* AL 1446/92 (K).

31. **Dendrobium smithianum**: SABAH: Pensiangan District, Batu Ponggol, 300–400 m: *Lamb* AL 385/85 (K). Kinabatangan District, Batu Urun: 400 m: *Lamb* AL 1115/89 (K). Sipitang District, Sungai Pa Sia, 1300 m: *Vermeulen* 555 (K, L).

32. **Dendrobium tridentatum**: SABAH: Mt. Alab, 1770 m: *Lamb* s.n. (K). Mt. Kinabalu: Mesilau Cave, 1900–2200 m: *Beaman* 9574 (K); Marai Parai, 1500 m: *Carr* SFN 27976 (SING); Tahubang River Head, 1500 m: *J. & M.S. Clemens* 40702 (BM); Layang-Layang/Mesilau Cave Route, 2100 m: *Collenette* 892 (K); Pig Hill: *Sutton* 17 (K).

33. **Dendrobium trullatum**: SABAH: Nabawan, 540 m: *Collenette* 2287 (K) & *Lamb* AL 841/87 (K).

34. **Dendrobium ventripes**: SABAH: Mt. Kinabalu: Kemburongoh, 2100 m: *Carr* SFN 27460 (SING); Minetuhan Spur, 2600 m: *J. & M.S. Clemens* 33792 (BM). Mt. Lumaku, 1740 m: *Lamb* SAN 92308 (K, SAN). Mt. Trus Madi, 1950 m: *Sands* 3675 (K) & 1920 m: *Wood* 879 (K).

35. **Dendrobium xiphophyllum**: SABAH: locality unknown, cult. Tenom Orchid Centre, *Chan* s.n. (K).

36. Dendrochilum anomalum: KALIMANTAN TIMUR: Apokayan, Kayan River between Long Ampung and Long Nawan, 550–600 m: *de Vogel* s.n., cult. Leiden no. 913346 (L). SABAH: Mt. Kinabalu, Hempuen Hill, 600 m: *Lamb & Surat* in *Lamb* AL 1365/91 (K). SARAWAK: Mt. Dulit, 800–1000 m: *Richards* 2497 (K, SING).

37. Dendrochilum auriculilobum: SABAH: Sipitang District, Long Pa Sia area: *Phillipps et al.* SNP 2948, 2969 & 2970 (Kinabalu Park Herbarium, Mount Kinabalu).

38. Dendrochilum crassum: SABAH: Mt. Alab, 900 m: *J.B. Comber & Lamb* in *Lamb* AL 345/85 (K). Kimanis road, 1500 m: *Vermeulen & Duistermaat* 693 (K, L) & 1200 m: *Wood* 584 (K). Mt. Kinabalu, Minitinduk Gorge, 900 m: *Carr* 3172, SFN 26668 (K, SING). Sinsuron road, 1500 m: *Wood* 733 (K). Mt. Trus Madi: *Joseph et al.* SAN 113509 (K, L, SAN, SAR).

39. Dendrochilum cruciforme var. cruciforme: SABAH: Sinsuron road, mile 27: *Bacon* 187 (E). Mt. Kinabalu: Penibukan, 1200–1500 m: *J. & M.S. Clemens* 30471 (E); Penataran Basin, 900–1200 m: *J. & M.S. Clemens* 34329 (AMES, BM, E, HBG, L); Tinekuk Falls, 1800 m: *J. & M.S. Clemens* 50278 (AMES, BM, K). Mt. Alab, 2000 m: *de Vogel* 8661 (K, L).

39a. Dendrochilum cruciforme var. longicuspe: SABAH: Known only from the type.

40. Dendrochilum cupulatum: SABAH: Kimanis road, 1400 m: *Kitayama* 893 (UKMS) & 1500–1600 m: *Vermeulen & Duistermaat* 667 (K, L). SARAWAK: Mt. Mulu, southern summit: 2100 m: *Argent & Coppins* 1126 (E, K). Mt. Mulu, west ridge, 1880 m: *Nielsen* 806 (AAU, K). Hose Mountains, ridge leading to Bukit Batu, 1600–1650 m: *de Vogel* 1013 (L).

41. Dendrochilum devogelii: SABAH: Known only from the type.

42. Dendrochilum exasperatum: SABAH: Tenom District, Mt. Anginon, 900 m: *H.F. Comber* 4037 (K). Mt. Kinabalu: Tenompok, 1600 m: *Carr* 3668, SFN 28029 (K, SING); Dallas, 900 m: *J. & M.S. Clemens* 26784A (BM) & 27256 (BM, K); Penibukan, 1200–1500 m: *J. & M.S. Clemens* 30640 (BM). Mt. Lumaku, 1200 m: *J.B. Comber* 122 (K). SARAWAK: Mt. Dulit, 1230 m: *Synge* S. 436 (K).

43. Dendrochilum galbanum: SARAWAK: Mt. Murud, 2250 m: *Yii* S. 44430 (K, L, SAR).

44. Dendrochilum geesinkii: KALIMANTAN TIMUR: Between Long Bawan and Panado, 1400 m: *Geesink* 8989 (L). SARAWAK: route to Mt. Batu Lawi, Bario, 1290 m: *Awa & Lee* S. 50586 (AAU, K, KEP, L, MEL, SING).

45. Dendrochilum gibbsiae: BRUNEI: Temburong District, Mt. Pagon, 1300 m: *de Vogel* 2043 (L). KALIMANTAN TIMUR: Apokayan, near Long Sungai Barang,

800 m: *de Vogel* 574 (L). SABAH: Sinsuron Road: *Argent* s.n., RBGE no. 801344 (E). Mt. Kinabalu: Pinosuk Plateau, golf course site, 1700–1800 m: *Beaman* 7239 (K); Kadamaian River, 1800 m: *Carr* 3709, SFN 28034 (SING); Kilembun River, 1400 m: *J. & M.S. Clemens* 32432 (BM, K); Marai Parai/Nungkek, 1400 m: *J. & M.S. Clemens* 32552 (BM); Park Headquarters, 1500 m: *Wood* 621 (K). SARAWAK: Mt. Mulu National Park, Mt. Api, 870 m: *Anderson* S. 30838 (A, E, K, L, SING) & 1050 m: *Chai* S. 30098 (A, BO, K, KEP, L, SAN, SING). Mt. Murud, 1600 m: *Yii* S. 44402 (K, L, SAR). Belaga District, Linau-Balui divide, Sungai Nawai, 780 m: *Burtt* B. 11460 (E).

46. **Dendrochilum grandiflorum**: SABAH: Mt. Kinabalu: Summit Trail, 2100–2925 m: *Bogle* 548 (AMES); Paka-Paka Cave, 3100 m: *Carr* 3476, SFN 27430A (K, SING); Gurulau Spur, 2400–2700 m: *J. & M.S. Clemens* 50661 (BM, K); Liwagu River Head, 2300 m: *Meijer* SAN 24131 (K, SAN); Layang-Layang, 2400–2700 m: *Sato* 699 (UKMS).

47. **Dendrochilum haslamii**: SABAH: Mt. Kinabalu: Paka-Paka Cave, 3100 m: *Carr* SFN 36565 (SING); Kinateki River Head, 2700 m: *J. & M.S. Clemens* 31830 (BM); Janet's Halt/Sheila's Plateau, 2900 m: *Collenette* 21535 (K); Summit Trail, 2700 m: *Gunsalam* 3 (K).

48. **Dendrochilum hologyne**: SABAH: Sipitang District, Meligan (Maligan) to Long Pa Sia trail, 1500 m: *Vermeulen & Duistermaat* 905 (K, L). Sipitang District, ridge between headwaters of Sungai Maga & Sungai Malabid, 1600 m: *Vermeulen & Duistermaat* 1011 (K, L). Sipitang District, ridge east of Sungai Maga, 1450 m: *de Vogel* 8339 (L). Sipitang District, Ulu Long Pa Sia, 1400 m: *Wood* 646 (K). SARAWAK: Mt. Dulit, Dulit Ridge, 1200 m: *Synge* S. 415 (K). Mt. Murud, 1900–2400 m: *Mjöberg* 65 (AMES).

49. **Dendrochilum hosei**: SARAWAK: Known only from the type.

50. **Dendrochilum imbricatum**: SABAH: Mt. Kinabalu: Tenompok, 1500 m: *Carr* 3653, SFN 27998 (SING); Kiau: *J. Clemens* 318 (AMES); Mamut River, 1400 m: *Collenette* 1042 (K); Park Headquarters, 1500 m: *Wood* 618 (K). SARAWAK: Sungai Pa Mario, Ulu Limbang, route to Mt. Batu Lawi, Bario, 1530 m: *Awa & Lee* S. 50704 (K, L, SAR, SING).

51. **Dendrochilum imitator**: SABAH: Sipitang District, Long Pa Sia to Long Samado trail, 1300 m: *Vermeulen & Duistermaat* 962 (K, L). Sipitang District, confluence of Maga and Pa Sia rivers, 1350 m: *de Vogel* 8426 (L).

52. **Dendrochilum johannis-winkleri**: BRUNEI: Bukit Retak, 1500 m: *Cantley* s.n., cult. Leiden no. 26836 (L). KALIMANTAN BARAT: Bukit Tilung, 800 m: *Winkler* 1495 (HBG).

53. **Dendrochilum kingii**: KALIMANTAN: Sungai Keribung: *Hallier* 1312 (BO, K, L). KALIMANTAN TIMUR: Sangkulirang, 200–300 m: *de Vogel* 919 (L).

SABAH: Lahad Datu District, Ulu Segama: *Ahmad* SAN 70988 (L, SAN). Nabawan, 420 m: *Lamb* AL 502/85 (K) & AL 1119/89 (K). Sandakan: *Madani & Sigin* SAN 107709 (K, L, SAN, SING). SARAWAK: Hose Mountains, Bukit Semako, 660 m: *Burtt & Martin* B. 4943 (E). Sungai Iban, Linau, Belaga: *Lee* S. 45514 (AAU, MEL, SAR). Bukit Sekiwa, Tubau, Bintulu: *Mohtar et al.* S. 53913 (K, L, SAR, SING). Mt. Temabok, Upper Baram Valley, 900 m: *Moulton* 6761 (K, SING). Ulu Lawas, near Puteh: *Paie* S. 31566 (A, BO, K, KEP, L, SAN, SAR, SING). Mt. Dulit, near Long Kapa, 300 m: *McLeod* in *Synge* S. 156 (K). Bukit Woen, Padawan, 350 m: *Yii* S. 61421 (AAU, L, SAR, SING).

54. Dendrochilum lacinilobum: SABAH: Kimanis road, 1220 m: *Bailes & Cribb* 632 (K). Tenom District, Mt. Anginon, 900 m: *H.F. Comber* 4032 (K). Tenom District, above Kallang Waterfall, 1000 m: *de Vogel* 8172 (L) & 8173 (L).

55. Dendrochilum lewisii: SARAWAK: Known only from the type.

56. Dendrochilum lumakuense: SABAH: Known only from the type.

57. Dendrochilum magaense: SABAH: Sipitang District, ridge east of Maga River, 1450 m: *de Vogel* 8350 (L).

58. Dendrochilum muluense: BRUNEI: Temburong District, Bukit Retak: *Wong* WKM 802 (BRUN, K, L, SING). KALIMANTAN TIMUR: Mt. Batu Harun, north of Long Bawan, Krayan, 1850–1950 m: *Kato et al.* 9758 (L). SABAH: Tambunan District, km 55 on Kota Kinabalu to Tambunan road, 1700–1800 m: *Beaman* 6876 (K, L) & 8031 (K). Mt. Alab, 1500 m: *Lamb* SAN 87459 (SAN). Crocker Range, Ulu Apin Apin, 1350 m: *Lamb* AL 1392/91 (K). Penampang District, path to kampong Longkogungan: *Madani & George* SAN 119278 (K, SAN). SARAWAK: Mt. Pagon Periuk, Limbang, 1400 m: *Awa & Lee* S. 47827 (AAU, K, KEP, L, SAR, SING). Ridge connecting to Mt. Batu Lawi, 1840 m: *Awa & Lee* S. 50937 (K, L, NY, SAR). Route from Bakelalan to Mt. Murud, 1680 m: *Burtt & Martin* B. 5244 (E). Mt. Mulu, 2040 m: *Lewis* 365 (K, SAR). Mt. Murud, 2200 m: *Yii* S. 44616 (K, L, SAR).

59. Dendrochilum ochrolabium: SABAH: Sipitang District, Long Pa Sia to Long Samado trail, near crossing with Malabid River, 1300 m: *Vermeulen & Duistermaat* 967 (L). SARAWAK: Kapit Division, Ulu Sungai Entulu: *Lee* S. 54797 (SAR). Mt. Penrissen, near Padawan, 800 m: *Schuiteman, Mulder & de Vogel* s.n., cult. Leiden no. 933108 (L).

60. Dendrochilum oxylobum: KALIMANTAN TIMUR: Mt. Malem, Long Bawan, Kerayan: 950–1050 m: *Ueda & Darnaedy* 11544 (L). SABAH: Nabawan, 450 m: *Lamb* AL 425/85 (K) & 700 m: *de Vogel* 8150 (L). Kimanis road, 900 m: *Lamb* AL 506/85 (K). SARAWAK: Kuching: *Hewitt* 46 (SAR) & *Hewitt* s.n. (K).

61. Dendrochilum pachyphyllum: SABAH: Sinsuron road, 1470 m: *Collenette* 49/79 (E). Kota Kinabalu to Tambunan road, 1700 m: *Vermeulen & Chan* 413 (L).

Sipitang District, Long Pa Sia to Long Samado trail, near crossing with Malabid River, 1300 m: *Vermeulen & Duistermaat* 965 (L). Mt. Alab, 1900 m: *de Vogel* 8646 (L). SARAWAK: route from Ulu Sungai Limbang to Bukit Buli, 1540 m: *Awa & Lee* S. 50774 (K, SAR).

62. **Dendrochilum pandurichilum**: BRUNEI: Belait District, Badas Forest Reserve, 10 m: *de Vogel* s.n., cult. Leiden no. 911260A (K, L). SABAH: Sipitang District, N.W. of Long Pa Sia to Long Samado trail, 1900 m: *de Vogel* 8569 (L). SARAWAK: Sungai Pa Mario, Ulu Limbang, route to Mt. Batu Lawi, 1530 m: *Awa & Lee* S. 50731 (K, L, SAR).

63. **Dendrochilum papillilabium**: SABAH: Beluran District, Bukit Monkobo, 1500 m: *Aban* SAN 95230 (K, L, SAN) & SAN 95243 (SAN). Crocker Range, mile 16 along Kimanis to Keningau road, 1290 m: *Dewol & Abas* SAN 89075 (SAN). Kimanis road, 1200 m: *Wood* 573 (K) & *Wood* 583 (K).

64. **Dendrochilum papillitepalum**: Known only from the type.

65. **Dendrochilum planiscapum**: SABAH: Mt. Alab, 1350 m: *Lamb* SAN 92253 (SAN). Sinsuron road, 1470 m: *Collenette* 46/79 (E). Mt. Kinabalu: Mamut: *Amin et al.* SAN 129373 (SAN); Mesilau Trail, 2100 m: *Chow & Leopold* SAN 74511 (K, SAN); Mesilau Cave Trail, 2400 m: *Meijer* SAN 48111 (K, SAN); Pinosuk Plateau, near Kundasang, 1600 m: *de Vogel* 8032 (L) & 8033 (L).

66. **Dendrochilum pubescens**: BRUNEI: Temburong River Valley, 50–250 m: *Johns* 7341 (BRUN, K). Belait Melilas, Sungai Ingei, 50–150 m: *Thomas* 197 (BRUN, K). Batu Melintang, 1–150 m: *de Vogel* 8888 (L). SARAWAK: Mt. Temabok, Upper Baram Valley, 900 m: *Moulton* 6763 (AMES, SING).

67. **Dendrochilum scriptum**: SABAH: Mt. Kinabalu: 3000 m: *Wong et al.* 93–43 (SAN).

68. **Dendrochilum stachyodes**: SABAH: Mt. Kinabalu: summit area, 3300 m: *Bogle* 532 (AMES); Summit Trail, 3000 m: *Chan & Gunsalam* 53/87 (SING); Paka-Paka Cave, 3000 m: *Carr* 3521, SFN 27531 (K, SING) & *Holttum* SFN 36570 (AMES, SING); Kilembun Basin, 2900 m: *J. & M.S. Clemens* 33177 (AMES, BM, E, HBG); Above Panar Laban, 3600 m: *Gardner* 104 (E); Summit area, 3500 m: *Gibbs* 4181 (BM, K); Summit Trail, 3400 m: *Wood* 605 (K).

69. **Dendrochilum suratii**: SABAH: Mt. Trus Madi, 2400 m: *Osman* SAN 87432 (SAN).

70. **Dendrochilum tenuitepalum**: Known only from the type.

71. **Dendrochilum trusmadiense**: SABAH: Mt. Trus Madi, 2300 m: *Suhaili et al.* TM17 (UKMS) & TM18 (UKMS).

72. **Epigeneium speculum**: KALIMANTAN: Bukit Kasian: *Nieuwenhuis* 274 (L). SABAH: Nabawan, 400–500 m: *Lamb* AL 447/85 (K) & *Wood* 597 (K). SARAWAK: Mt. Dulit, Dulit Ridge, 900 m: *Synge* S. 499 (K).

73. **Epigeneium tricallosum**: BRUNEI: Temburong District, Mt. Pagon East: *Coode* 7532 (BRUN, K). SABAH: Mt. Alab: 1650 m: *Cribb* 89/47 (K) & 1800 m: *Wood* 773 (K). Mt. Kinabalu: Tenompok, 1600 m: *Carr* SFN 26955 (K, SING); Mt. Nungkek, 800 m: *J. & M.S. Clemens* 32906 (BM); Dallas, 900 m: *J. & M.S. Clemens* s.n. (BM).

74. **Eria caricifolia var. caricifolia**: SARAWAK: Known only from the type.
Eria caricifolia var. glabra: SABAH: Sipitang District, Maligan to Long Pa Sia trail, 1500 m: *Vermeulen & Duistermaat* 914 (K, L).

75. **Eria grandis**: SABAH: Mt. Kinabalu: Summit Trail, 3000 m: *Beaman* 8307 (K); Marai Parai, 2100 m: *Carr* SFN 27432 (SING); Paka-Paka Cave, 3000 m: *Carr* SFN 27534 (SING); Kemburongoh, 2700 m: *J. & M.S. Clemens* 27165 (BM); Kilembun River Head, 2700–2900 m: *J. & M.S. Clemens* 33962 (BM); Paka-Paka Cave, 2900–3200 m: *Gibbs* 4268 (BM, K) & 3000 m: *Meijer* SAN 29293 (K, SAN); Eastern Shoulder, Camp 4, 2900 m: RSNB 1121 (K).

76. **Eria lanuginosa**: SABAH: Sipitang District, Pa Sia River, near Long Pa Sia, 930 m: *Wood* 725 (K). SARAWAK: Bario, Sungai Menalio, Pa Dappur, 950 m: *Awa* S. 51069 (K, SAR).

77. **Hapalochilus lohokii**: SABAH: Mt. Alab, 2000 m: *Vermeulen* 651 (K, L, UKMS). Mt. Kinabalu, Pinosuk Plateau, 1590 m: *Lamb* AL 820/87 (K).

78. **Liparis aurantiorbiculata**: SABAH: Crocker Range, km 59 on Kota Kinabalu to Tambunan road, 1400 m: *Beaman* 7378 (K). Mt. Kinabalu: Penibukan, 1200 m: *J. & M.S. Clemens* 30698 (BM); Tenompok Orchid Garden, 1500 m: *J. & M.S. Clemens* 50255 (BM).

79. **Liparis kinabaluensis**: SABAH: Mt. Kinabalu: Lohan River, 800 m: *Bailes & Cribb* 658 (K); West Mesilau River: *Brentnall* 124 (K); Tinekuk Falls, 1800 m: *J. & M.S. Clemens* 40908 (BM, K); Tahubang River, 1100 m: *J. & M.S. Clemens* 51462 (BM, K).

80. **Liparis lobongensis**: SABAH: Mt. Kinabalu: Bundu Tuhan, 1200 m: *Carr* SFN 27420 (K); Park Headquarters, 1600 m: *Chan* 121/89 (K); Marai Parai, 1500 m: *J. & M.S. Clemens* 33100 (BM); Mesilau Cave/Janet's Halt, 2300 m: *Collenette* 21574 (K).

81. **Liparis pandurata**: SABAH: Mt. Kinabalu: Mesilau Cave Trail, 1700–1900 m: *Beaman* 8011 (K); West Mesilau River, 1600–1700 m: *Beaman* 8623 (K) & 1600 m: 8998 (K); Bundu Tuhan, 1200 m: *Carr* 3041, SFN 27416 (K, SING); Minirinteg Cave, 2100 m: *J. & M.S. Clemens* 29285 (BM, K); Liwagu River, 1600 m: *Darnton* 556 (BM); Mesilau Trial, 1700 m: *Fuchs & Collenette* 21498 (K). Sipitang District, Ulu Long Pa Sia, near Maga River, 1260 m: *Wood* 660 (K).

82. **Malaxis punctata**: SABAH: Mt. Kinabalu: Penataran Basin, 1200 m: *J. & M.S. Clemens* 34059 (BM); Tahubang River, 1400 m: *J. & M.S. Clemens* 40824 (BM); Tinekuk Falls, 1800 m: *J. & M.S. Clemens* 40909 (BM, K); Kinateki River, 800 m: *Collenette* A113 (BM); Mesilau Cave Trail, 1800 m: *Wood* 846 (K). Crocker Range, Moyog side of Mt. Alab, 1200 m: *Lamb* AL 555/86 (K); Sipitang District, Ulu Long Pa Sia, near Maga River, 1320 m: *Wood* 638 (K).

83. **Neuwiedia zollingeri var. javanica**: SABAH: Mt. Kinabalu: Hempuen Hill, 800–1000 m: *Beaman* 7416 (K) & *Madani* SAN 89375 (K, SAN); Melangkap Kapa, 700–1000 m: *Beaman* 8782 (K). Bukit Tangkunan, Telupid: *Rahim et al.* SAN 93285 (K, L, SAN, SAR).

84. **Paphiopedilum kolopakingii**: KALIMANTAN: locality withheld.

85. **Paphiopedilum lowii**: KALIMANTAN, SABAH & SARAWAK: localities withheld.

86. **Paphiopedilum stonei**: SARAWAK: locality withheld.

87. **Pilophyllum villosum**: SABAH: Mt. Kinabalu: Penibukan, 1200 m: *J. & M.S. Clemens* 30696 (BM); Gurulau Spur, 1500 m: *J. & M.S. Clemens* 50377 (BM) & 51023 (BM, K); Mt. Nungkek, 1200 m: *Darnton* 482 (BM). Crocker Range, Moyog area, 900 m: *Lamb* AL 372/85 (K). SARAWAK: Bario, Batu Buli: sight record by *Lamb*.

88. **Plocoglottis gigantea**: SABAH: Mt. Kinabalu: Dallas, 900 m: *J. & M.S. Clemens* s.n. (BM); Langanan River: *Lohok* 10 (K).

89. **Plocoglottis hirta**: SARAWAK: Bau District, Seburan, 60 m: *Anderson* 8991 (K, SAR). Mt. Api, 600 m: *Anderson* S. 24020 (K, SAR). Mt. Mulu National Park, base of cliffs of Mt. Api: *Collenette* 2354 (K). Mt. Jambusan: *Lee* S. 38609 (K, L, SAR, SING). Mt. Mulu, Hidden Valley, 450–500 m: *Nielsen* 967 (AAU). Mt. Mulu National Park, Medalam River, 100 m: *Yii* S. 39914 (K, SAR).

90. **Porpax borneensis**: SABAH: Mt. Kinabalu, Melangkap Tomis, 900–1000 m: *Beaman* 8994a (K). Sandakan District, Mt. Tawai (Tavai), 900 m: *Lamb* AL 651/86 (K, sketch only).

91. **Porrorhachis galbina**: Kimanis road, 1080 m: *Lamb* AL 849/87 (K) & 1200 m: AL 1130/89 (K). Mt. Alab, 1500 m: *Lamb* AL 869/87 (K) & 1560 m: AL 1133/89 (K). Mt. Kinabalu: Bundu Tuhan, 1400 m: *Darnton* 199 (BM); Tenompok, 1500 m: *J. & M.S. Clemens* 29580 (BM) & 30162 (K).

92. **Pteroceras biserratum**: KALIMANTAN BARAT: Sungai Pinang, near Samarinda, *Schlechter* 13348 (fide Schlechter). SABAH: Lahad Datu: *Vermeulen & Lamb* in *Lamb* AL 1147/89 (K).

93. Pteroceras erosulum: SABAH: Sipitang District, 8 km NW of Long Pa Sia, 1260 m: *Wood* 676 (K). SARAWAK: Photographic record by *Vogel*.

94. Robiquetia crockerensis: SABAH: Mt. Alab, 1050 m: *Lamb* AL 31/82 (K). Mt. Kinabalu: Golf Course Site, 1600 m: *Bailes & Cribb* 707 (K); Hempuen Hill, 800–1200 m: *Beaman* 7717 (K); Tenompok, 1500 m: *J. & M.S. Clemens* 28554 (BM); Marai Parai Spur, 1900 m: *Collenette* A65 (BM).

95. Robiquetia pinosukensis: SABAH: Mt. Kinabalu: East Mesilau River, 1600–1700 m: *Bailes & Cribb* 743 (K); Golf Course Site, 1700–1800 m: *Beaman* 7227 (K) & 1700 m: *Brentnall* 156 (K); Tenompok, 1500 m: *J. & M.S. Clemens* 28141 (BM) & 29252 (BM); Mesilau, 1500 m: *Cockburn* SAN 70120 (K, SAN).

96. Robiquetia transversisaccata: SABAH: Sinsuron Road, 900–1200 m: *Lamb* AL 196/84 (K). Mt. Kinabalu: Kiau, 900 m: *J. Clemens* 67 (AMES); Marai Parai Spur: *J. Clemens* 246 (AMES); Penibukan, 1500 m: *J. & M.S. Clemens* 50340 (BM).

97. Spathoglottis microchilina: BRUNEI: Belait District, Ulu Ingei, 50 m: *Boyce* 315 (K). Sungai Ingei, 50–150 m: *Thomas* 194 (K). KALIMANTAN: Mt. Njapa, Kelai River, 1000 m: *Kostermans* 21494 (BO, K, L). SABAH: Crocker Range, Sinsuron road, 1350 m: *Chan* 127 (K). Mongaya River, 900 m: *J.B. Comber* 114 (K). Mt. Kinabalu: Marai Parai Spur, 1700 m: *Bailes & Cribb* 828 (K); Dallas, 900 m: *J. & M.S. Clemens* 26767 (BM); Kiau, 1200 m: Darnton 599 (BM); Park Headquarters, 1500 m: *Tan & Phillipps* SNP 579 (K). Crocker Range, Kimanis road, 1350 m: *Wood* 820 (K).

98. Tainia vegetissima: SABAH: Maliau Basin, 1000 m: *Lamb* AL 1428/92 (K). SARAWAK: Mt. Dulit, Dulit Ridge, 1230 m: *Synge* S. 434 (K).

99. Trichotosia brevipedunculata: Mt. Kinabalu: Golf Course Site, 1700–1800 m: *Beaman* 7235 (K); Mesilau Cave Trail, 1800 m: *Beaman* 7491 (K); Marai Parai Spur: *J. Clemens* 291 (AMES); Sediken River/Marai Parai, 1500 m: *J. & M.S. Clemens* 32452 (BM); Tinekuk Falls, 2000 m: *J. & M.S. Clemens* 40916 (BM, K); Park Headquarters, 1500 m: *Wood* 619 (K). Penampang District, km 49.5 on Kota Kinabalu to Tambunan road, 1350–1500 m: *Beaman* 10486 (K). Mt. Trus Madi, above Kidukarok, 1560 m: *Wood* 872 (K).

100. Tropidia connata: Known only from the type.

Opposite: *Dendrobium cinnabarinum* Rchb.f. Sabah, Maliau Basin.
(Photo: R.S. Beaman)

COLOUR PLATES

PLATE 1

A. *Acanthephippium eburneum* Kraenzl. Sarawak, Forest Research Institute. *(Photo: P.J. Cribb)* (1)

B. *Agrostophyllum laterale* J.J. Sm. Kalimantan Timur, Apokayan. *(Photo: P.J. Cribb)* (3)

C. *Appendicula longirostrata* Ames & C. Schweinf. Sabah, Kinabalu Park Headquarters. *(Photo: C.L. Chan)* (5)

D. *Appendicula longirostrata* Ames & C. Schweinf. Close-up of inflorescence. Sabah, Kinabalu Park Headquarters. *(Photo: C.L. Chan)* (5)

E. *Arachnis grandisepala* J.J. Wood. Sabah, growing in the Crocker Range. *(Photo: A. Lamb)* (7)

F. *Arachnis grandisepala* J.J. Wood. Sabah, cult. Poring Orchid Centre. *(Photo: C.L. Chan)* (7)

A
Photo: P.J. Cribb
B
Photo: P.J. Cribb
C
Photo: C.L. Chan
D
Photo: C.L. Chan
E
Photo: A. Lamb
F
Photo: C.L. Chan

PLATE 2

A. *Arachnis longisepala* (J.J. Wood) Shim & A. Lamb. Sabah, *Bailes & Cribb* 654 (K), cult. R.B.G. Kew. *(Photo: C.L. Chan)* (8)

B. *Ascochilopsis lobata* J.J. Wood & A. Lamb. Sabah, *Lamb & Surat* in *Lamb* AL 1252/90 (holotype K). *(Photo: A. Lamb)* (9)

C. *Brachypeza zamboangensis* (Ames) Garay. Habit. Sabah, cult. Tenom Orchid Centre. *(Photo: C.L. Chan)* (10)

D. *Brachypeza zamboangensis* (Ames) Garay. Close-up of flower, front view. Sabah, cult. Tenom Orchid Centre. *(Photo: C.L. Chan)* (10)

E. *Brachypeza zamboangensis* (Ames) Garay. Close-up of flower, side view. Sabah, cult. Tenom Orchid Centre. *(Photo: C.L. Chan)* (10)

A
B
C
D
E
Photo: C.L. Chan
Photo: A. Lamb
Photo: C.L. Chan
Photo: C.L. Chan
Photo: C.L. Chan

PLATE 3

A. *Bulbophyllum beccarii* Rchb. f. Sabah, showing leaf litter trapped by the leaf. *(Photo: P.J. Cribb)* (11)

B. *Bulbophyllum beccarii* Rchb. f. Sabah, cult. Tenom Orchid Centre, showing pollinating bluebottles. *(Photo: Photo: A. Lamb)* (11)

C. *Bulbophyllum beccarii* Rchb. f. Close-up of inflorescence. *(Photo: P. Jongejan)* (11)

D. *Bulbophyllum kemulense* J.J. Sm. Sabah. *(Photo: A. Lamb)* (12)

E. *Bulbophyllum polygaliflorum* J.J. Wood. Sabah, cult. The Netherlands. *(Photo: P. Jongejan)* (13)

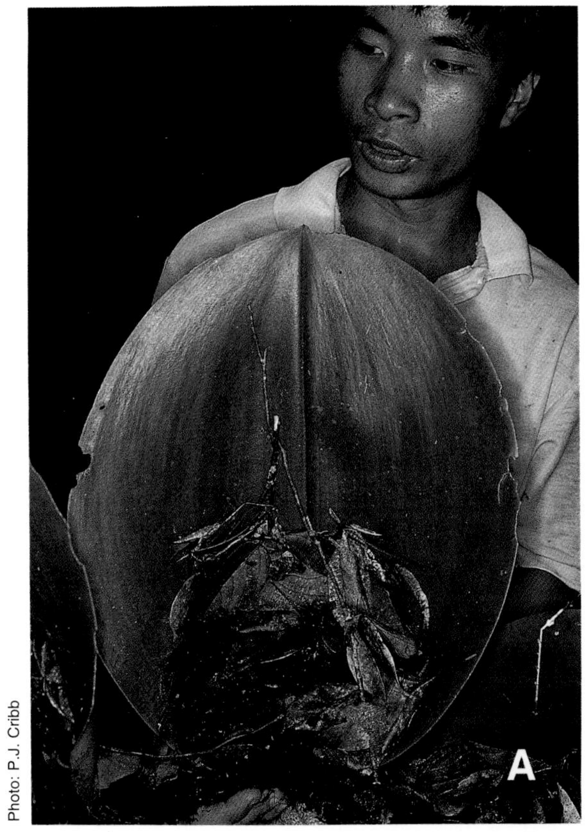

Photo: P.J. Cribb

Photo: A. Lamb

Photo: P. Jongejan

Photo: A. Lamb

Photo: P. Jongejan

PLATE 4

A. *Calanthe undulata* J.J. Sm. Kalimantan Timur, Apokayan, *de Vogel & Cribb* 9140 (K). *(Photo: P.J. Cribb)* (14)

B. *Cleisocentron merrillianum* (Ames) Christenson. Sabah, cult. R.B.G. Kew. *(Photo: J.B. Comber)* (15)

C. *Cleisocentron merrillianum* (Ames) Christenson. Sabah, cult. R.B.G. Kew. *(Photo: C.L. Chan)* (15)

D. *Corybas muluensis* J. Dransf. Sabah, *Wood* 823 (K). *(Photo: J.B. Comber)* (16)

E. *Corybas muluensis* J. Dransf. Sabah, *Wood* 823 (K). *(Photo: J.B. Comber)* (16)

Photo: P.J. Cribb
A
Photo: J.B. Comber
B
C
Photo: C.L. Chan
Photo: J.B. Comber
D
E
Photo: J.B. Comber

PLATE 5

A. *Cymbidium borneense* J.J. Wood. Close-up of flower. Sabah, Sepilok Forest Reserve. *(Photo: C.L. Chan)* (17)

B. *Cymbidium borneense* J.J. Wood. Habit. Sabah, Sepilok Forest Reserve. *(Photo: C.L. Chan)* (17)

C. *Dendrobium alabense* J.J. Wood. Habit. Sabah, *Wood* 851 (K). *(Photo. J.B. Comber)* (18)

D. *Dendrobium alabense* J.J. Wood. Close-up of flower. Sabah, *Wood* 776 (K). *(Photo: J.B. Comber)* (18)

Photo: C.L. Chan

Photo: C.L. Chan

Photo: J.B. Comber

Photo: J.B. Comber

PLATE 6

A. *Dendrobium cymboglossum* J.J. Wood. Sabah, cult. Poring Orchid Centre. *(Photo: C.L. Chan)* (20)

B. *Dendrobium cymboglossum* J.J. Wood. Sabah, cult. Poring Orchid Centre. *(Photo: A. Lamb)* (20)

C. *Dendrobium hamaticalcar* J.J. Wood & Dauncey. Sabah, *Bacon* 142 (holotype E), cult. R.B.G. Edinburgh. *(Photo: R.B.G. Edinburgh)* (21)

D. *Dendrobium kiauense* Ames & C. Schweinf. Habit. Sabah, Mt. Kinabalu. *(Photo: C.L. Chan)* (22)

E. *Dendrobium kiauense* Ames & C. Schweinf. Close-up of flower. Sabah, Mt. Kinabalu. *(Photo: C.L. Chan)* (22)

A

Photo: C.L. Chan

Photo: R.B.G. Edinburgh

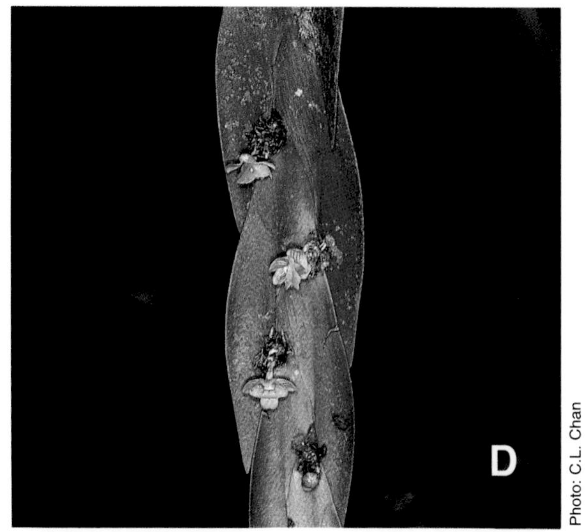

Photo: C.L. Chan

Photo: A. Lamb

Photo: C.L. Chan

PLATE 7

A. *Dendrobium lambii* J.J. Wood. Sabah, *Lamb* s.n. (holotype K). *(Photo. S. Collenette)* (23)

B. *Dendrobium lancilobum* J.J. Wood. Sabah, *Wood* 684 (K). *(Photo: J.B. Comber)* (24)

C. *Dendrobium lowii* Lindl. Curtis's Botanical Magazine, Plate 5303 (1862). *(Photo: R.B.G. Kew)* (26)

D. *Dendrobium lowii* Lindl. (var. *pleiotrichum*). Cult. South Africa. *(Photo. L. Vogelpoel)* (26)

E. *Dendrobium nabawanense* J.J. Wood & A. Lamb. Sabah, cult. Tenom Orchid Centre. *(Photo: C.L. Chan)* (27)

Photo: S. Collenette

Photo: J.B. Comber

Photo: L. Vogelpoel

Photo: R.B.G. Kew

Photo: C.L. Chan

PLATE 8

A. *Dendrobium panduriferum* Hook. f. Sabah, cult. Poring Orchid Centre. *(Photo. P.J. Cribb)* (29)

B. *Dendrobium patentilobum* Ames & C. Schweinf. Close-up of flower, side view. Sabah, Mt. Kinabalu, *Bailes & Cribb* 800, cult. R.B.G. Kew. *(Photo: R.B.G. Kew)* (30)

C. *Dendrobium patentilobum* Ames & C. Schweinf. Close-up of flower, front view. Sabah, Mt. Kinabalu, *Bailes & Cribb* 800, cult. R.B.G. Kew. *(Photo: R.B.G. Kew)* (30)

D. *Dendrobium trullatum* J.J. Wood & A. Lamb. Sabah, *Collenette* 2287 (K). *(Photo: S. Collenette)* (33)

E. *Dendrobium ventripes* Carr. Sarawak. *(Photo: J. Dransfield)* (34)

A

Photo: P.J. Cribb

B

Photo: R.B.G. Kew

C

Photo: R.B.G. Kew

D

Photo: S. Collenette

E

Photo: J. Dransfield

PLATE 9

A. *Dendrobium xiphophyllum* Schltr. Sabah, cult. Tenom Orchid Centre. *(Photo: C.L. Chan)* (35)

B. *Dendrochilum anomalum* Carr. Sabah, cult. Poring Orchid Centre. *(Photo. C.L. Chan)* (36)

C. *Dendrochilum crassum* Ridl. Sabah, *Wood* 584 (K). *(Photo: J.B. Comber)* (38)

D. *Dendrochilum crassum* Ridl. Close-up of flowers. Sabah, Crocker Range, Kimanis road. *(Photo: C.L. Chan)* (38)

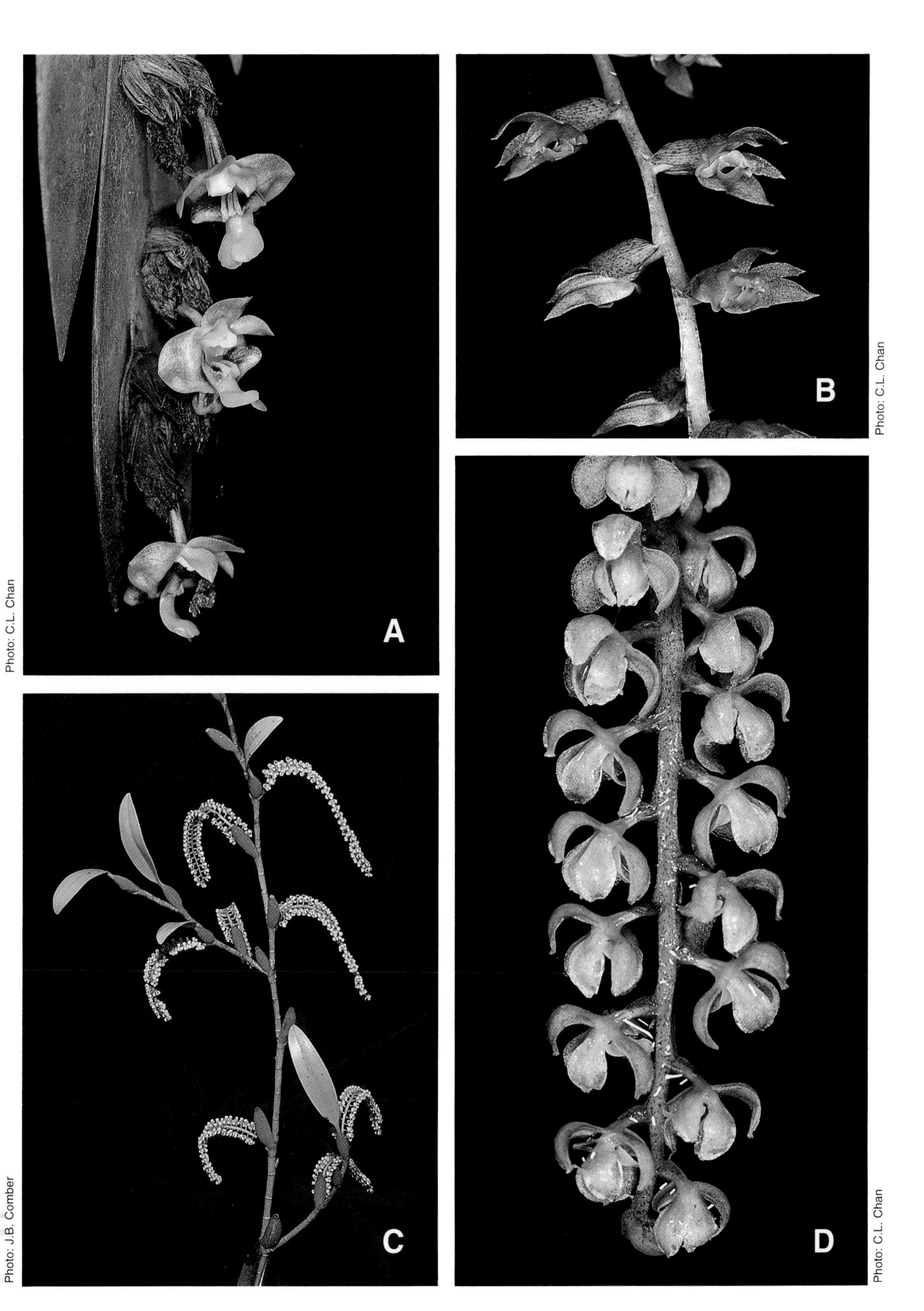
A
Photo: C.L. Chan
B
Photo: C.L. Chan
C
Photo: J.B. Comber
D
Photo: C.L. Chan

PLATE 10

A. *Dendrochilum cruciforme* J.J. Wood var. *cruciforme*. Sabah, Maliau Basin. *(Photo. C.L. Chan)* (39)

B. *Dendrochilum cupulatum* J.J. Wood. Sabah, Kimanis road. *(Photo: T.J. Barkman)* (40)

C. *Dendrochilum gibbsiae* Rolfe. Habit. Sabah, Kimanis road. *(Photo. A. Lamb)* (45)

D. *Dendrochilum gibbsiae* Rolfe. Sabah, *Wood* 621 (K). *(Photo. J.B. Comber)* (45)

A
Photo: C.L. Chan
B
Photo: T.J. Barkman
C
Photo: A. Lamb
D
Photo: J.B. Comber

PLATE 11

A. *Dendrochilum grandiflorum* (Ridl.) J.J. Sm. Sabah, Mt. Kinabalu. *(Photo: P.J. Cribb)* (46)

B. *Dendrochilum grandiflorum* (Ridl.) J.J. Sm. Sabah, Mt. Kinabalu. *(Photo: A. Lamb)* (46)

C. *Dendrochilum haslamii* Ames. Sabah, Mt. Kinabalu. *(Photo: T.J. Barkman)* (47)

D. *Dendrochilum hologyne* Carr. Sabah, *Wood* 646 (K). *(Photo: J.B. Comber)* (48)

A

Photo: P.J. Cribb

Photo: A. Lamb

Photo: T.J. Barkman

Photo: J.B. Comber

PLATE 12

A. *Dendrochilum imbricatum* Ames. Habit. Sabah, Mt. Kinabalu. *(Photo: T.J. Barkman)* (50)

B. *Dendrochilum imbricatum* Ames. Close-up of flowers. Sabah, *Wood* 618 (K). *(Photo: J.B. Comber)* (50)

A
B
Photo: T.J. Barkman

PLATE 13

A. *Dendrochilum kingii* (Hook. f.) J.J. Sm. Sabah, Crocker Range. *(Photo: A. Lamb)* (53)

B. *Dendrochilum lacinilobum* J.J. Wood & A. Lamb. Sabah. *(Photo: A. Lamb)* (54)

C. *Dendrochilum muluense* J.J. Wood. Close-up of flowers. Sabah, Crocker Range. *(Photo: C.L. Chan)* (58)

D. *Dendrochilum muluense* J.J. Wood. Habit. Sabah, Crocker Range. *(Photo: C.L. Chan)* (58)

Photo: A. Lamb

Photo: C.L. Chan

Photo: C.L. Chan

Photo: A. Lamb

PLATE 14

A. *Dendrochilum ochrolabium* J.J. Wood. Sabah, Mt. Alab. *(Photo: A. Lamb)* (59)

B. *Dendrochilum oxylobum* Schltr. Close-up of flowers. Sabah, cult. Tenom Orchid Centre, ex Nabawan. *(Photo: A. Lamb)* (60)

C. *Dendrochilum oxylobum* Schltr. Close-up of flower. Cult. Hortus Botanicus, Leiden, The Netherlands. *(Photo: Ed de Vogel)* (60)

D. *Dendrochilum pachyphyllum* J.J. Wood & A. Lamb. Close-up of flowers. Sabah, Crocker Range. *(Photo: C.L. Chan)* (61)

A
B
C
D
Photo: A. Lamb
Photo: A. Lamb
Photo: Ed de Vogel
Photo: C.L. Chan

PLATE 15

Dendrochilum pachyphyllum J.J. Wood & A. Lamb. Habit. Sabah, Crocker Range. *(Photo: C.L. Chan)* (61)

Photo: C.L. Chan

PLATE 16

A. *Dendrochilum planiscapum* Carr. Sabah. *(Photo: A. Lamb)* (65)

B. *Dendrochilum pubescens* L.O. Williams. Habit. Brunei, *Thomas* 197 (K). *(Photo. J.B. Comber)* (66)

C. *Dendrochilum pubescens* L.O. Williams. Close-up of flower. Brunei, *Thomas* 197 (K). *(Photo. J.B. Comber)* (66)

D. *Dendrochilum scriptum* Carr. Close-up of flower. Sabah, Mt. Kinabalu. *(Photo. T.J. Barkman)* (67)

Photo: A. Lamb
A
D
C
B
Photo: T.J. Barkman
Photo: J.B. Comber
Photo: J.B. Comber

PLATE 17

A. *Dendrochilum scriptum* Carr. Habit. Sabah, Mt. Kinabalu. *(Photo: T.J. Barkman)* (67)

B. *Dendrochilum stachyodes* (Ridl.) J.J. Sm. Habitat. Sabah, Mt. Kinabalu. *(Photo: A. Lamb)* (68)

Photo: T.J. Barkman

Photo: A. Lamb

PLATE 18

A. *Dendrochilum stachyodes (*Ridl.) J.J. Sm. Sabah, Mt. Kinabalu. *(Photo: A. Lamb)* (68)

B. *Epigeneium speculum* (J.J. Sm.) Summerh. Close-up of flower. Sabah, Crocker Range. (*Photo: C.L. Chan)* (72)

C. *Epigeneium tricallosum* (Ames & C. Schweinf.) J.J. Wood. Close-up of flowers. Sabah, Mt Kinabalu. *(Photo: C.L. Chan)* (73)

D. *Eria grandis* Ridl. Habit. Sabah, Mt. Kinabalu, 3000 metres. *(Photo: C.L. Chan)* (75)

E. *Eria grandis* Ridl. Close-up of flowers. Sabah, Mt. Kinabalu, 3000 metres. *(Photo: C.L. Chan)* (75)

A
B
C
D
E
Photo: A. Lamb
Photo: C.L. Chan
Photo: C.L. Chan
Photo: C.L. Chan
Photo: C.L. Chan

PLATE 19

A. *Eria lanuginosa* J.J. Wood. Habit. Sabah, *Wood* 725 (holotype K). *(Photo: J.B. Comber)* (76)

B. *Eria lanuginosa* J.J. Wood. Close-up of flower. Sabah, *Wood* 725 (holotype K). *(Photo: J.B. Comber)* (76)

C. *Hapalochilus lohokii* (J.J. Verm. & A. Lamb) Garay, Hamer & Siegerist. Habit. Sabah, Mt. Alab. *(Photo. P.J. Cribb)* (77)

D. *Hapalochilus lohokii* (J.J. Verm. & A. Lamb) Garay, Hamer & Siegerist. Close-up of flower. Sabah, Mt. Alab. *(Photo. P.J. Cribb)* (77)

E. *Liparis aurantiorbiculata* J.J. Wood & A. Lamb. Habit. Sabah, Mt. Kinabalu. *(Photo: C.L. Chan)* (78)

F. *Liparis aurantiorbiculata* J.J. Wood & A. Lamb. Close-up of flower. Sabah, Mt. Kinabalu. *(Photo: C.L. Chan)* (78)

Photo: J.B. Comber
A
B
Photo: J.B. Comber
Photo: P.J. Cribb
C
D
Photo: P.J. Cribb
Photo: C.L. Chan
E
F
Photo: C.L. Chan

PLATE 20

A. *Liparis kinabaluensis* J.J. Wood. Habit. Sabah, *Brentnall* 124 (K), cult. R.B.G. Kew. *(Photo: R.B.G. Kew)* (79)

B. *Liparis kinabaluensis* J.J. Wood. Close-up of flower. Sabah, *Brentnall* 124 (K), cult. R.B.G. Kew. *(Photo: R.B.G. Kew)* (79)

C. *Liparis lobongensis* Ames. Close-up of flower. Sabah, Mt. Kinabalu. *(Photo: C.L. Chan)* (80)

D. *Liparis lobongensis* Ames. Habit. Sabah, Mt. Kinabalu. *(Photo: C.L. Chan)* (80)

E. *Liparis pandurata* Ames. Sabah, *Wood* 660 (K). *(Photo: J.B. Comber)* (81)

A
B
C
D
E
Photo: R.B.G. Kew
Photo: R.B.G. Kew
Photo: C.L. Chan
Photo: C.L. Chan
Photo: J.B. Comber

PLATE 21

A. *Malaxis punctata* J.J. Wood. Sabah, *Sands* 4011 (holotype K), plant showing spotted foliage, cult. R.B.G. Kew. *(Photo: R.B.G. Kew)* (82)

B *Malaxis punctata* J.J. Wood. Sabah, *Sands* 4011 (holotype K), close-up of inflorescence, cult. R.B.G. Kew. *(Photo: R.B.G. Kew)* (82)

C. *Malaxis punctata* J.J. Wood. Sabah, *Wood* 638 (K), purple-flowered form, cult. R.B.G. Kew. *(Photo: J.B. Comber)* (82)

D. *Malaxis punctata* J.J. Wood. Sabah, *Wood* 638 (K), close-up of inflorescence of purple-flowered form, cult. R.B.G. Kew. *(Photo: J.B. Comber)* (82)

Photo: R.B.G. Kew

A

Photo: J.B. Comber

Photo: J.B. Comber

PLATE 22

A. *Neuwiedia zollingeri* Rchb. f. var. *javanica* (J.J. Sm.) de Vogel. Sabah, cult. Poring Orchid Centre. *(Photo: A. Lamb)* (83)

B. *Neuwiedia zollingeri* Rchb. f. var. *javanica* (J.J. Sm.) de Vogel. Sabah. *(Photo: A. Lamb)* (83)

C. *Paphiopedilum kolopakingii* Fowlie. Cult. R.B.G. Kew. *(Photo: P.J. Cribb)* (84)

D. *Paphiopedilum kolopakingii* Fowlie. Cult. R.B.G. Kew. *(Photo: J.B. Comber)* (84)

Photo: A. Lamb

Photo: A. Lamb

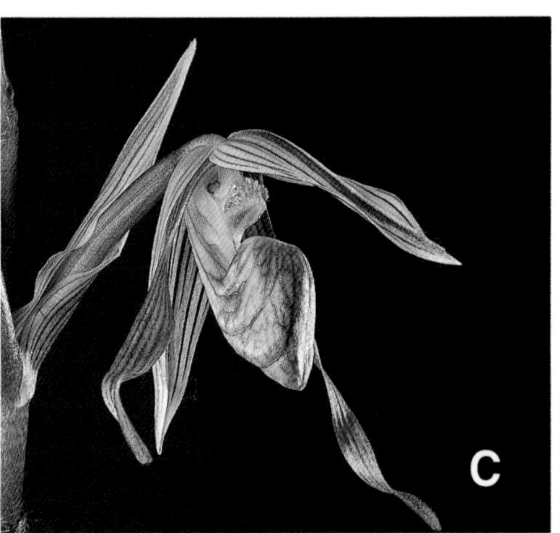

Photo: P.J. Cribb

Photo: J.B. Comber

PLATE 23

A. *Paphiopedilum lowii* (Lindl.) Stein. Sabah, Mt. Kinabalu. *(Photo: C.L. Chan)* (85)

B. *Paphiopedilum lowii* (Lindl.) Stein. Close-up of flower showing staminode. *(Photo: R. Zabeau)* (85)

C. *Paphiopedilum lowii* (Lindl.) Stein forma *aureum* P.J. Cribb. Sarawak. *(Photo: Yii Puan Ching)* (85)

D. *Paphiopedilum stonei* (Hook. f.) Stein. Sarawak. *(Photo: P.J. Cribb)* (86)

E. *Pilophyllum villosum* (Blume) Schltr. Cult. Tenom Orchid Centre. *(Photo: A. Lamb)* (87)

A
Photo: C.L. Chan
B
Photo: R. Zabeau
C
Photo: Yii Puan Ching
D
Photo: P.J. Cribb
E
Photo: A. Lamb

PLATE 24

A. *Plocoglottis gigantea* (Hook. f.) J.J. Sm. Sabah, cult. Poring Orchid Centre. *(Photo: C.L. Chan)* (88)

B. *Plocoglottis hirta* Ridl. Sarawak. *(Photo: J. Dransfield)* (89)

C. P*orpax borneensis* J.J. Wood & A. Lamb. Sabah, Tawai. *(Photo: A. Lamb)* (90)

D. *Porrorhachis galbina* (J.J. Sm.) Garay. Habit. Sabah, Crocker Range. *(Photo: C.L. Chan)* (91)

E. *Porrorhachis galbina* (J.J. Sm.) Garay. Close-up of flowers. Sabah, Crocker Range. *(Photo: C.L. Chan)* (91)

A

Photo: C.L. Chan

B

Photo: J. Dransfield

C

Photo: A. Lamb

D

Photo: C.L. Chan

E

Photo: C.L. Chan

PLATE 25

A. *Pteroceras biserratum* (Ridl.) Holttum. Sabah. *(Photo: A. Lamb)* (92)

B. *Pteroceras erosulum* H. Ae. Peders. Close-up of flower. Sabah, *Wood* 676 (holotype K). *(Photo: J.B. Comber)* (93)

C. *Robiquetia crockerensis* J.J. Wood & A. Lamb. Sabah, cult. R.B.G. Kew. *(Photo. J.B. Comber)* (94)

D. *Robiquetia transversisaccata* (Ames & C. Schweinf.) J.J. Wood. Sabah. *(Photo: A. Lamb)* (96)

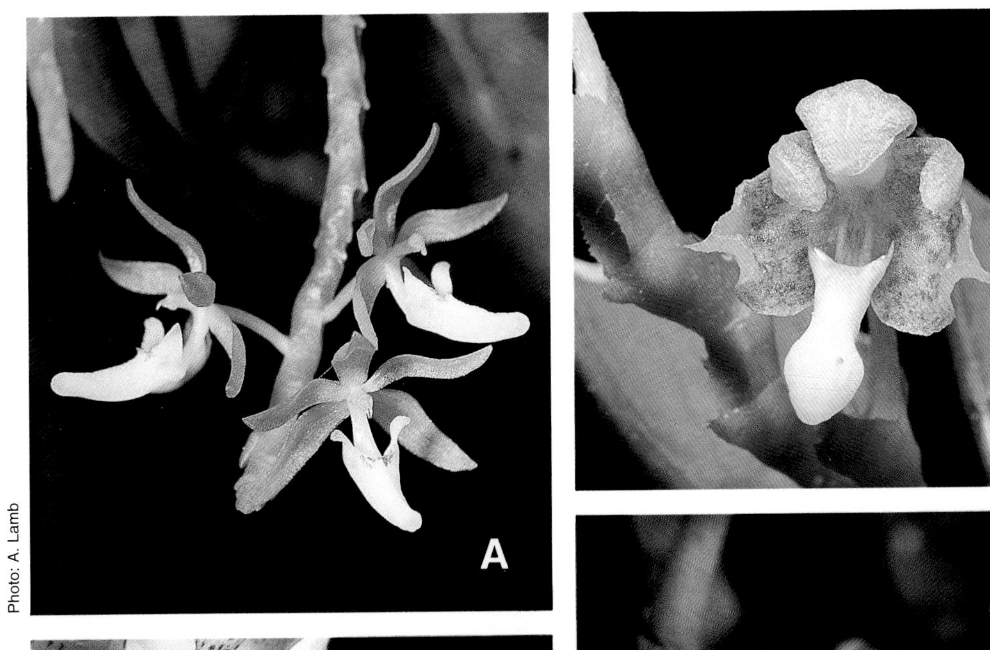

Photo: A. Lamb

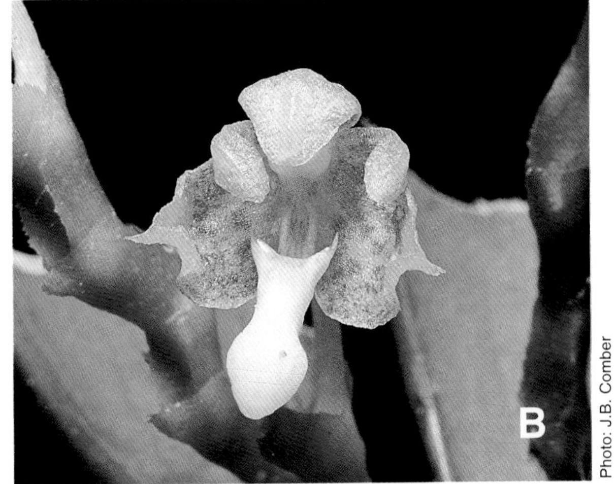

Photo: J.B. Comber

Photo: J.B. Comber

Photo: A. Lamb

PLATE 26

A. *Spathoglottis microchilina* Kraenzl. Sabah, Mt. Kinabalu. *(Photo: C.L. Chan)* (97)

B. *Tainia vegetissima* Ridl. Peninsular Malaysia. *(Photo: J. Dransfield)* (98)

C. *Trichotosia brevipedunculata* (Ames & C. Schweinf.) J.J. Wood. Sabah, *Wood* 619 (K). *(Photo: J.B. Comber)* (99)

D. *Tropidia connata* J.J. Wood & A. Lamb. Sabah, *Lamb, Surat & Lim* in *Lamb AL* 1512/92 (holotype K). *(Photo: C.L. Chan)* (100)

A
Photo: C.L. Chan
B
Photo: J. Dransfield
C
Photo: J.B. Comber
D
Photo: C.L. Chan

INDEX TO ORCHID SCIENTIFIC NAMES

(Accepted names appear in roman type. Synonyms appear in *italics*. Numbers in **bold** type indicate pages with detailed treatment. Numbers within brackets indicate pages with illustrations).